T0395292

Island Power Systems

Island Power Systems

Lukas Sigrist
Enrique Lobato
Francisco M. Echavarren
Ignacio Egido
Luis Rouco

CRC Press
Taylor & Francis Group
Boca Raton London New York

CRC Press is an imprint of the
Taylor & Francis Group, an **Informa** business

CRC Press
Taylor & Francis Group
6000 Broken Sound Parkway NW, Suite 300
Boca Raton, FL 33487-2742

Printed on acid-free paper
Version Date: 20161101

International Standard Book Number-13: 978-1-4987-4636-6 (Hardback)

Library of Congress Cataloging-in-Publication Data

Names: Sigrist, Lukas, author. | Lobato, Enrique, author. | Echavarren, Francisco M., author. | Egido, Ignacio, author. | Rouco, Luis, author.
Title: Island power systems / Lukas Sigrist, Enrique Lobato, Francisco M. Echavarren, Ignacio Egido and Luis Rouco.
Description: Boca Raton, FL : Taylor & Francis Group, LLC, CRC Press is an imprint of the Taylor & Francis Group, an Informa Business, [2017] | Includes bibliographical references and index.
Identifiers: LCCN 2016024596| ISBN 9781498746366 (hardback : acid-free paper) | ISBN 9781498746380 (ebook)
Subjects: LCSH: Island networks (Electricity)
Classification: LCC TK3101 .S54 2017 | DDC 621.31/7--dc23
LC record available at https://lccn.loc.gov/2016024596

Visit the Taylor & Francis Web site at
http://www.taylorandfrancis.com

and the CRC Press Web site at
http://www.crcpress.com

Printed and bound in the United States of America by
Edwards Brothers Malloy on sustainably sourced paper

Contents

Preface

This book is concerned with understanding the technical and economic problems of island power systems. Specifically, we are interested in island systems that exhibit two distinct features: they are small and isolated.

Isolated power systems cannot count on the technical and economic support of external systems as interconnected power systems do. Small isolated systems are not only more vulnerable from the frequency stability point of view, but they also carry higher generation costs.

Frequency stability is the most relevant technical issue in small isolated systems because of the fact that a single generating unit can comprise a major fraction of the total online generation capacity, and frequency excursions due to its tripping can lead to cascade tripping of other units if load-shedding schemes have not been implemented. The cost of electricity supply in island power systems is higher than in large systems due to the lack of scale economy in power generation and the high cost of fuel. Island power systems are based on smaller generating units than large systems: typically, diesel engines and gas turbines. Smaller units use more expensive fuel and exhibit lower efficiency than larger units.

Island power systems are not only challenged by technical and economic issues; they also represent a unique opportunity. Thanks to their isolation, island power systems have the potential to become places of demonstration, as test beds for solutions whose value is difficult to show in larger systems. They represent microcosms of larger power systems in which initiatives aimed at building "smart island power systems" can be tested quickly and effectively.

The book brings together in a single source much of the knowledge we have developed over 20 years while addressing the technical and economic problems of the Spanish island power systems (Balearics and Canary Islands) in a great number of research and consultancy projects developed for Endesa and Red Eléctrica de España. Endesa produces and distributes electric power in the Spanish island systems, while Red Eléctrica de España is their transmission system operator.

The book is organized in two parts. The first part, comprising Chapters 2 through 5, is devoted to the technical problems of island systems. We address both frequency and voltage stability problems. Frequency stability fundamentals are provided together with protections for frequency stability. Moreover, advanced control devices such as ultracapacitors and flywheels for frequency control are presented. The second part, comprising Chapters 6 and 7, is interested in the economic issues of island systems. Economic dispatch models incorporating specific constraints of island systems are presented first. Then, the application of economic dispatch models for assessing initiatives aimed at building smart island power systems is presented. The book starts with an introductory chapter that highlights the features of the island systems and the nature of the problems they face.

Lukas Sigrist, Enrique Lobato, Francisco Echavarren, Ignacio Egido, and Luis Rouco
Universidad Pontificia Comillas

Acknowledgments

We owe much gratitude to many people in Endesa: José Ramón Diago, Miguel González, José Antonio Torres, Jordi Magriñá, Alberto Barrado, Pablo Fontela, Eloisa Porras, Salvador Rubio, Alberto Morón, and José Luis Ruiz.

We are also indebted to Santiago Marín, Jesús Rupérez, Alfredo Rodríguez, Cristóbal Castro, and René Ascanio in Red Eléctrica de España.

Authors

Lukas Sigrist earned his MSc in electrical and electronics engineering from École Polytechnique Fédérale de Lausanne in 2007 and his PhD from Universidad Pontificia Comillas de Madrid in 2010. Dr. Sigrist is a research associate at the Instituto de Investigación Tecnológica of Universidad Pontificia Comillas. His areas of interest are power system stability and control.

Enrique Lobato earned his MS and PhD from Universidad Pontificia Comillas, Madrid, Spain, in 1998 and 2002, respectively. He is an associate professor in the School of Engineering of Universidad Pontificia Comillas. Dr. Lobato develops his research activities at the Instituto de Investigación Tecnológica of Universidad Pontificia Comillas. His areas of interest are power system planning and operation.

Franscisco Echavarren earned his MS and PhD degrees from Universidad Pontificia Comillas, Madrid, Spain, in 2001 and 2006, respectively. He is a research associate at the Instituto de Investigación Tecnológica of Universidad Pontificia Comillas. His areas of interest are power system modeling and analysis including voltage.

Ignacio Egido earned his MS and PhD degrees from Universidad Pontificia Comillas, Madrid, Spain, in 2000 and 2005, respectively. He is an assistant professor in the School of Engineering of Universidad Pontificia Comillas. He develops his research activities at the Instituto de Investigación Tecnológica of Universidad Pontificia Comillas. His areas of interest are power system stability and control.

Luis Rouco obtained his MS and PhD from Universidad Politécnica de Madrid, Spain, in 1985 and 1990, respectively. He is a professor in the School of Engineering of Universidad Pontificia Comillas in Madrid, Spain. He served as head of the Department of Electrical Engineering from 1999 to 2005. Professor Rouco develops his research activities at the Instituto de Investigación Tecnológica. His areas of interest are modeling, analysis, simulation, control, and identification of electric power systems. He has supervised a number of research projects for Spanish and European public administrations and Spanish and foreign companies. Dr. Rouco has published extensively in international journals and conferences. He is a member of the Institute of Electrical and Electronics Engineers, president of the Spanish chapter of the Institute of Electrical and Electronics Engineers

Power Engineering Society, and a member of the Executive Committee of the Spanish National Committee of the International Council on Large Electric Systems. He has been a visiting scientist at Ontario Hydro (Toronto, Canada), Massachusetts Institute of Technology (Cambridge, Massachusetts), and ABB Power Systems (Vasteras, Sweden).

1

Introduction

This chapter highlights the features of island systems and the nature of the issues they face. In addition, a review of recent initiatives aimed at overcoming the technical and economic issues of island power systems is provided.

1.1 Features of Island Power Systems

This book is interested in island systems that exhibit two specific features: they are small and isolated. Isolated systems are much smaller than interconnected systems. In addition, they cannot count on the support of neighboring systems.

The Continental European system and the Eastern and Western systems of the United States and Canada are examples of large interconnected systems. Large interconnected systems are the result of the interconnection of regional systems. For example, the Continental European system comes from the interconnection of the national European systems in Continental Europe.

There are many instances of large isolated power systems. Some of them are island systems, such as Japan, Great Britain, and Ireland. Others are continental systems, such as the Nordic and the Baltic countries in Europe; Texas in North America; Argentina, Chile, and Brazil in South America; and China and South Korea in Asia.

The separation between small and large island systems is a difficult problem. Hence, we propose to collect demand information for many systems and to identify island prototypes using clustering techniques. In addition to demand, generation information has been collected to investigate the capacity factors of the island systems. Grid configuration and economic and technical regulation frameworks will be reviewed as well.

1.1.1 Demand

We have collected information on 40 non-EU island systems and 19 EU island systems. Tables 1.1 and 1.2 provide the total annual demand (in gigawatt hours) of the non-EU island systems and the EU island systems, respectively. Non-EU island system demand data have been taken from the International Renewable Energy Agency (IRENA) report (IRENA 2012). Complementary

TABLE 1.1

Demand in Non-EU Island Systems

Island	Area	Demand (GWh)
Cuba	Caribbean	17,700
Dominican Republic	Caribbean	15,000
Trinidad and Tobago	Caribbean	7,700
Jamaica	Caribbean	5,500
Bahamas	Caribbean	2,139
Barbados	Caribbean	1,068
Guyana	Caribbean	723.1
Haiti	Caribbean	721
Saint Lucia	Caribbean	363
Grenada	Caribbean	203
Saint Kitts and Nevis	Caribbean	142
Saint Vincent and the Grenadines	Caribbean	140
Antigua and Barbuda	Caribbean	119
Dominica	Caribbean	92.7
Sao Tome and Principe	Central Africa	48.9
Belize	Central America	264.5
Mauritius	Eastern Africa	2689
Seychelles	Eastern Africa	275.7
Comoros	Eastern Africa	43
Fiji	Melanesia	793.5
Papua New Guinea	Melanesia	486
Salomon Islands	Melanesia	84.3
Vanuatu	Melanesia	64.7
Palau	Micronesia	84.9
Federated States of Micronesia	Micronesia	69
Marshall Islands	Micronesia	66
Nauru	Micronesia	21.2
Kiribati	Micronesia	19.4
Bahrain	Persian Gulf	12,100
Samoa	Polynesia	108
Tonga	Polynesia	51.6
Cook Islands	Polynesia	32.8
Tuvalu	Polynesia	6.4
Niue	Polynesia	3
Maldives	Sourthern Asia	296
Suriname	South America	1,618
Singapore	South-Eastern Asia	41,800
Timor-Leste	South-Eastern Asia	131.7
Cape Verde	Western Africa	296
Guinea-Bissau	Western Africa	14.3
Mean		2,826.97
Standard deviation		7,544.58
Maximum		41,800
Minimum		3

TABLE 1.2
Demand in EU Island Systems

Island	Area	Demand (GWh)	Peak Demand (MW)	Peak/ Average
Tenerife	Atlantic	3,625	593	1.43
Gran Canaria	Atlantic	3,653	598	1.43
Lanzarote–Fuerteventura	Atlantic	1,458.7	256.5	1.54
La Palma	Atlantic	254.8	48.4	1.66
La Gomera	Atlantic	66.7	12.1	1.59
El Hierro	Atlantic	35.7	7	1.72
Ibiza–Formentera	Mediterranean	851	215	2.21
Mallorca–Menorca	Mediterranean	4,913	1,062	1.89
Malta	Mediterranean	2,278	429	1.65
Cyprus	Mediterranean	5,305.8	862	1.42
Corsica	Mediterranean	2,130	463	1.9
Sardinia	Mediterranean	11,661	1,800	1.35
Guadeloupe	Caribbean	1,692	256	1.33
Martinique	Caribbean	1,576	241	1.34
La Reunion	Indian	2,750	442	1.41
Isle of Man	Atlantic	426	90	1.85
Guernsey, Channel Islands	Atlantic	381.2	78.3	1.8
Jersey, Channel Islands	Atlantic	651	154	2.07
Faroe	North Sea	273.8	48	1.54
Santa Maria	Atlantic	20	5	2.19
Sao Miguel	Atlantic	414	72	1.52
Terceira	Atlantic	195	39	1.75
Graciosa	Atlantic	13	2	1.35
Sao Jorge	Atlantic	28	5	1.56
Pico	Atlantic	43	9	1.83
Faial	Atlantic	47	9	1.68
Flores	Atlantic	11	2	1.59
Corvo	Atlantic	1	0.2	1.75
Madeira	Atlantic	880.28	145.48	1.45
Porto Santo	Atlantic	29.47	7.97	2.37
Mean		1,522.15	265.07	1.67
Standard deviation		2,439.75	398.82	0.28
Maximum		11,661	1,800	2.37
Minimum		1	0.2	1.33

information can be found on the website of the US Energy Information Agency; 2016. Table 1.2 shows the total demand (in gigawatt hours) and the peak demand (in megawatts) of 30 EU island systems. EU island system demand data have been taken from the Eurelectric report (Eurelectric). Complementary information can be found on the websites of a numbers of companies

(EDF en Corse, 2016; EDF en La Reunion, 2016; EDF en Martinique, 2016; EDF en Guadeloupe, 2016; Electricidade dos Azores, 2016; Electricidade do Madeira, 2016; Electricity Authority of Cyprus, 2016; Enemalta, 2016; Manx Electricity Authority, 2016; Jersey Electricity, 2016; Guernsey Electricity) and regulatory bodies (Comisión Nacional de Energía 2012). Table 1.2 also shows the ratio between peak demand and average demand (computed as Equation 1.1).

$$\text{Ratio\_peak\_average} = \frac{\text{Peak\_demand}(\text{MW})}{\frac{\text{Demand}(\text{GWh}) \cdot 1000}{8760}} \quad (1.1)$$

It is very interesting to note that the average ratio between peak demand and average demand is 1.64 with a standard deviation of 0.26.

It is also interesting to investigate the relationship between gross domestic product (GDP) per capita and demand per capita. Table 1.3 shows population GDP, GDP per capita, demand, and demand per capita in non-EU island systems. Figure 1.1 displays the GDP per capita versus demand per capita and a linear regression. Table 1.4 shows population GDP, GDP per capita, demand, and demand per capita in EU island systems. Figure 1.2 displays the GDP per capita versus demand per capita and a linear regression. Figures 1.1 and 1.2 confirm the correlation between GDP per capita and demand per capita.

1.1.2 Generation

Table 1.5 shows the generation and demand data of non-EU island systems. Table 1.6 shows the generation and data of EU island systems. The capacity factor defined according to Equation 1.2 is also provided.

$$\text{Capacity\_factor} = \frac{\frac{\text{Installed\_capacity}(\text{MW}) \cdot 8760}{1000}}{\text{Demand}(\text{GWh})} \quad (1.2)$$

1.1.3 Grid

Two key features of the grid of island systems are the highest voltage level and the existence of an interconnector with a large system. The highest voltage level is determined by the distances between generation and load centers and the amount of power transmitted. Interconnectors can be at either high-voltage alternating current (HVAC) or high-voltage direct current (HVDC) depending on the distance from the continent and the amount of power transmitted. Table 1.7 shows the grid features of the EU island systems. Among 30 EU island systems, there are three HVDC links and three HVAC links connecting the island systems to the continent.

TABLE 1.3

Population, GDP, and Demand in Non-EU Island Systems

Island	Area	Population	GDP (Billion €)	GDP per Capita (€)	Demand (GWh)	Demand per Capita (kWh)
Cuba	Caribbean	11,300,000	49.581	4,388	17,700	1,566
Dominican Republic	Caribbean	99,00,000	39.562	3,996	15,000	1,515
Trinidad and Tobago	Caribbean	13,00,000	15.614	12,011	7,700	5,923
Jamaica	Caribbean	27,00,000	10.661	3,949	5,500	2,037
Bahamas	Caribbean	3,43,000	5.98	17,434	2,139	6,236
Barbados	Caribbean	2,73,000	3.157	11,564	1,068	3,912
Guyana	Caribbean	7,54,000	1.737	2,304	723.1	959
Haiti	Caribbean	10,000,000	5.108	511	721	72
Saint Lucia	Caribbean	1,74,000	0.922	5,299	363	2,086
Grenada	Caribbean	1,04,000	0.6	5,769	203	1,952
Saint Kitts and Nevis	Caribbean	52,000	0.514	9,885	142	2,731
Saint Vincent and the Grenadines	Caribbean	1,09,000	0.518	4,752	140	1,284
Antigua and Barbuda	Caribbean	89,000	0.891	10,011	119	1,337
Dominica	Caribbean	68,000	0.364	5,353	92.7	1,363
Sao Tome and Principe	Central Africa	1,65,000	0.154	933	48.9	296
Belize	Central America	3,45,000	1.079	3,128	264.5	767
Mauritius	Eastern Africa	13,00,000	7.582	5,832	2,689	2,068
Seychelles	Eastern Africa	87,000	0.745	8,563	275.7	3,169
Comoros	Eastern Africa	7,35,000	0.416	566	43	59
Papua New Guinea	Melanesia	6,90,000	11.841	17,161	3,501	5,074
Fiji	Melanesia	32,00,000	9.076	2,836	793.5	248
Salomon Islands	Melanesia	5,38,000	0.522	970	84.3	157

(Continued)

TABLE 1.3 (CONTINUED)

Population, GDP, and Demand in Non-EU Island Systems

Island	Area	Population	GDP (Billion €)	GDP per Capita (€)	Demand (GWh)	Demand per Capita (kWh)
Vanuatu	Melanesia	2,40,000	0.531	2,213	64.7	270
Palau	Micronesia	20,000	0.129	6,450	84.9	4,245
Federated States of Micronesia	Micronesia	1,11,000	0.228	2,054	69	622
Marshall Islands	Micronesia	54,000	0.125	2,315	66	1,222
Nauru	Micronesia	10,000	0.041	4,100	21.2	2,120
Kiribati	Micronesia	1,00,000	0.117	1,170	19.4	194
Bahrain	Persian Gulf	13,00,000	18.184	13,988	12,100	9,308
Samoa	Polynesia	1,83,000	0.457	2,497	108	590
Tonga	Polynesia	1,04,000	0.275	2,644	51.6	496
Cook Islands	Polynesia	24,000	0.191	7,958	32.8	1,367
Tuvalu	Polynesia	10,000	0.024	2,400	6.4	640
Niue	Polynesia	1,600	0.006	3,750	3	1,875
Maldives	Southern Asia	3,16,000	1.597	5,054	296	937
Suriname	South America	5,25,000	3.349	6,379	1,618	3,082
Singapore	South-Eastern Asia	51,00,000	164.718	32,298	41,800	8,196
Timor-Leste	South-Eastern Asia	11,00,000	0.659	599	131.7	120
Cape Verde	Western Africa	4,96,000	1.276	2,573	296	597
Guinea-Bissau	Western Africa	15,00,000	0.636	424	14.3	10
Mean		13,85,515	8.98	5,902	2,902	2,018
Standard deviation		27,92,032	27.27	6,119	7,536	2,227
Maximum		11,300,000	164.718	32,298	41,800	9,308
Minimum		1,600	0.006	424	3	10

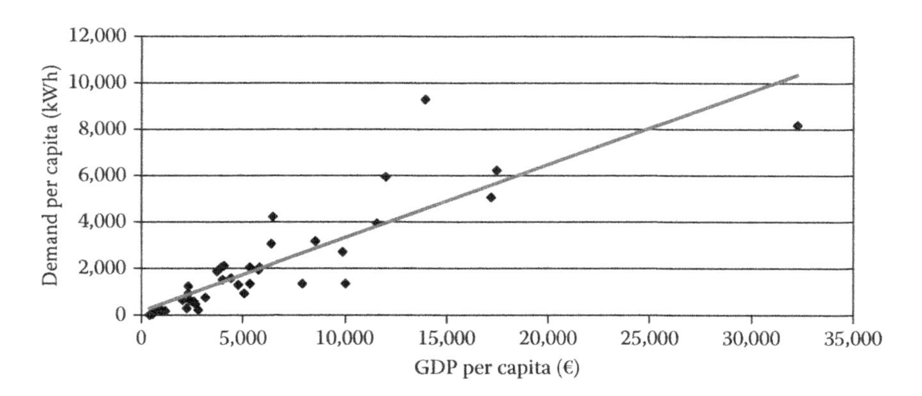

FIGURE 1.1
GDP per capita and demand per capita in non-EU island systems.

1.1.4 Island Prototypes

Island prototypes or island representative instances can be identified using clustering techniques. Cluster analysis forms part of data mining. Data mining techniques emerged in the early 1990s, and they allow relevant information to be processed and extracted from large databases. Cluster analysis refers to the partitioning of a data set into clusters, so that the data in each subset ideally share some common trait and differ from the data in other subsets. Different methods (Abonyi and Feil 2007, Michie et al. 1994, Rojas 1996), based on advanced statistical and modeling techniques such as K-Means, Fuzzy C-Means, and Kohonen Self Organizing Map (KSOM), are commonly employed.

The island clusters will be investigated considering as variables the demand (in gigawatt hours) and the installed capacity (in megawatts). It is quite reasonable to state that installed capacity and demand should be correlated in some way.

Before attempting to identify the island clusters, it may be helpful to plot each island in the demand–installed capacity plane. Figure 1.3 displays the islands, assuming a logarithmic scale for both demand and installed capacity. Demand and installed capacity are well correlated on the logarithmic scale.

The K-Means method is used to determine the island clusters. A key step of the application of the method is to determine a suitable number of clusters. For this purpose, the objective function of the K-Means method is represented as a function of the number of clusters. The knee of the objective function would provide the most suitable number of clusters. Figure 1.4 shows that the most suitable number of clusters is five.

Figure 1.5 is the same as Figure 1.3 but includes the cluster representatives. Table 1.8 shows the demand and the installed capacity of the representatives of island clusters.

TABLE 1.4

Population, GDP, and Demand in EU Island Systems

Island	Area	Population	GDP	GDP per Capita	Demand (GWh)	Demand per Capita (kWh)
Canary Islands	Atlantic	21,27,000	40.34	18,966	9,093.9	4,275
Baleares	Mediterranean	11,13,114	26.041	23,395	5,764	5,178
Malta	Mediterranean	4,17,608	4.7	11,255	2,278	5,455
Cyprus	Mediterranean	8,04,435	17.8	22,127	5,305.8	6,596
Corsica	Mediterranean	3,11,000	7	22,508	2,130	6,849
Sardinia	Mediterranean	16,37,193	33.823	20,659	11,661	7,123
Guadeloupe	Caribbean	4,04,000	7.75	19,183	1,692	4,188
Martinique	Caribbean	4,03,000	7.9	19,603	1,576	3,911
La Reunion	Indian	8,24,000	14.7	17,840	2,750	3,337
Isle of Man	Atlantic	80,000	4.14	51,750	426	5,325
Guernsey, Channel Islands	Atlantic	62,451	1.93	30,904	381.2	6,104
Jersey, Channel Islands	Atlantic	91,000	4.26	46,813	651	7,154
Faroe	North Sea	48,372	1.7	35,144	273.8	5,660
Azores	Atlantic	2,46,000	3.7	15,041	772	3,138
Madeira	Atlantic	2,67,785	5.224	19,508	909.75	3,397
Mean		5,89,131	12.07	24,980	3,044	5,179
Standard deviation		6,16,060	12.17	11,445	3,443	1,409
Maximum		21,27,000	40.34	51,750	11,661	7,154
Minimum		48,372	1.7	11,255	273.8	3,138

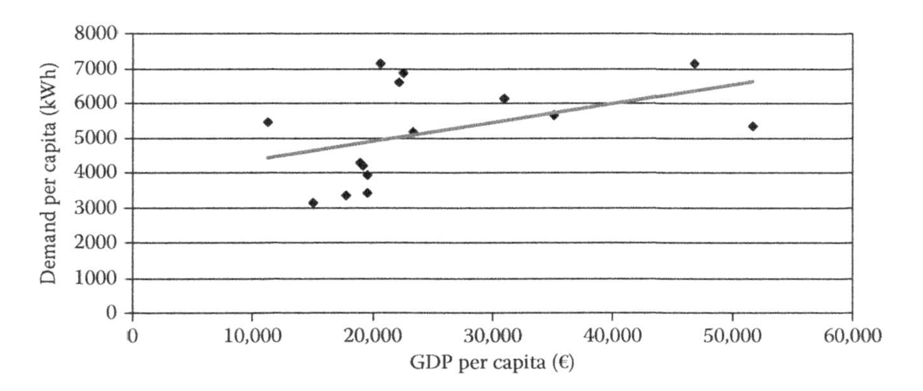

FIGURE 1.2
GDP per capita and demand per capita in EU island systems.

1.2 Issues of Island Power Systems

The size and the lack of external support make isolated systems more vulnerable than interconnected systems in the case of disturbances due to stability issues. Such features also affect the cost of electricity supply. This section introduces the key technical and economic features of island power systems.

1.2.1 Stability

Power system stability is concerned with its ability either to remain in a stable equilibrium point or to arrive at a new equilibrium point after a disturbance. The power system stability problem is a very difficult one. Its study is facilitated by separating it into three subproblems: angle, frequency, and voltage stability problems (Kundur 1994).

Rotor angle stability is concerned with the ability of generators to run in synchronism. Running in synchronism means that generator rotors rotate at exactly the same electrical speed (it is proportional to frequency), and therefore, the difference of the generator rotor angles remains constant. Large-disturbance (transient) rotor angle stability investigates the ability of synchronous generators to remain in synchronism after large disturbances such as faults. The key question is whether generators can withstand secured cleared faults: in other words, whether the critical clearing times of the faults are higher than the protection and circuit breaker times. Small-disturbance (small-signal) rotor angle stability is concerned with the damping of the rotor oscillations. Poorly damped or even undamped rotor oscillations can arise after a line switching or a change of generation pattern. Rotor angle stability problems arise mainly in long transmission systems under stressed conditions. Hence, rotor angle stability problems are not frequent in island power systems, since distances are not large.

TABLE 1.5

Generation and Demand in Non-EU Island Systems

Island	Area	Demand (GWh)	Generation Capacity (MW)	Capacity Factor
Cuba	Caribbean	17,700	5,500	2.72
Dominican Republic	Caribbean	15,000	2,973	1.74
Trinidad and Tobago	Caribbean	7,700	1,429	1.63
Jamaica	Caribbean	5,500	1,198	1.91
Bahamas	Caribbean	2,139	493	2.02
Barbados	Caribbean	1,068	239	1.96
Guyana	Caribbean	723.1	343	4.16
Haiti	Caribbean	721	240	2.92
Saint Lucia	Caribbean	363	76	1.83
Grenada	Caribbean	203	33.2	1.43
Saint Kitts and Nevis	Caribbean	142	22	1.36
Saint Vincent and the Grenadines	Caribbean	140	41	2.57
Antigua and Barbuda	Caribbean	119	27	1.99
Dominica	Caribbean	92.7	24.3	2.3
Sao Tome and Principe	Central Africa	48.9	14	2.51
Belize	Central America	264.5	102	3.38
Mauritius	Eastern Africa	2,689	740	2.41
Seychelles	Eastern Africa	275.7	95	3.02
Comoros	Eastern Africa	43	6	1.22
Fiji	Melanesia	793.5	215	2.37
Papua New Guinea	Melanesia	486	722	13.01
Salomon Islands	Melanesia	84.3	36	3.74
Vanuatu	Melanesia	64.7	30.7	4.16
Palau	Micronesia	84.9	39.3	4.05
Federated States of Micronesia	Micronesia	69	28	3.55
Marshall Islands	Micronesia	66	17	2.26
Nauru	Micronesia	21.2	4.85	2
Kiribati	Micronesia	19.4	5.8	2.62
Bahrain	Persian Gulf	12,100	3,168	2.29
Samoa	Polynesia	108	41.5	3.37
Tonga	Polynesia	51.6	12.075	2.05
Cook Islands	Polynesia	32.8	8.08	2.16
Tuvalu	Polynesia	6.4	3.9	5.34
Niue	Polynesia	3	2.45	7.15
Maldives	Sourthern Asia	296	62	1.83
Suriname	South America	1,618	389	2.11
Singapore	South-Eastern Asia	41,800	10,700	2.24
Timor-Leste	South-Eastern Asia	131.7	44	2.93
Cape Verde	Western Africa	296	75	2.22

(Continued)

TABLE 1.5 (CONTINUED)

Generation and Demand in Non-EU Island Systems

Island	Area	Demand (GWh)	Generation Capacity (MW)	Capacity Factor
Guinea-Bissau	Western Africa	14.3	15	9.19
Mean		2,826.97	730.38	3.09
Standard deviation		7,544.58	1,945.04	2.22
Maximum		41,800	10,700	13.01
Minimum		3	2.45	1.22

Frequency stability is concerned with the capacity of generators to supply the loads at acceptable frequency ranges in the case of generator tripping. Frequency results from the generator rotor speeds. Generator rotor speeds result from the equilibrium between the power supplied by their prime movers (either turbines or engines) and the power consumed by the loads. Frequency stability is governed by the inertia of the rotating masses of prime movers and generators and the gain and time constant of the primary frequency regulation of prime movers in such a way that

- After a generator trips, frequency decays with a rate of change that depends on the inertia of prime mover–generator rotating masses and the magnitude of the generation lost.
- Prime mover primary frequency regulation reacts to the frequency decay increasing the output of the power supplied by the prime movers.
- Frequency stabilizes if two conditions are fulfilled: the remaining online generators have enough reserve to supply the generation lost, and they are also able to increase the power output fast enough to avoid frequency falling below the settings of generator underfrequency protection, to avoid generator cascade tripping.

Frequency stability is at risk in isolated power systems because of the fact that the frequency rate of change in case of generator tripping is higher than in an interconnected power system. The inertia or the kinetic energy of the rotating masses of an interconnected system is much higher than the inertia of the rotating masses of an isolated system. In addition, the magnitude of the generation that can be tripped compared with the total rotating generation is much larger in an isolated system than in an interconnected one. Figure 1.6a compares frequency excursions arising in interconnected systems and island systems in the case of normal disturbances. Figure 1.6b compares frequency excursions arising in the case of normal disturbance and severe disturbances in an island system.

Frequency stability can only be preserved by appropriate load-shedding schemes that disconnect fractions of the load to prevent the system from

TABLE 1.6

Generation and Demand in EU Island Systems

Island	Area	Demand (GWh)	Generation Capacity (MW)	Capacity Factor	Peak Demand (MW)	Peak/Average
Tenerife	Atlantic	3,625	1,084.28	2.62	593	1.43
Gran Canaria	Atlantic	3,653	1,111.8	2.67	598	1.43
Lanzarote–Fuerteventura	Atlantic	1,458.7	377.97	2.27	256.5	1.54
La Palma	Atlantic	254.8	105.52	3.63	48.4	1.66
La Gomera	Atlantic	66.7	20.1	2.64	12.1	1.59
El Hierro	Atlantic	35.7	11.31	2.78	7	1.72
Ibiza–Formentera	Mediterranean	851	305.1	3.14	215	2.21
Mallorca–Menorca	Mediterranean	4,913	1,943.95	3.47	1,062	1.89
Malta	Mediterranean	2,278	700	2.69	429	1.65
Cyprus	Mediterranean	5,305.8	1,733.7	2.86	862	1.42
Corsica	Mediterranean	2,130	524	2.16	463	1.9
Sardinia	Mediterranean	11,661	2767	2.08	1,800	1.35
Guadeloupe	Caribbean	1,692	472.3	2.45	256	1.33
Martinique	Caribbean	1,576	402.3	2.24	241	1.34
La Reunion	Indian	2,750	629.9	2.01	442	1.41
Isle of Man	Atlantic	426	244.6	5.03	90	1.85

Guernsey, Channel Islands	Atlantic	381.2	125.35	2.88	78.3	1.8
Jersey, Channel Islands	Atlantic	651	290	3.9	154	2.07
Faroe	North Sea	273.8	83.63	2.68	48	1.54
Santa Maria	Atlantic	20	5.7	2.5	5	2.19
Sao Miguel	Atlantic	414	98	2.07	72	1.52
Terceira	Atlantic	195	35.8	1.61	39	1.75
Graciosa	Atlantic	13	3.45	2.32	2	1.35
Sao Jorge	Atlantic	28	7.03	2.2	5	1.56
Pico	Atlantic	43	13.3	2.71	9	1.83
Faial	Atlantic	47	15.7	2.93	9	1.68
Flores	Atlantic	11	2.902	2.31	2	1.59
Corvo	Atlantic	1	0.64	5.61	0.2	1.75
Madeira	Atlantic	880.28	354.5	3.53	145.48	1.45
Porto Santo	Atlantic	29.47	20.73	6.16	7.97	2.37
Mean		1,522.15	449.69	2.94	265.07	1.67
Standard deviation		2,439.75	666.55	1.05	398.82	0.28
Maximum		11,661	2767	6.16	1800	2.37
Minimum		1	0.64	1.61	0.2	1.33

TABLE 1.7
High Voltage Level of the Grid of EU Island Systems

Island	Area	Highest Voltage Level (kV)	Interconnector
Tenerife	Atlantic	220	
Gran Canaria	Atlantic	220	
Lanzarote-Fuerteventura	Atlantic	66	
La Palma	Atlantic	66	
La Gomera	Atlantic	20	
El Hierro	Atlantic	20	
Ibiza–Formentera	Mediterranean	132	
Mallorca–Menorca	Mediterranean	220	HVDC 250 kV, 400 MW, Mainland Spain
Malta	Mediterranean	132	
Cyprus	Mediterranean	132	
Corsica	Mediterranean	150	HVDC 200 kV, 100 MW, Corsica-Sardinia-Mainland Italy HVDC 200 kV, 200 MW, Corsica-Sardinia-Mainland Italy
Sardinia	Mediterranean	380	HVDC 500 kV, 1000 MW, Sardinia-Mainland Italy
Guadeloupe	Caribbean	63	
Martinique	Caribbean	63	
La Reunion	Indian	63	
Isle of Man	Atlantic	90	HVAC 90 kV, 104 km, 65 MW Isle of Man-England
Guernsey, Channel Islands	Atlantic	90	HVAC 90 kV, 37 km, 50 MVA, Guernesey-Jersey-France
Jersey, Channel Islands	Atlantic	90	HVAC 90 kV, 27 km, 85 MVA, Guernesey-Jersey-France
Faroe	North Sea	60	
Santa Maria	Atlantic	10	
Sao Miguel	Atlantic	60	
Terceira	Atlantic	30	
Graciosa	Atlantic	15	
Sao Jorge	Atlantic	15	
Pico	Atlantic	30	
Faial	Atlantic	15	
Flores	Atlantic	15	
Corvo	Atlantic	0.4	
Madeira	Atlantic	60	
Porto Santo	Atlantic	30	

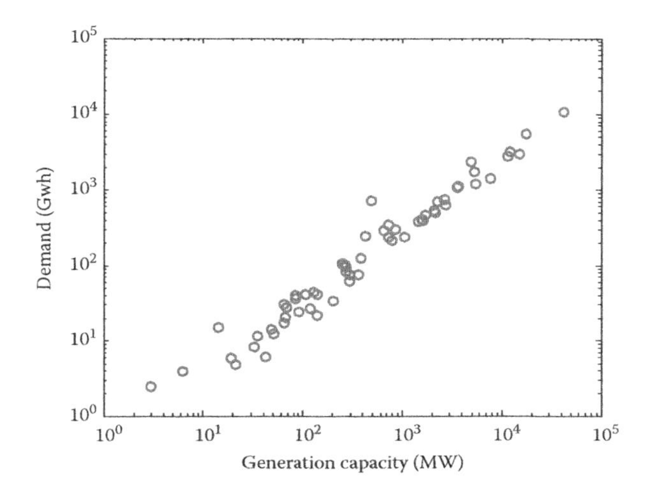

FIGURE 1.3
Generation capacity (MW) versus demand (MW).

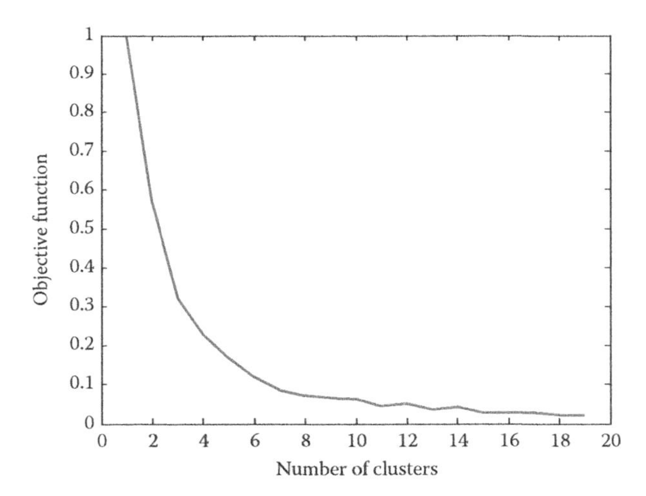

FIGURE 1.4
Objective function of the K-Means clustering algorithm in terms of the number of clusters.

collapsing either in the case of lack of reserve or in the case of extreme disturbances. Load-shedding schemes are controlled by underfrequency and rate of change of frequency protections.

Voltage stability is concerned with the capacity to supply the loads at acceptable voltage ranges. Voltage stability problems are of an electromagnetic nature, in contrast to angle and frequency stability problems, which are mainly of an electromechanical nature. Voltage stability problems arise from the limited capability of an alternating current (ac) circuit to deliver active

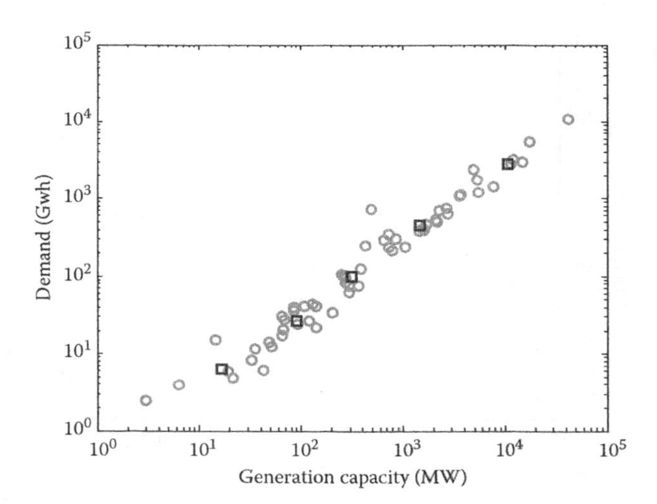

FIGURE 1.5
Generation capacity (MW) versus demand (MW) including the cluster representative.

TABLE 1.8
Cluster Representatives

Demand (GWh)	Generation Capacity (MW)
10627	2802.7
1445.7	457.4
309.8	99.2
90.5	26.7
16.6	6.2

power to a load. They are also affected by the availability of reactive power resources. Voltage stability problems arise in island systems when remote loads are fed through high-impedance paths and with lack of reactive power resources at the load side.

1.2.2 Economic Operation

The cost of electricity supply in island power systems is higher than in large systems due to the lack of scale economy in power generation and the high cost of fuel. Island power systems are based on smaller generating units than large systems: typically diesel engines and gas turbines. Coal-fired stations can be found in very large islands. Combined-cycle power plants are becoming usual in large islands. Smaller units make use of costlier fuel and exhibit lower efficiency than larger units.

Figure 1.7 shows the average marginal cost as a function of generating unit rated output in island systems. The units under consideration are in the

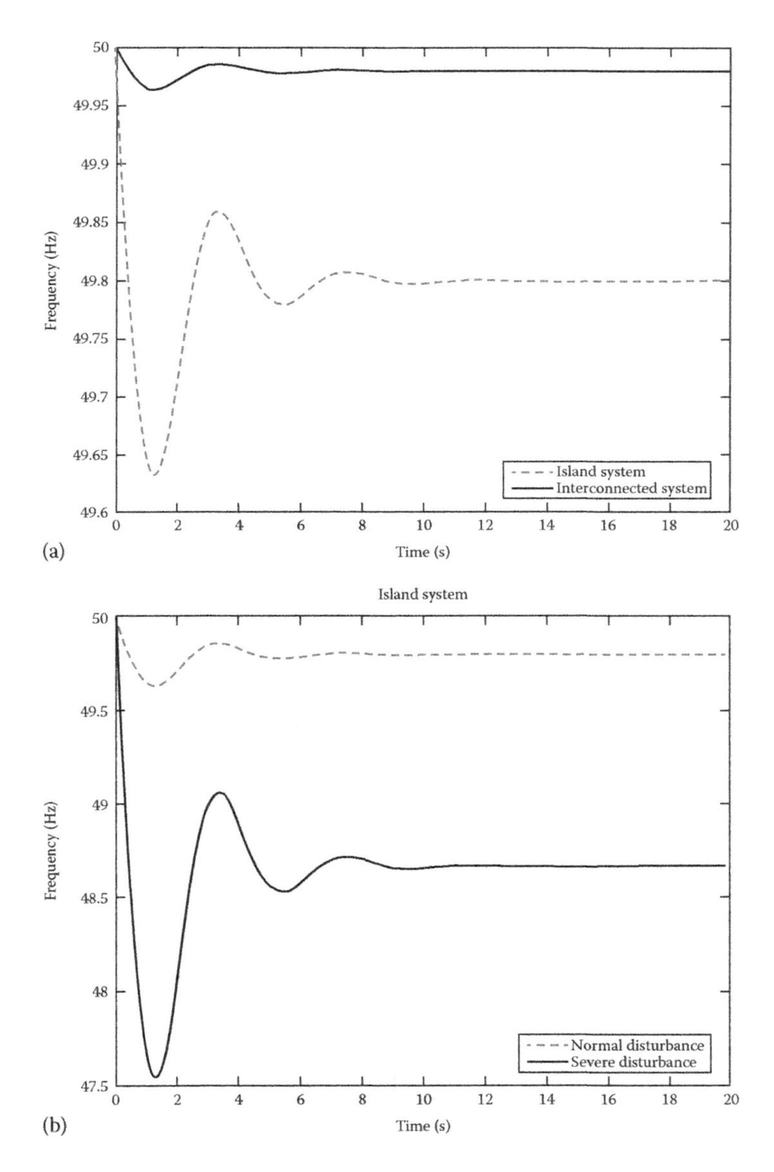

FIGURE 1.6
Comparison of frequency excursions. Interconnected system versus island system (a) and normal disturbance versus severe disturbance in an island system (b).

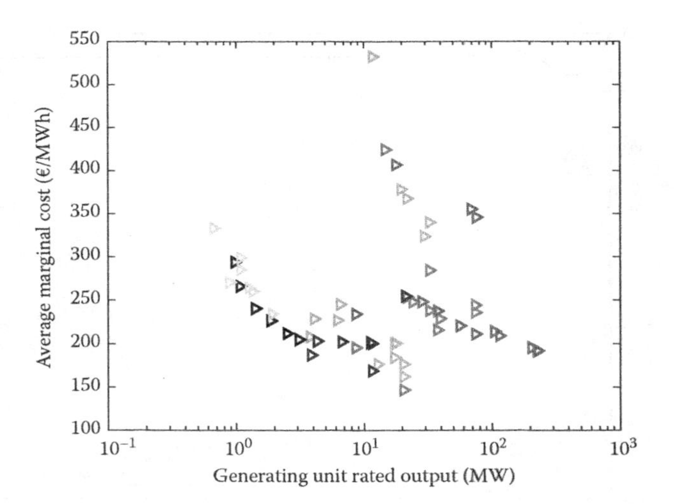

FIGURE 1.7
Average marginal cost as a function of generating unit rated output.

range between 1 and 200 MW. The average marginal costs are between 150 and 500 €/MWh. Marginal costs in island systems could be compared with marginal costs in very large interconnected systems, such as the Continental European one, which frequently exhibit a marginal cost around 50 €/MWh.

1.3 Regulation

Regulation is needed in economic sectors where perfect competition does not work or where competition has been legally excluded by a political decision. Regulation must balance obligations to both customers and regulated companies and also the costs and benefits of the regulatory system itself. The electric utility industry has been regulated almost from its beginning due to the importance of the electricity supply for a nation's economy and social welfare together with the monopolistic nature of some of its activities. In the electricity sector, there is generally good potential for competition in the generation and retailing businesses, but transmission and distribution activities remain regulated. A comprehensive review on regulation of the power sector can be found in Pérez-Arriaga (2013).

This section highlights some key aspects of the economic and technical regulation of island power systems.

1.3.1 Economic Regulation

Large power systems are usually market-driven systems with different degrees of competition. Small isolated power systems are operated either under a classical centralized scheme or under a market-driven scheme. However, most of them operate under a classical centralized scheme (Pertrov 2013).

In a classical centralized scheme, generating units are programmed according to economic dispatch rules. Generation is remunerated according to the so-called standard costs of different generation technologies. Generation expansion is also planned in a centralized way. Competitive bids can be organized to select new generation capacity.

In a market-driven scheme, generating units can be dispatched by either the clearing of a spot market or long-term contracts. In the case of a spot market scheme, all generating units are remunerated by market clearing price. Generation expansion is decided by market agents. Security of supply payments is usually incorporated into generation remuneration to encourage generation investment.

As an example, the principles of the economic operation of the Spanish isolated power systems are (Ministerio de Economía 2003; Ministerio de Industria, Turismo y Comercio 2006a)

- The remuneration of the generation businesses is done according to the remuneration of the regulated activities in the Spanish mainland power system.
- The basis is the rate of return of transmission system assets in the Spanish mainland power system.

The generation remuneration comprises two components:

- Energy component (€)
- Security of supply component (€)

$$CG_{g,h} = \underbrace{E_{g,h} \cdot \left[PMP + PRF_{g,h} \right]}_{\text{Energy component}} + \underbrace{GPOT_{g,h} \cdot PDIS_{g,h}}_{\text{Security of supply component}}$$

where:

$CG_{g,h}$ is the generation cost of generator g in hour h (€)
$E_{g,h}$ is the generation of generator g in hour h (MWh)
PMP is the average price in the Spanish mainland power system (€/MWh)
$PRF_{g,h}$ is the premium of generator g in hour h to cover the fuel costs (€/MWh)
$GPOT_{g,h}$ is the payment due to security of supply of generator g in hour h (€/MWh)
$PDIS_{g,h}$ is the available generation of generator g in hour h (MWh)

The price of power purchase (€/MWh) is computed as

$$PFG_h = \frac{\sum_g CG_{g,h}}{\sum_g E_{g,h}}$$

The generation expansion is done according to the following criteria:

- The installed capacity should allow a loss of load probability of 1 day in 10 years.
- New capacity will be assigned through a competitive bid.
- The clearance of the bid will be based on the required generation technologies and the offers of security of supply payments.

1.3.2 Technical Regulation

Frequency stability is the most relevant issue in the operation of isolated power systems (Centeno et al. 2004). Moreover, the increasing penetration of non-synchronous generation is challenging frequency stability of isolated power systems (Egido et al. 2007). Frequency stability is guaranteed by assigning enough reserves and ensuring a fast primary frequency regulation system and a proper tuned frequency load-shedding scheme (Egido et al. 2009 and Sigrist 2012).

Technical regulation of island power systems sets

- The amount of primary reserves
- The speed of response of the primary frequency regulation
- The principles of the design of the underfrequency and rate-of-change-of-frequency load-shedding schemes

As an example, the principles of the technical regulation of the Spanish isolated power systems are (Ministerio de Industria, Turismo y Comercio 2006b)

- The amount of primary reserve must be at least 50% of the actual output of the unit with greatest scheduled generation.
- The primary frequency reserves must be effectively provided in less than 30 s.
- The load-shedding schemes must prevent frequency from falling below 47.5 Hz for more than 3 s.

1.4 Initiatives toward Smarter Island Power Systems

The high cost of electricity supply in island power systems may make initiatives economically feasible that are not feasible in large systems. A review of recent initiatives would help to understand their value. The Euroelectric report (Euroelectric 2012) is a mandatory starting point for any review on initiatives to increase island power system sustainability. Euroelectric suggests looking at the following solutions:

- Oil to gas
- Flexible generation
- Interconnectors
- Renewable energy sources (RES)
- Storage
- Demand-side management (DSM)

Hence, the review has been oriented to these six categories. In addition, our review has considered a seventh category, which is a combination of the three last categories considered.

The analysis of each initiative informs about

- Location
- Area (oil to gas, flexible generation, interconnectors, RES, storage, DSM)
- Status (feasibility study, project, in operation, results reported)
- Technology
- Objectives
- Actors (generation, transmission and distribution, others)
- Funding
- Budget

Twenty-nine initiatives have identified and analyzed. Table 1.9 summarizes the initiatives reported according to the area (oil to gas, flexible generation, interconnectors, RES, storage, DSM). The components of hybrid initiatives have been highlighted. We have found an initiative on oil to gas that has taken place in Spain (Cantalellas 2012 and Pescador 2013) and one on flexible generation that was implemented in Malta (Grima 2013). We have found two initiatives on interconnectors: one in Spain (Granadino 2012; de la Torre 2013b) and one in Malta (Enemalta 2016b). Six initiatives on RES are worthy of being reported: one has taken place in Denmark (Nielsen 2013), three in Australia (Hydro Tasmania 2016), one in Portugal (European Investment Bank 2016), and one in Spain (Gobierno de Canarias 2006, 2011). A review of applications of energy storage systems can be found in IRENA 2012b. However, the applications reported are on a very small scale. We are aware of six initiatives on storage worthy of being reported. Three have taken place in the Canary Islands (Arrojo 2010; Margina 2011; Rodriguez 2011; Egido 2016), two in the French Islands (Pons 2012; Barlier 2012; Buriez et al. 2012), and one in Germany (Schütt 2012). We are aware of three initiatives on DSM worthy of being reported. Two have taken place in Denmark (Faroe [Nielsen and Windolf 2010] and Bornholm [Kumagai 2013], [Ecogrid Project 2016]) and one in the French Islands (Pons 2012; Barlier 2012; Buriez et al. 2012). Many of the initiatives (nine) are hybrid ones. Two have taken place in the Canary Islands,

TABLE 1.9

Summary of Initiatives: Area

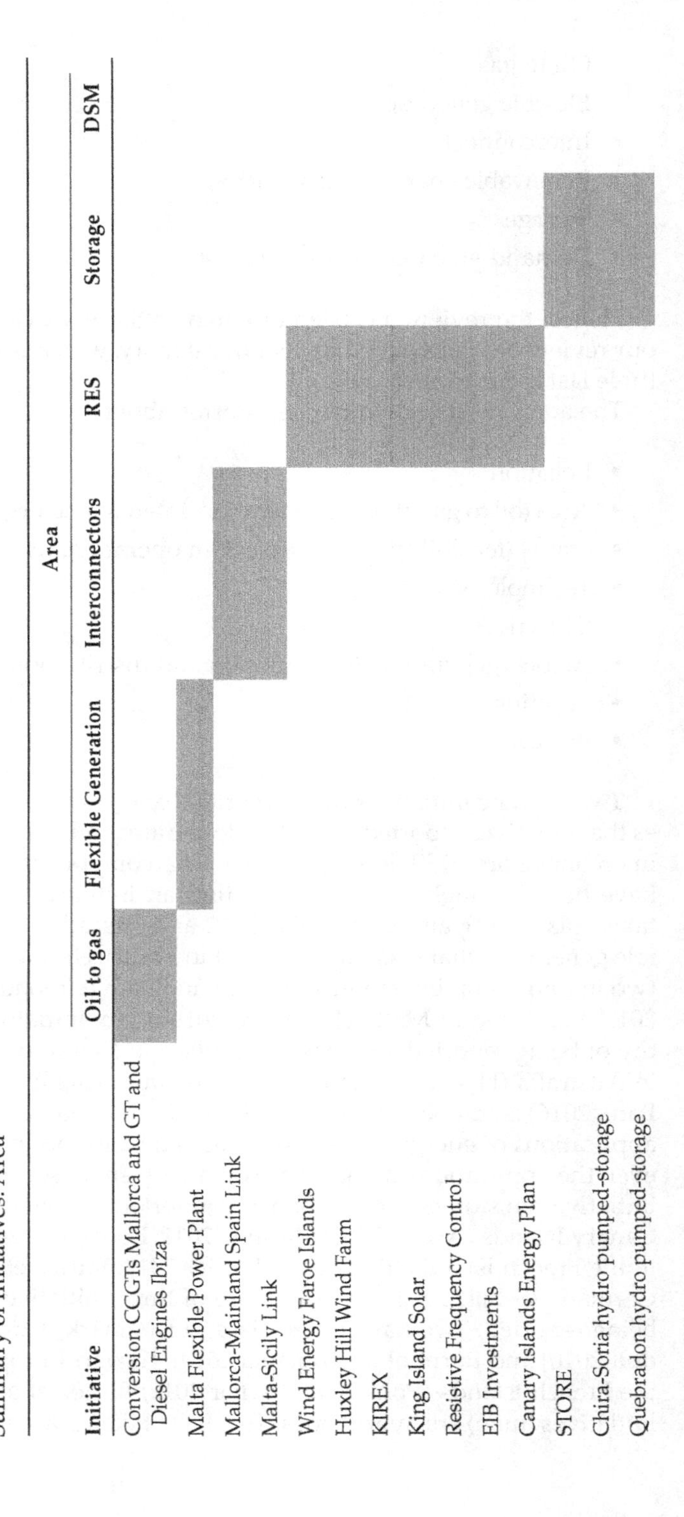

Initiative	Area					
	Oil to gas	**Flexible Generation**	**Interconnectors**	**RES**	**Storage**	**DSM**
Conversion CCGTs Mallorca and GT and Diesel Engines Ibiza	■					
Malta Flexible Power Plant		■				
Mallorca-Mainland Spain Link			■			
Malta-Sicily Link			■			
Wind Energy Faroe Islands				■		
Huxley Hill Wind Farm				■		
KIREX				■		
King Island Solar				■		
Resistive Frequency Control				■		
EIB Investments				■		
Canary Islands Energy Plan				■		
STORE					■	
Chira-Soria hydro pumped-storage					■	
Quebradon hydro pumped-storage					■	

1 MW NaS Battery La Reunion
SEPMERI
Smart Region Pellworm
GRANI
Bornholm Test Site Ecogrid
MILLENER
El Hierro 100% Renovable
STORIES
JeJu Consortium
HECO Wind Integration Project
KIREIP
Flywheels
Younicos Project
Eco Island
DAFNI network

Single-area initiative
Multi-area initiative

TABLE 1.10
Summary of Initiatives: Status

Initiative	Status			
	Feasibility Study	Project	Operation	Results Reported
Conversion CCGTs Mallorca and GT and Diesel Engines Ibiza			■	
Malta Flexible Power Plant			■	
Mallorca-Mainland Spain Link			■	
Malta-Sicily Link		■		
Wind Energy Faroe Islands			■	
Huxley Hill Wind Farm				■
KIREX				■
King Island Solar				■
Resistive Frequency Control				■
EIB Investments		■		
Canary Islands Energy Plan		■		
STORE		■		
Chira-Soria hydro pumped-storage		■		
Quebradon hydro pumped-storage		■		
1 MW NaS Battery La Reunion			■	
SEPMERI	■			
Smart Region Pellworm		■		
GRANI			■	
Bornholm Test Site Ecogrid		■		
MILLENER			■	
El Hierro 100% Renovable			■	
STORIES	■			
JeJu Consortium		■		
HECO Wind Integration Project				■
KIREIP		■		
Flywheels				■
Younicos Project			■	
Eco Island		■		
DAFNI network		■		

(Quintero 2013) two in the Azores (Botelho 2013; Da Costa 2012), and one each in Korea (Geun 2013), Hawaii (Kaneshiro 2013), Australia (Hydro Tasmania), the Isle of Wright (Eco-Island project 2016), and Greece (Dafni Network 2016).

Table 1.10 provides a map of the initiative status (feasibility study, project, operation, and results reported). Table 1.11 also shows a map of the initiative actors (generation, T&D, and others).

Figures 1.8 through 1.10 show in a polygonal graph the dominant initiatives according to the area, status, and actors, respectively.

TABLE 1.11
Summary of Initiatives: Actors

Initiative	Actors		
	Generation	T&D	Other
Conversion CCGTs Mallorca and GT and Diesel Engines Ibiza	■		
Malta Flexible Power Plant	■		
Mallorca-Mainland Spain Link		■	
Malta-Sicily Link		■	
Wind Energy Faroe Islands	■		
Huxley Hill Wind Farm	■		
KIREX	■		
King Island Solar	■		■
Resistive Frequency Control	■		
EIB Investments			■
Canary Islands Energy Plan			■
STORE	■		
Chira-Soria hydro pumped-storage	■		
Quebradon hydro pumped-storage	■		
1 MW NaS Battery La Reunion	■		
SEPMERI	■	■	
Smat Region Pellworm	■		■
GRANI			■
Bornholm Test Site Ecogrid			■
MILLENER	■	■	
El Hierro 100% Renovable	■		■
STORIES			■
JeJu Consortium	■		■
HECO Wind Integration Project	■		
KIREIP	■		
Flywheels	■		
Younicos Project	■		■
Eco Island			■
DAFNI network	■	■	■

The review of 29 initiatives provides the following conclusions:

- Most initiatives are on storage and RES.
- Storage initiatives are driven by generator-side actors.
- Storage devices are used to comply with security constraints.
- RES initiatives are driven by generator and T&D-side actors.
- RES generation is used to reduce operational costs.
- Security constraints prevent RES integration.
- Fewer initiatives deal with DSM.
- Most initiatives are in project (storage) or results reported (RES) status.

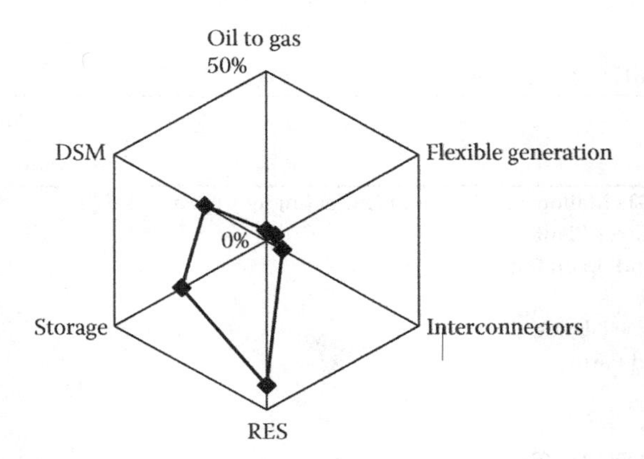

FIGURE 1.8
Summary of initiatives: Area.

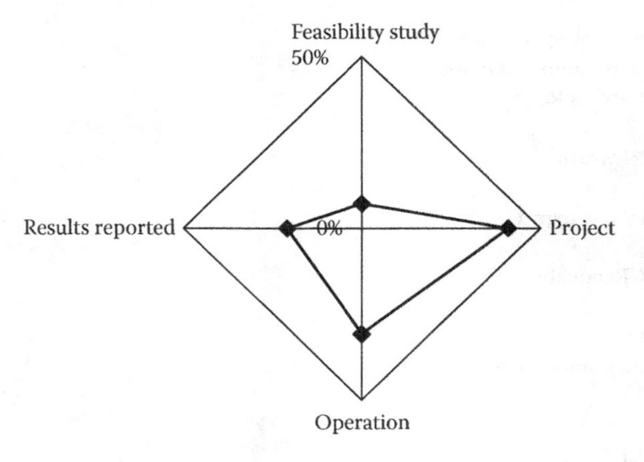

FIGURE 1.9
Summary of initiatives: Status.

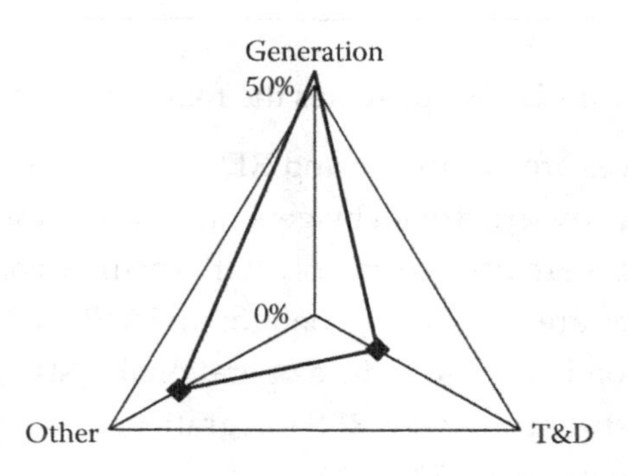

FIGURE 1.10
Summary of initiatives: Actors.

References

Abonyi, J., and B. Feil. 2007. *Cluster Analysis for Data Mining and System Identification*. Basel: Birkhäuser.

Arrojo, J. 2010. Storage as a strategic tool to manage RES intermittency, adding value to the grid, DSM and change mobility. In *3rd NESSIS Euroelectric Conference toward a Sustainable Energy Future for Island Systems*, Las Palmas de Gran Canaria, available from www.catedraendesared.ulpgc.es/index.php/descargas/doc_details/ 31-session-iii-jose-arrojo. accessed September 2016.

Barlier, Y. 2012. How to manage intermittency on islands? Pritemps de la Recherche, Electricité de France, Clamart, France.

Botelho, F. 2013. The Azores and the renewables. GENERA 2013, Session on Island Power Systems, Madrid.

Buriez, N., J. Mairem, Y. Barlier, and S. Lascaud. 2012. Will small-scale PV contribute to French insular grid operation? In *Integration of Renewables into the Distribution Grid, CIRED 2012 Workshop*. Lisbon: IET, Paper 261.

Cantarellas, A. 2012. El gas natural en las centrales térmicas de Baleares. Primeras Jornadas para un Sistema Eléctrico Sostenible en Les Illes Balears: de Sistema Aislado a Sistema Interconectado, Palma de Mallorca.

Centeno, P., F. Fernández, L. Rouco, M. González, J. M. Rojo, and J. R. Diago. 2004. Estabilidad de los sistemas eléctricos insulares. *Anales de Mecánica y Electricidad* LXXXI (Fascículo IV):33 –40.

Comisión Nacional de Energía. 2012. Informe marco sobre la demanda de energía elléctrica y gas de natural, y su cobertura, available from www.cne.es/cne/doc/publicaciones/PA006_12.pdf.

Da Costa, P. J. 2012. Case studies for sustainable energy on islands: EDA experience. Euroelectric, Workshop on EU Islands: Towards a Sustainable Energy Future, Brussels, available from www.eurelectric.org/media/64767/da%20Costa.pdf.

Dafni Network, available from www.dafni.net.gr/en/home.htm.

de la Torre, J. M. 2013b. HVDC-250 kV Morvedre-Santa Ponsa Spanish mainland-Balearic Islands electric interconnection. GENERA 2013, Session on Island Power Systems, Madrid.

Ecogrid Project, available from www.eu-ecogrid.net/. Accessed September 2016.

Eco-Island, available from www.eco-island.org/. Accessed September 2016.

EDF. Développer Les Énergies Renouvelables, available from http://reunion.edf.com/edf-a-la-reunion/nos-energies/developper-les-energies-renouvelables-49334.html. Accessed September 2016.

EDF en Corse, available from http://corse.edf.com/edf-en-corse-47855.html. Accessed September 2016.

EDF en Guadeloupe, available from http://guadeloupe.edf.com/. Accessed September 2016.

EDF en La Reunion, available from http://reunion.edf.com/. Accessed September 2016.

EDF en Martinique, available from http://martinique.edf.com/. Accessed September 2016.

Egido, I., F. Fernández-Bernal, P. Centeno, and L. Rouco. 2009. Maximum frequency deviation calculation in small isolated power systems. *IEEE Transactions on Power Systems* 24 (4):1731–1738.

Egido, I., L. Rouco, F. Fernández-Bernal, A. Rodríguez, J. Rupérez, F. Rodríguez-Bobada, and S. Marín. 2007. Máxima generación eólica en sistemas eléctricos aislados. XII ERIAC (Encuentro Regional Iberoamericano de Cigré), Foz de Iguazú.
Electricidade do Madeira. Caracterização da Rede de Transporte e Distribuicao em AT e MT, available from www.eem.pt, accessed September 2016.
Electricity Authority of Cyprus, available from www.eac.com.
Egido I., Lobato E., Rouco L., Sigrist L., Barrado A., Fontela P., Magriñá J., "Store: a comprehensive research and demostration project on the application of energy storage systems in island power systems", 46 CIGRE Session - CIGRE 2016, Paris, France, 21–26 August 2016, Paper C4-109.
Enemalta 2016a, available from www.enemalta.com.mt, accessed September 2016
Enemalta. Electricity interconnection Malta-Sicily, available from www.enemalta.com.mt/index.aspx?cat=1&art=118&art1=175. Accessed September 2016.
Eurelectric. EU Islands: Towards a Sustainable Energy Future, available from www.eurelectric.org/media/38999/eu_islands_-_towards_a_sustainable_energy_future_-_eurelectric_report_final-2012-190-0001-01-e.pdf.
Eurelectric. Workshop on EU Islands: Towards a Sustainable Energy Future, Brussels, available from www.eurelectric.org/media/64743/Fastelli.pdf.
European Investment Bank. Portugal: Azores receive funding for renewable energy and transmission, available from www.eib.org/projects/press/2013/2013-052-portugal-azores-receive-funding-for-renewable-energy-and-transmission.htm.
Fastelli, I. 2012. Energy storage on islands: A sustainable energy future for islands.
Geun, P. M. 2013. Experience of Jeju smart grid test-bed and national strategy. GENERA 2013, Session on Island Power Systems, Madrid.
Gobierno de Canarias. 2006. Plan Energético de Canarias 2006–2015, available from www.gobiernodecanarias.org/industria/pecan/pecan.pdf.
Gobierno de Canarias. 2011. Revisión Plan Energético de Canarias 2006–2015, available from www.gobiernodecanarias.org/energia/doc/planificacion/pecan/DOCUMENTO_REVISION_PECAN2006.pdf.
Granadino, R. 2012. Rómulo: Interconexión eléctrica Península-Baleares. Primeras Jornadas para un Sistema Eléctrico Sostenible en Les Illes Balears: De sistema aislado a sistema interconectado, Palma de Mallorca.
Grima, P. 2013. Flexible power plant. GENERA 2103, Session on Island Power Systems, Madrid.
Guernsey Electricity, available from www.electricity.gg/.
Hydro Tasmania. King Island: Towards a sustainable, renewable energy future, available from www.hydro.com.au/system/files/documents/King_Island_Renewable_Energy_PK_2008.pdf.
IRENA. 2012. Renewable energy country profiles: Special edition on the occasion of the Renewables and Islands Global Summit, available from www.irena.org/DocumentDownloads/Publications/Country_profiles_special_edition-islands .pdf.
IRENA. Electricity storage and renewables for island power: A guide for decision makers, available from www.irena.org/DocumentDownloads/Publications/Electricity%20Storage%20and%20RE%20for%20Island%20Power.pdf.
Jersey Electricity, available from www.jec.co.uk.
Kaneshiro, R. S. 2013. Hawaii Island (Big Island) integration of renewable energy. GENERA 2013, Session on Island Power Systems, Madrid.
Kumagai, J. 2013. The smartest, greenest grid. *IEEE Spectrum* 50 (5). 42–47.
Kundur, P. 1994. *Power System Stability and Control*. New York: McGraw Hill.

Manx Electricity Authority, available from www.gov.im/mea/.
Margina, J. 2011. Almacenamiento de Energía: Proyecto Store. Cátedra Endesa Red, Universidad de Las Palmas de Gran Canaria, available from www.catedraendesared.ulpgc.es/index.php/descargas/cat_view/73-ciclo-de-conferencias-en-el-area-de-ingenieria-electrica.
Michie, D., D. J. Spiegelhalter, et al. 1994. *Machine Learning, Neural and Statistical Classification*. Upper Saddle River, NJ: Prentice Hall.
Ministerio de Economía, Real Decreto 1747/2003 de 19 de diciembre por el que se regulan los sistemas eléctricos insulares y extrapeninsulares. *Boletín Oficial de Estado* 311:46316–46322.
Ministerio de Industria, Turismo y Comercio. 2006a. Orden ITC/913/2006, de 30 de marzo, por la que se aprueban el método de cálculo del coste de cada uno de los combustibles utilizados y el procedimiento de despacho y liquidación de la energía en los sistemas eléctricos insulares y extrapeninsulares. *Boletín Oficial del Estado* 77 (31 de marzo):12484–12556.
Ministerio de Industria, Turismo y Comercio. 2006b. Resolución de 28 de abril de 2006, de la Secretaría General de Energía, por la que se aprueba un conjunto de procedimientos de carácter técnico e instrumental necesarios para realizar la adecuada gestión técnica de los sistemas eléctricos insulares y extrapeninsulares. *Boletín Oficial del Estado* 129 (31 de mayo) (129): 1–168.
Nielsen, T. 2013. Faroe Islands. GENERA 2103, Session on Island Power Systems, Madrid.
Nielsen, T., and M. Windolf. 2010. GRANI: A full scale "Live Lab" for new intelligent solutions within renewable energy. In *3rd NESSIS Euroelectric Conference toward a Sustainable Energy Future for Island Systems*, Las Palmas de Gran Canaria, available from www.catedraendesared.ulpgc.es/index.php/descargas/doc_download/27-session-i-windolf-a-nielsen.
Pérez-Arriaga, I. J., ed. 2013. *Regulation of the Power Sector*. London: Springer.
Pescador, I. 2013. A sustainable energy future for islands: Insights from EURELECTRIC report. GENERA 2013, Session on Island Power Systems, Madrid.
Petrov, K. 2013. Regulation of small isolated systems: Normative principles and practical issues. GENERA 2013, Session on Island Power Systems, Madrid.
Piernavieja, G. 2010. El Hierro wind-pumped-hydro power station penetration Estoril.
Pons, T. 2012. How to manage intermittency on islands? EDF case studies. In *Eurelectric Annual Convention and Conference*, Malta, available from www.eurelectric.org/media/50462/PONS%20new.pdf.
Rodriguez, R. 2011. La central de bombeo de Soria-Chira. Cátedra Endesa Red, Universidad de Las Palmas de Gran Canaria, available from www.catedraendesared.ulpgc.es/index.php/descargas/cat_view/73-ciclo-de-conferencias-en-el-area-de-ingenieria-electrica.
Rojas, R. 1996. *Neural Networks*. Berlin: Springer.
Schütt, R. 2012. *Decentralized Energy Supply Using the Example of Pellworm*. Eilat, Israel: SEEEI Electricity 2012.

2

Frequency Stability

This chapter describes the modeling and simulation of frequency stability in island power systems. It further provides a sensitivity study on the impact of different model parameters and, in particular, of renewable energy sources on frequency dynamics and frequency stability.

Frequency stability refers to the ability of a power system to maintain frequency within an acceptable range following a severe system upset resulting in a significant imbalance between real power generation and load demand (IEEE/CIGRE 2004). Large imbalances cause the frequency to deviate from its nominal value. The underlying frequency dynamics rely to a great extent on the responses of the generating units, be they conventional generators (CGs), renewable energy sources, or any other device such as energy storage systems responding in a controlled manner to frequency deviations. Load dependency on frequency and voltage also affects frequency dynamics on the final customer side, but to a much lower extent than generating units. However, underfrequency load shedding (UFLS) influences frequency dynamics, although in a very discontinuous way, by curtailing load blocks.

Island power systems are especially sensitive to power imbalances, usually originating from outages of a single generating unit and leading to significant frequency deviations. Comparable frequency deviations in large interconnected power systems would only occur if the system were split into several separated islands. From a frequency stability point of view, an island power system can be described as a system that is not interconnected to any other power system and in which any individual generating unit infeed presents a substantial portion of the total demand (Horne et al. 2004). The latter indirectly implies that the rotating system inertia is much smaller than in a large interconnected system. The increasing penetration of nondispatchable renewable energy sources, usually connected to the system by a stage of power converters, further reduces rotating inertia by substituting CGs, unless inertia is emulated within the converters' control scheme. Analogously, nondispatchable renewable energy sources currently do not provide primary frequency reserve, with the side effect that additional CGs or energy storage systems might be needed to satisfy reserve criteria.

The rest of the chapter is dedicated to the modeling and simulation of frequency stability in island power systems. First, power system models suitable to reflect frequency dynamics are presented, and common simulation approaches are exposed, before the impact of model parameters on frequency dynamics is determined. Second, the impact of nondispatchable renewable energy sources on frequency dynamics and frequency stability is shown with the help of simplified and actual island power systems.

2.1 Modeling and Simulation of Frequency Stability in Island Systems

Power systems are highly complex systems consisting of many individual elements. The interconnections between these elements give rise to a large set of possible dynamic interactions. Normally, power system dynamics are divided into four groups: wave, electromagnetic, electromechanical, and thermodynamic phenomena. Short-term frequency dynamics are generally classified as slower electromechanical phenomena. In fact, rotor dynamics still play an important role in short-term frequency dynamics, but the dynamics associated with the turbine-governor system have a much greater influence (Machowski et al. 1997).

2.1.1 Power System Models

The response of a power system to a power imbalance can be described in four stages (Machowski et al. 1997). The transition between these stages is smooth, and a real distinction does not exist, but this differentiation helps with studying and determining the principal elements participating in frequency dynamics. Figure 2.1 illustrates these four stages.

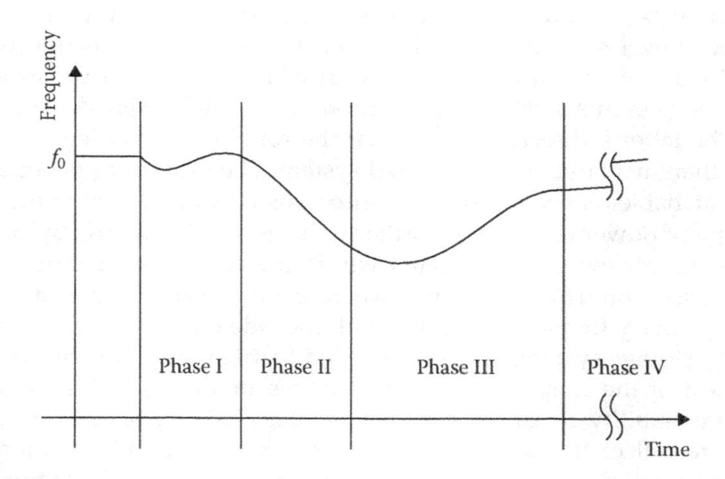

FIGURE 2.1
Four stages of frequency dynamics.

In the first stage, during the first few seconds, the system's behavior is dominated by rotor swings due to rotor angle oscillations. The share of each generator in meeting the power loss depends on its electrical distance from the disturbance. The following two stages, lasting from a few seconds to several seconds, consecutively describe the frequency droop (with constant rate of change of frequency [ROCOF]) and the action of the turbine-governor system. In fact, it has been tacitly supposed that generators remain in synchronism during the first stage. Thus, after the first stage, the power imbalance instigates an increase or a decrease in frequency depending on whether load or generation has been lost. During the second stage, frequency increases or decreases with a constant slope, since still no primary frequency control action takes place. The participation of each generator during this stage depends on its inertia. The lower the inertia, the faster the frequency decay or increase. Subsequent to the second stage, primary controls of turbine-governor systems intervene and try to balance power generation and load demand. Turbine-governor systems provide a means of controlling power and frequency. The contribution of each generator is basically a function of its governor speed drop, the speediness of the turbine-governor system response, and the amount of available spinning reserve. In fact, the amount of available spinning reserve is not unlimited, which might have a destabilizing effect on short-term frequency dynamics. Finally, within a time frame of one or several minutes, secondary control action and energy supply system dynamics prevail in frequency dynamics.

Short-term frequency dynamics are affected by the aforementioned first three stages: the rotor swings, the inertial response, and the turbine-governor system response. Figure 2.2 portrays the functional relationship between the elements involved in short-term frequency dynamics. The principal elements involved are the turbine-governor system; the generator, particularly its inertia; and the electrical system, including other generator units, loads, and so on. The secondary control system and the energy supply system are

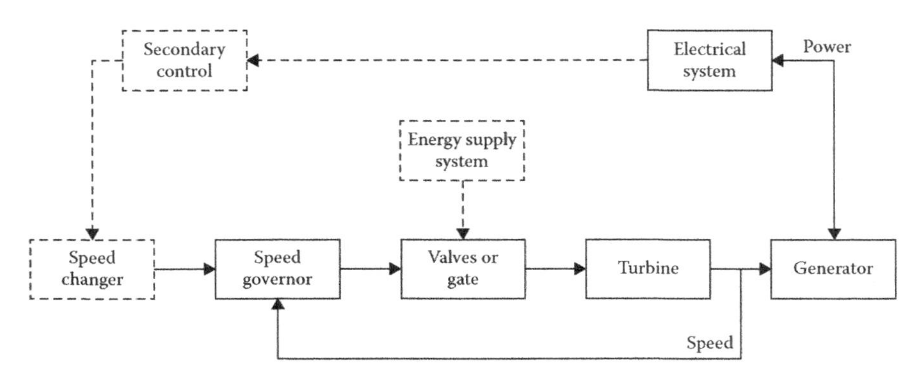

FIGURE 2.2
Elements involved in short-term frequency dynamics.

not involved in short-term frequency dynamics, as they are too slow, and are therefore neglected (Chan et al. 1972; Anderson 1999).

It will be shown in the following sections that for short-term frequency stability analysis, generators could actually be represented by their turbine-governor systems. Further, it is reasonable for an island power system to neglect the transmission network and therefore to assume uniform frequency, disregarding oscillations between generating units. A uniform frequency directly implies that all rotating inertias are unified in a single equivalent inertia. Finally, loads can be considered voltage (and in some cases even frequency) independent, leading to a rather conservative representation of loads with respect to frequency dynamics. These assumptions on frequency, generating unit, and load modeling converge to a so-called system frequency dynamics (SFD). SFD models simplify power system models significantly, but they retain the essential components to reflect frequency dynamics in a sufficiently accurate manner.

2.1.1.1 Generator Representation

Due to the nature of frequency stability phenomena, it seems reasonable to assume that short-term frequency dynamics are mainly affected by the turbine-governor system and the generating unit's inertia. Moreover, the models of the excitation system and the synchronous machine can be neglected, since their associated dynamics are usually too fast and therefore, hardly influence the response of the generating unit. Actually, the response of the excitation systems restores voltage in a relatively short time to its pre-fault value in comparison with the response time of turbine-governor systems (Elgerd 1982). In addition, frequency dynamics influence generators, transformers, or excitation systems to a much lower extent.

The assumption that the generator can be represented by its turbine-governor system is verified by considering typical turbine-governor systems for island power systems. In particular, diesel engines, gas turbines, and steam turbines are analyzed. Figure 2.3 shows and compares the simulated response in terms of mechanical power of the complete model of the generating unit (synchronous machine, excitation system, and turbine-governor system) and of the detailed turbine-governor system model only to a rise in active power. It can be inferred that there exist only small differences between the model of the full generating unit and the detailed model of the turbine governor–rotor system.

From Figure 2.3, it can be further inferred that the generating unit responses resemble the responses of a second-order or third-order closed-loop (CL) system. This CL system is actually formed by the lower loop of Figure 2.2 or, in other words, by the interaction of the speed-governor and the shaft rotation by means of the equation of motion. A CL second-order model implies a first-order open-loop (OL) model, since the retroaction, governed by the relationship between angular speed and power described by

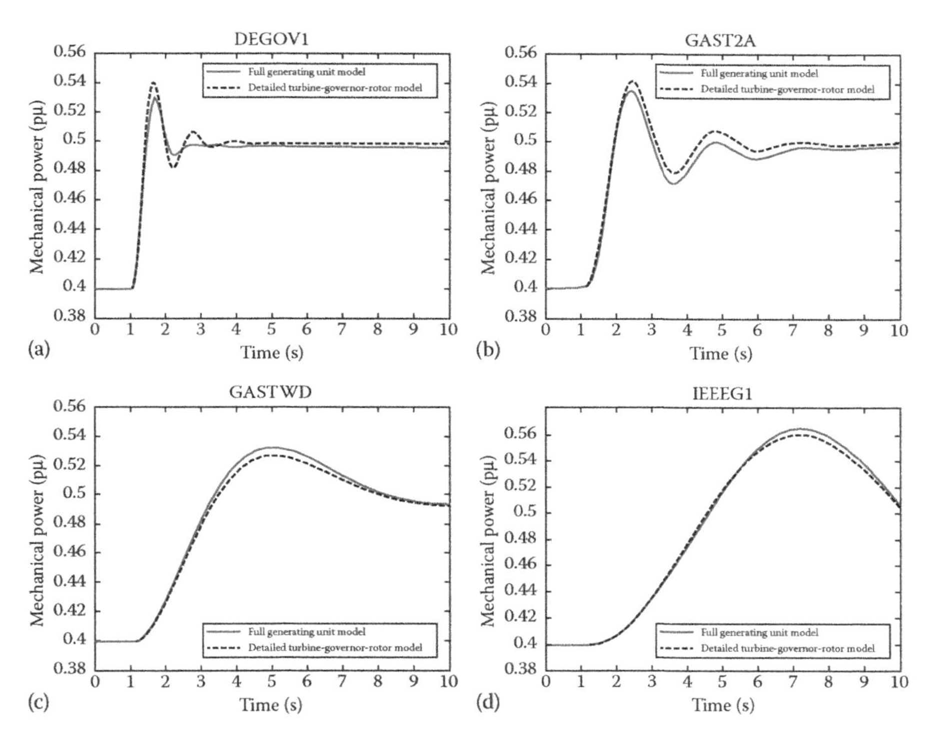

FIGURE 2.3
Comparison of the responses of (a) a diesel-based power plant (DEGOV1), (b) a gas-based power plant (GAST2A), (c) a gas-based power plant (GASTWD), and (d) a steam-based power plant (IEEEG1).

the equation of motion of a rigid body (inertia), is a first-order loop. Similarly, a second-order OL model leads to a third-order CL model. Typically, steam-driven turbines can be modeled by first-order transfer functions, whereas gas-driven or diesel-driven combustion turbines might require second-order models (Anderson and Mirheydar 1990). SFD models in one form or another simplify turbine-governor systems to first- or second-order OL systems (see Subsection 2.1.1.4), reducing computational cost but sacrificing a certain level of accuracy.

2.1.1.2 Network Representation

Island power systems are usually of comparatively small geographic dimensions, implying that electrical distances are rather short in comparison with large interconnected power systems. The assumption on neglecting the network is verified by analyzing the impact of representing the network on frequency dynamics.

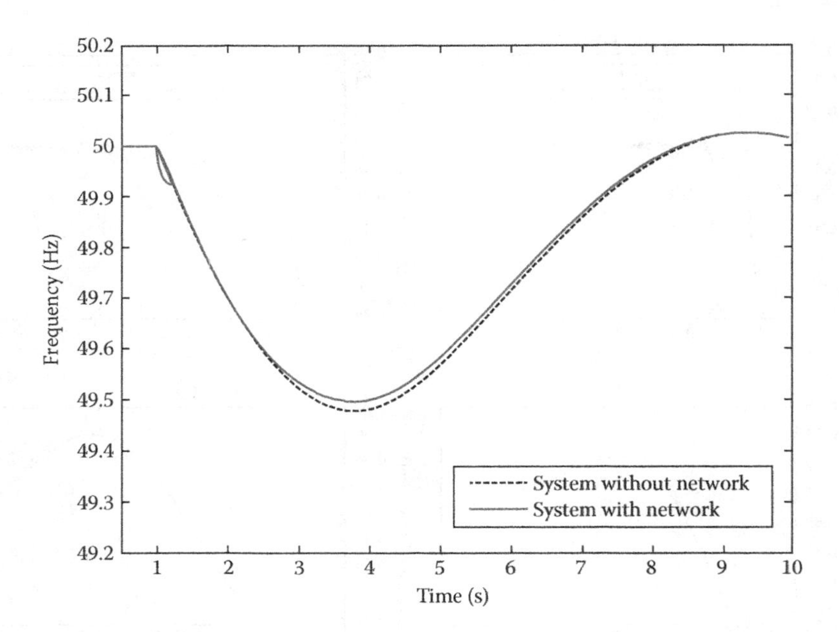

FIGURE 2.4
Comparison of the frequency variation of a system with and without a transmission network.

Figure 2.4 compares the power system frequency when modeling the network and when omitting the network by concentrating load demand and power generation on a single bus. Frequencies at different buses are shown for the case including the network. Results show that that the average frequency does not significantly differ between the two cases, and thus, the network has a negligible impact on the system frequency. Furthermore, without the network, the connections between the generators (and the loads) are absolutely rigid, and consequently, as shown in Figure 2.4, the intermachine oscillations in frequency tend to disappear.

Disregard of intermachine oscillations in frequency dynamics by neglecting to synchronize power and transmission performance results in an average or uniform frequency. In fact, frequency dynamics have been simulated using the concept of uniform frequency (or “dynamic energy balance” [Crevier and Schweppe 1975]) to reduce the complexity of the model of a large power system and to eliminate the effect of intermachine oscillations (Stanton 1972). Actually, intermachine oscillations are avoided by setting up dynamic models that allow relative movement between generator rotors without representing the transient slow-down–speed-up process associated with oscillatory transfer of kinetic energy from rotor to rotor. The key feature for eliminating intermachine oscillations is the assumption that the variation of speed of each generating unit with respect to the variation of speed of the center of inertia is zero, and therefore, the equation of motion of each generating unit reduces to an energy balance. Consequently, the center of inertia for the system and the associated frequency of the center of inertia can then be used. Frequency dynamics are described by an aggregate equation of motion of the center of inertia, reducing the total number of equations of motion of each generator. This is the key feature of SFD models. The complexity can be further reduced by developing dynamic equivalents of different generating units by applying delay or canonical models (Chan et al. 1972) or by identifying coherent generating units (Germond and Podmore 1978).

2.1.1.3 Load Representation

There exist many different types of loads (motors, refrigerators, lamps, furnaces, etc.), and load composition changes depending on many factors, including time, weather conditions, and the state of the economy. Commonly, instead of representing each individual load, aggregated load models as seen from the transmission network delivery points are used. Load models can be classified into static and dynamic load models. Static models represent load dependency on voltage and frequency with an algebraic equation: voltage dependency is modeled by an exponential or polynomial (ZIP) model, whereas dependency on frequency is assumed to be linear. The use of static models is justified under modest amplitudes of voltage and frequency variations. Load dynamics are mostly influenced by electric motors, consuming around 60%–70% of the total energy supplied to the system. Other load components exhibiting dynamic attributes, but of usually much slower response than frequency dynamics, include the operation protection relays, thermostatically controlled loads, and underload tapping changing transformers.

Figure 2.5 shows the response of an induction machine to voltage and frequency swings typical of an island power system after the loss of a generating unit. The power consumption of the induction machine has been plotted for three cases: the stator is fed with the given voltage and frequency swings; the stator is fed with the given voltage swing but at constant frequency; and the stator is fed with the given frequency swing but at constant voltage. The power consumption under constant voltage and varying frequency matches the actual power consumption very well. It is noteworthy that power consumption is somewhat proportional to frequency deviation after the initial transient, but during the first 500 ms, power consumption is rather dependent on the ROCOF.

Figure 2.6 compares the impact of modeling the load voltage dependency on the response in terms of frequency of an island power system to a loss of a generating unit with a constant load power modeling. Active power has been assumed to be proportional to the voltage, whereas reactive power has been assumed to be proportional to the square of the voltage. A maximum difference of 0.02 Hz is observable at the instant of minimum frequency. SFD models can easily reflect frequency dependency of loads.

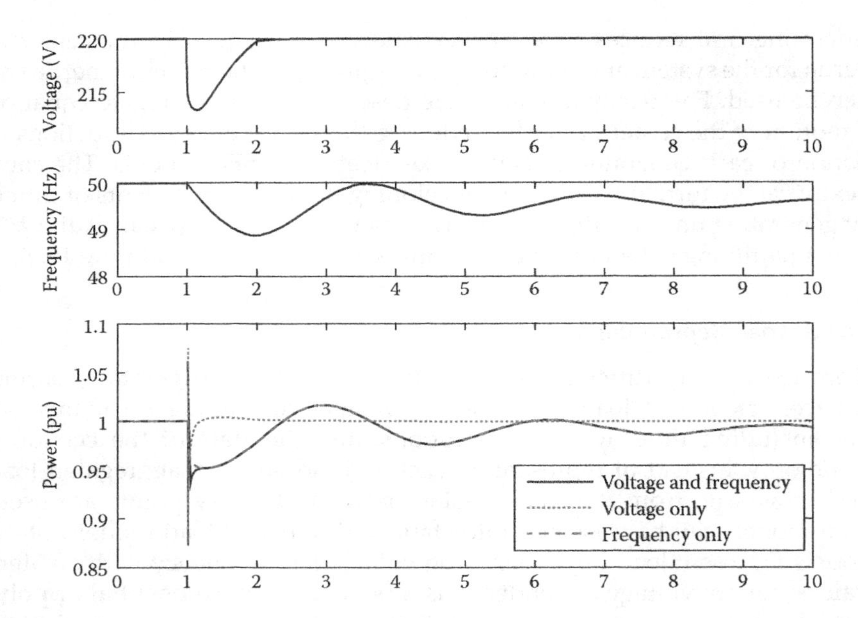

FIGURE 2.5
Comparison of responses to frequency and voltage swings of an induction machine.

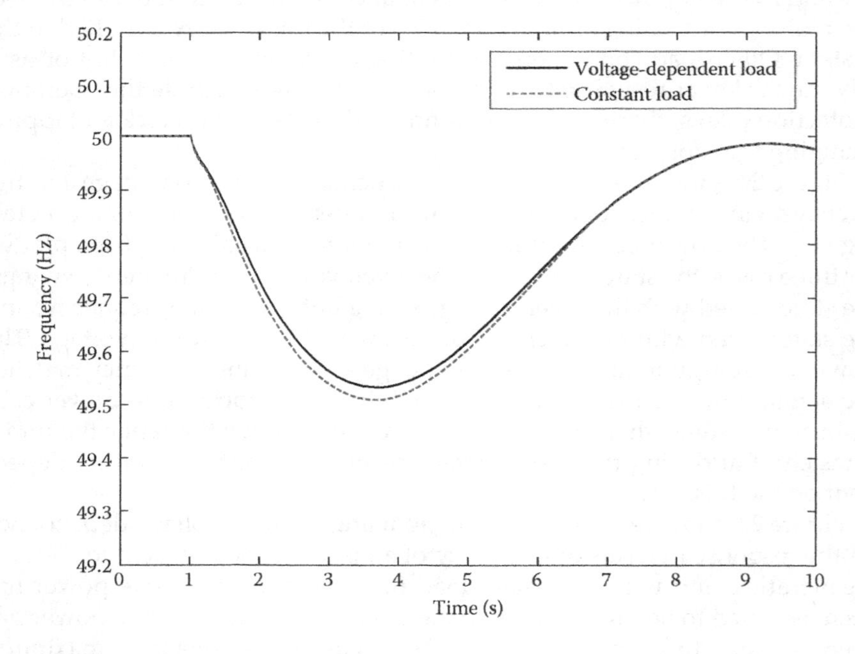

FIGURE 2.6
Comparison of frequency variation of a system with constant and voltage-dependent ($P_{load} \sim V$ and $Q_{load} \sim V^2$) loads.

2.1.1.4 System Frequency Dynamics Models

The principal idea of SFD models is to provide low-order models that retain the essential frequency shape of a system with typical time constants and active turbine-governor systems (IEEE 1973, 1991). An important feature of SFD models is the assumption of uniform frequency.

Several SFD models have been proposed in the technical literature. The first SFD models only considered the impact of the load-damping factor on short-term frequency dynamics (New 1977). Therefore, a simple linear single-bus SFD model has been developed, in which the generating units are represented by a single first-order model of the turbine-governor systems (Anderson and Mirheydar 1990). This model was based on the assumption that the generating units are predominantly driven by steam turbines. In Denis Lee Hau (2006), a general linear SFD model is outlined, allowing the representation of several first-order models of turbine-governor systems. A similar SFD model is presented in Egido et al. (2010). However, these SFD models are all linear and do not take into account the generator output limitations. For short-term frequency dynamics in small isolated power systems, it is important to be aware of the finite reserve. In fact, the assumption of a linear model is only valid for small disturbances and is therefore questionable for the design of an UFLS scheme for a small isolated power system. In addition, the first-order models of the generating units adopted in Anderson and Mirheydar (1990) and Denis Lee Hau (2006) correspond to steam turbines. In general, the generation mix also contains gas-driven and diesel-driven turbines. Thus, a nonlinear SFD model needs to be defined that enables different turbine types to be implemented. Mainly, delays in the response of generating units demand a higher-order model. A first step in this direction is given in Egido et al. (2010) by adjusting the first-order models in the function of different turbine-governor systems. Finally, the order of the simplified model of each generating unit may vary in the function of the turbine type and its parameters, and therefore, the model adopted to represent a particular generating unit should be the one that best represents the detailed turbine-governor system. A generic second-order model is used here to represent turbine-governor systems.

A complete SFD model consists of n_g generators with n_g equation of motions and n_g different turbine-governor models. The equation of motion of the *i*th generator is given by

$$2H_i\dot{\omega}_i = p_{G,i} - p_{D,i} \tag{2.1}$$

or uniquely considering deviations:

$$2H_i\Delta\dot{\omega}_i = \Delta p_{G,i} - \Delta p_{D,i} \tag{2.2}$$

where:

H_i is the inertia (s)
$\Delta\dot{\omega}_i$ is the frequency deviation (pu)
$\Delta p_{G,i}$ is the mechanical power deviation (pu)
$\Delta p_{D,i}$ is the load power deviation (pu)

Equation 2.2 is given on the base of the rating of the ith machine $M_{\text{base},i}$. The dynamics of the generic second-order turbine-governor system of the ith generating unit are described by the following set of nonlinear equations:

$$\begin{bmatrix} \Delta\dot{x}_{1tg,i} \\ \Delta\dot{x}_{2tg,i} \end{bmatrix} = \begin{bmatrix} 0 & 1 \\ -1/a_{1,i} & -a_{1,i}/a_{2,i} \end{bmatrix} \begin{bmatrix} \Delta x_{1tg,i} \\ \Delta x_{2tg,i} \end{bmatrix} + K_i \begin{bmatrix} 0 \\ 1/a_{2,i} \end{bmatrix} \Delta\omega$$

$$\Delta p'_{G,i} = \begin{bmatrix} 1 - \dfrac{b_{2,i}}{a_{2,i}} & b_{1,i} - \dfrac{a_{1,i} \cdot b_{2,i}}{a_{2,i}} \end{bmatrix} \begin{bmatrix} \Delta x_{1tg,i} \\ \Delta x_{2tg,i} \end{bmatrix} + K_i \cdot \frac{b_{2,i}}{a_{2,i}} \Delta\omega$$

$$\Delta p_{G,i} = \max\left(\Delta p_{i,\min}, \min\left(\Delta p_{i,\max}, \Delta p'_{G,i}\right)\right)$$

where:

$a_{1,i}$ and $a_{2,i}$, and $b_{1,i}$ and $b_{2,i}$ are the poles and zeros of the generic second-order system
K_i is the inverse of the droop in pu on $M_{\text{base},i}$
$\Delta p_{i,\max}$ and $\Delta p_{i,\min}$ represent the maximum and minimum generator output

Figure 2.7 shows the model of the ith generating unit for the purpose of the analysis of frequency dynamics.

Assuming that the equations are normalized to the size of the isolated power system, the system base, S_{base}, is the sum of the ratings of all generating units. Thus,

$$2H_i \frac{M_{\text{base}.i}}{S_{\text{base}}} \Delta\dot{\omega}_i = \frac{M_{\text{base}.i}}{S_{\text{base}}} \left(\Delta p_{G,i} - \Delta p_{D,i}\right) \tag{2.3}$$

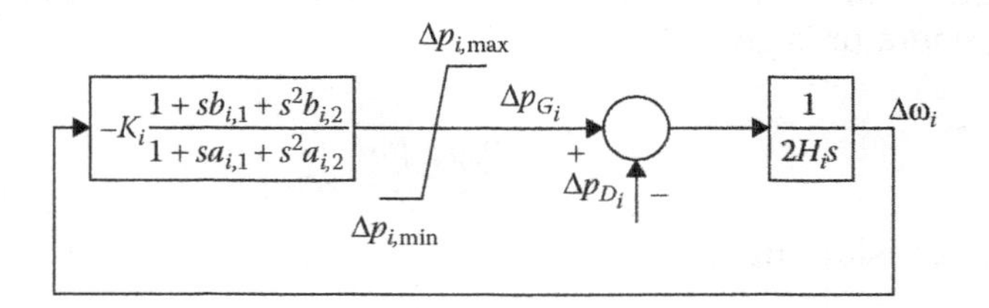

FIGURE 2.7
Modeling of the ith generating unit.

with

$$S_{\text{base}} = \sum_{i=1}^{n} M_{\text{base}.i}$$

Summing the n equations of motion yields

$$2\sum_{i=1}^{n_g} H_i \frac{M_{\text{base}.i}}{S_{\text{base}}} \Delta\dot{\omega}_i = \sum_{i=1}^{n_g} \frac{M_{\text{base}.i}}{S_{\text{base}}} \Delta p_{G,i} - \sum_{i=1}^{n_g} \frac{M_{\text{base}.i}}{S_{\text{base}}} \Delta p_{D,i} \tag{2.4}$$

At this stage, the equivalent inertia H can be defined as

$$H = \sum_{i=1}^{n} \frac{H_i \cdot M_{\text{base},i}}{S_{\text{base}}} \tag{2.5}$$

Combining Equations 2.4 and 2.5, the uniform frequency deviation is described by the average frequency deviation $\Delta\dot{\omega}$ given in Equation 2.6:

$$2H\Delta\dot{\omega} = \Delta p_{G,\text{tot}} - \Delta p_{D,\text{tot}} \tag{2.6}$$

where $\Delta p_{G,\text{tot}}$ and $\Delta p_{D,\text{tot}}$ are the total power generation and total load demand deviation, respectively.

Writing the equation of motion of the center of inertia on the base of the system rating S_{base} requires the gain of the generic turbine-governor system model K_i to be expressed according to S_{base}:

$$k_i = \frac{M_{\text{base},i} \cdot K_i}{S_{\text{base}}} \tag{2.7}$$

Finally, Figure 2.8 details the power system model proposed to design robust and efficient UFLS schemes of small isolated power systems of n generating units. This is an SFD model with equivalent inertia in which the whole generation and load demand is assumed to be connected to the same bus. Generators can be represented by the generic second-order model shown in Figure 2.7. Load is considered to be frequency dependent according to the load-damping factor D. The UFLS scheme additionally acts on the available load. The UFLS scheme permits both underfrequency relays and ROCOF relays to be implemented.

The SFD model of Figure 2.8 can be readily extended to renewable energy sources. Currently, this kind of generation essentially suffers from two drawbacks: its limited primary frequency control capacity and its negligible inertial response. Both the lack of primary frequency control and the lack

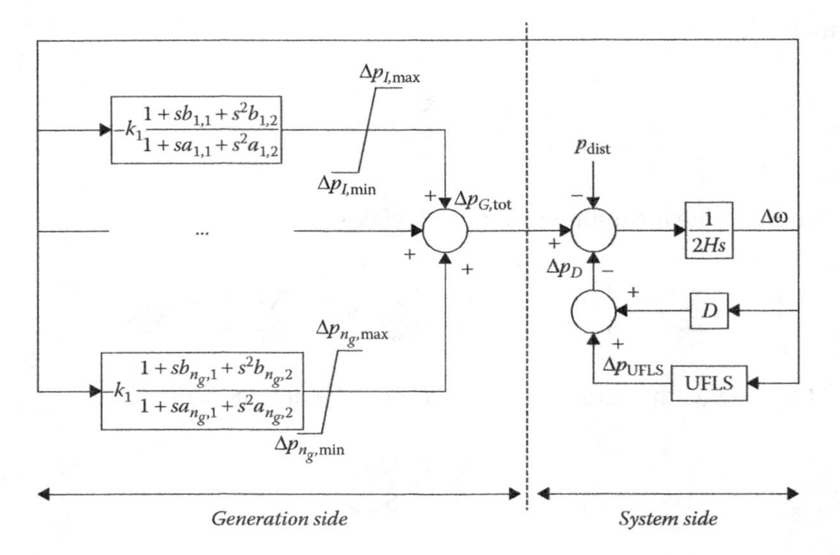

FIGURE 2.8
Nonlinear multigenerator SFD model of a small isolated power system.

of inertial response have a negative impact on frequency stability. In fact, the initial ROCOF increases with decreasing inertia. Since renewable energy sources substitute conventional generating units, the spinning reserve provided by the latter takes more time to be fully activated. Thus, the smaller the inertia and the slower the spinning reserve that is provided, the larger are the frequency deviations.

The differences between the characteristics of conventional and renewable energy sources and the provision of an increasing amount of installed capacity of renewable energy sources require us to contemplate scenarios with different penetration levels. Due to their specific characteristics and in terms of the nonlinear multigenerator SFD model of Figure 2.8, renewable energy sources are modeled like a generating unit, but with zero inertia H_i and zero gain K_i, unless they emulate inertia or operate at an operating point below the maximum power point.

2.1.2 Simulation Approaches

The frequency stability of an island power system is assessed by analyzing the underlying frequency dynamics. SFD models are able to reflect frequency dynamics in a sufficiently accurate manner. Three simulation approaches are possible: analytical expression of the power system's response, time-domain simulation, and prediction of the power system's response. The first and the third approaches are faster than the second one, but they either require further simplification of the power system model or might lead to inaccurate results if not updated continuously.

2.1.2.1 Analytical Expression

In Anderson and Mirheydar (1990), a CL expression for an islanded power system is presented. This analytical expression assumes that the generating units within the islanded system can be represented by a single equivalent generating unit. The generation mix is supposed to be dominated by steam turbines. In addition, power output limitations have been neglected, and the UFLS scheme has been omitted Figure 2.9 shows the corresponding SFD model.

By the analysis of the block diagram of the SFD model of Figure 2.9, the frequency deviation $\Delta\omega$ due to power imbalance p_{dist} can be computed as

$$\Delta\omega(s) = \frac{R\omega_n^2}{DR+1} \cdot \frac{(1+T_R s)}{s^2 + 2\varsigma\omega_n + \omega_n^2} p_{\text{dist}}(s) \tag{2.8}$$

where

$$\omega_n^2 = \frac{DR+1}{2HRT_R}$$

$$\varsigma = \frac{2HR + (DR+F_H)T_R}{2(DR+1)} \omega_n^2 \tag{2.9}$$

If the disturbance represents a sudden load imbalance such as the loss of a generating unit, then Equation 2.8 becomes

$$\Delta\omega(s) = \frac{R\omega_n^2}{DR+1} \cdot \frac{(1+T_R s)}{s^2 + 2\varsigma\omega_n + \omega_n^2} \frac{\Delta p}{s} \tag{2.10}$$

The application of the inverse Laplace transform results in

$$\Delta\omega(t) = \frac{R\Delta p}{DR+1}\left(1+\alpha e^{-\varsigma\omega_n t}\sin(\omega_r t+\phi)\right) \tag{2.11}$$

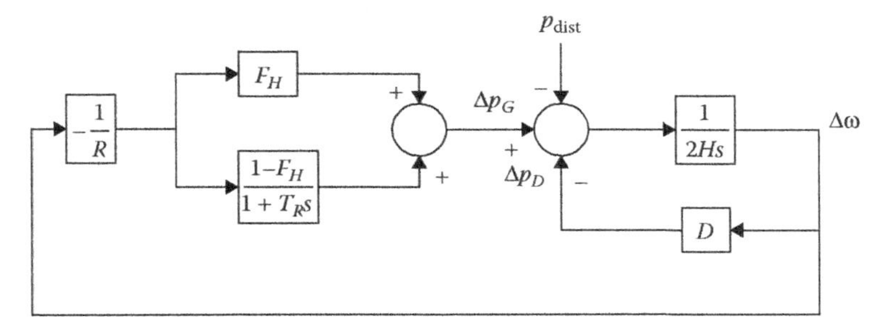

FIGURE 2.9
SFD model with one single equivalent generator.

where

$$\alpha = \sqrt{\frac{1 - 2T_R\varsigma\omega_n + (T_R\omega_n)^2}{1 - \varsigma^2}}$$

$$\omega_r = \omega_n\sqrt{1 - \varsigma^2}$$

and

$$\phi = \tan^{-1}\left(\frac{T_R\omega_r}{1 - \varsigma T_R\omega_n}\right) - \tan^{-1}\left(\frac{\sqrt{1 - \varsigma^2}}{-\varsigma}\right)$$

The system response in terms of frequency to a sudden power imbalance represents a damped sinusoidal. Minimum and maximum frequency deviation can be easily deduced from Equation 2.11.

2.1.2.2 Time-Domain Simulations

By using time-domain simulations, the SFD model shown in Figure 2.8 can be simulated without any further simplification. The system of nonlinear differential equations describing the SFD model is nonstiff; that is, there are no terms in the differential equations that can lead to rapid variations in the solution (Hoffman 2001). Several software packages, such as Matlab, offer powerful simulation tools that implement different methods of numerical integration of nonlinear differential equations. One can distinguish between explicit and implicit integration methods. Explicit methods can be further divided into variable and fixed-step methods: the former adapt the integration step according to the underlying dynamics of the set of nonlinear differential equations and some tolerances, whereas the latter keep the integration step constant. For island power systems with turbine-governor systems with similar time constants, variable methods are not necessarily superior in terms of computational speed. In any case, once the UFLS scheme intervene, small integration steps are required to accurately determine frequency crossing instants and opening times.

Heun's integration method, also known as Modified Euler's method, is an explicit, fixed-step integration method. It can be seen as an extension of the Euler method into a two-stage second-order Runge–Kutta method (Kendall et al. 2009). In fact, Heun's method divides the numerical integration into a predictor and a corrector step. The PSS/E software package applies Heun's numerical integration method to solve nonlinear differential equations (Siemens 2005).

Consider, for example, the system of differential equations given in Equation 2.12:

$$\dot{\mathbf{x}}_s(t) = \mathbf{f}(t, \mathbf{x}_s(t)), \quad \mathbf{x}_s(t_0) = \mathbf{x}_{s,0} \tag{2.12}$$

where:

- $\mathbf{x}_s$ is a vector of state variables
- $\mathbf{x}_{s,0}$ are its initial values
- $\mathbf{f}(\bullet)$ is a vector of functions of time t and $\mathbf{x}_s$

Here, the vector of functions $\mathbf{f}(\bullet)$ implements the system of nonlinear differential equations describing the SFD model. Heun's method resolves the initial value problem of Equation 2.12 as follows:

$$\begin{aligned} \tilde{\mathbf{x}}'_{s,k+1} &= \mathbf{x}_{s,k} + h \cdot \mathbf{f}(t_k, \mathbf{x}_{s,k}) \\ \mathbf{x}_{s,k+1} &= \mathbf{x}_{s,k} + \frac{h}{2} \cdot \left(\mathbf{f}(t_k, \mathbf{x}_{s,k}) + \mathbf{f}(t_{k+1}, \mathbf{x}'_{s,k+1})\right) \end{aligned} \tag{2.13}$$

where:

- h is the integration step
- k is the iteration number

First, a predictor $\mathbf{x}'_{s,k}$ is calculated by using Euler's method, and subsequently the corrector $\mathbf{x}_{s,k}$ is determined by means of an explicit trapezoidal method.

Figures 2.10 and 2.11 show the system frequency obtained by the SFD model of two real power systems and compare it with the system frequency obtained by using the detailed power system model. In Figure 2.10, the loss of 66.68 MW (corresponding to 21.9% of total demand) of the Gran Canaria power system has been simulated, whereas in Figure 2.11, the loss of 2.35 MW (corresponding to 13% of the total demand) of the La Palma power system has been simulated. The responses of the SFD models fit very well the responses of the detailed power system models.

In Denis Lee Hau (2006), a method to solve the SFD model without simulating the whole transient has been proposed and applied. With regard to Anderson and Mirheydar (1990), the SFD model represents each generating unit as well as the UFLS scheme, but it still neglects power output limitations. The response of the SFD model including UFLS actions is obtained by evaluating the analytical expressions of steady-state frequency deviation, minimum frequency deviation, and time instants of load shedding in a sequential order. Figure 2.12 shows the proposed SFD model.

The CL transfer function between $\Delta\omega$ and p_{dist} can be expressed as a sum of partial fractions (Denis Lee Hau 2006):

$$\Delta\omega(s) = \sum_{j=0}^{M} \Delta P_j \cdot e^{-\tau_j s} \cdot \sum_{i=1}^{n_g+1} \frac{A_i}{p_i} \cdot \left(\frac{1}{s} + \frac{1}{s - p_i}\right) \tag{2.14}$$

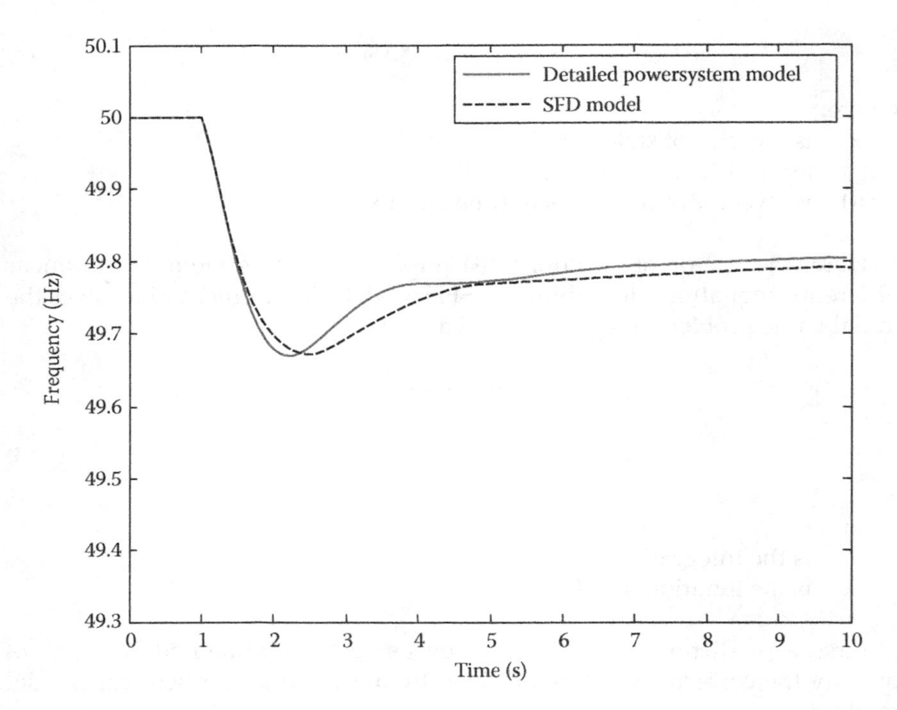

FIGURE 2.10
Comparison of the responses of the detailed Gran Canaria power system model and the SFD model.

where:

A_i is real or complex a

p_i is a pole of the denominator of the CL transfer function between $\Delta\omega$ and p_{dist} and may be real or a complex conjugate-pair

ΔP_j is the amount of load shed at instant τ_j except for $j=0$, where $\Delta P_{j=0}$ is equal to p_{dist} and $\tau_{j=0}$

Equation 2.14 can be written in the time domain as

$$\Delta\omega(t) = \sum_{j=0}^{M} \Delta P_j \cdot \sum_{i=1}^{n_g+1} \frac{A_i}{p_i} \cdot \left(1 - e^{p_i(t-\tau_j)}\right) \cdot U\left(t - \tau_j\right) \tag{2.15}$$

The steady-state frequency deviation can be computed as

$$\Delta\omega_{ss} = \lim_{t\to\infty} \Delta\omega(t) = \sum_{j=0}^{M} \Delta P_j \cdot \sum_{i=1}^{n_g+1} \frac{A_i}{p_i} \tag{2.16}$$

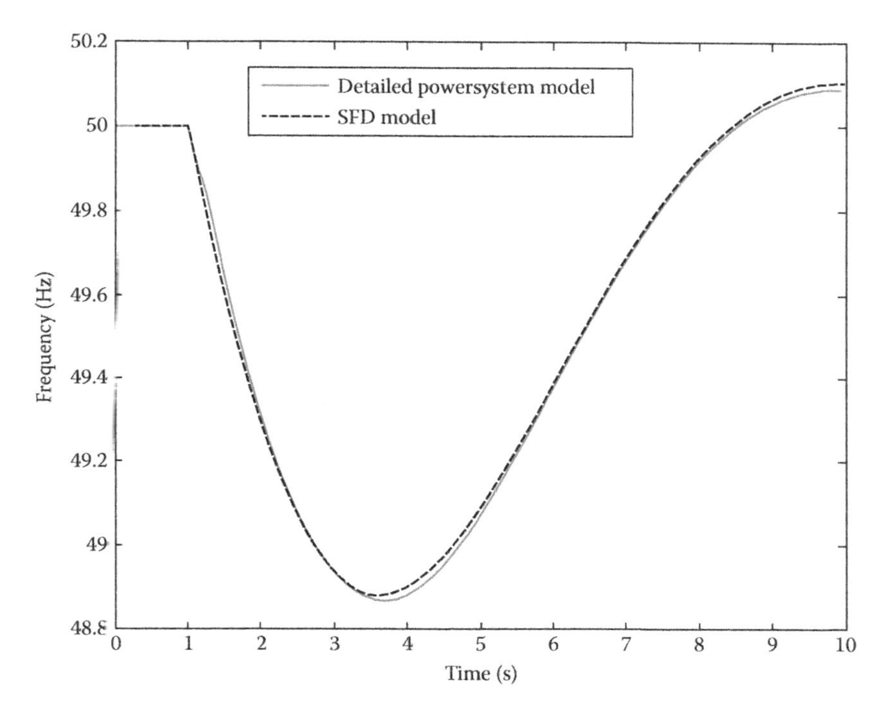

FIGURE 2.11
Comparison of the responses of the detailed La Palma power system model and the SFD model.

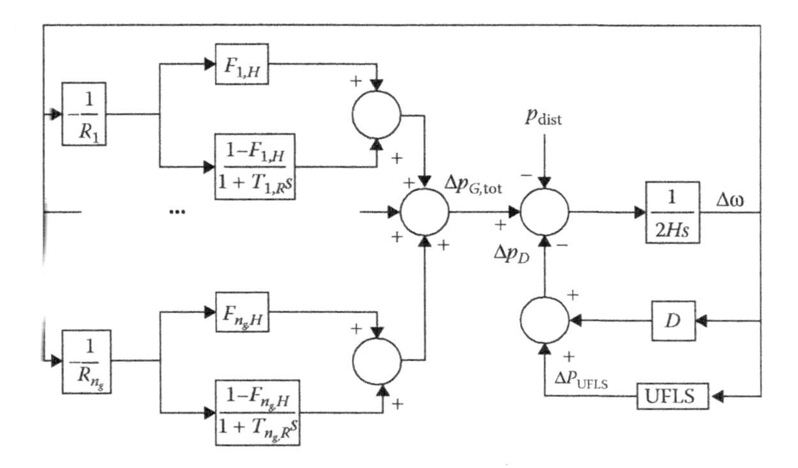

FIGURE 2.12
Multigenerator SFD model with UFLS scheme but without generator power output limitations.

Differentiating Equation 2.15 yields Equation 2.17, which allows the instant of minimum frequency and consequently, by means of Equation 2.15, the minimum frequency deviation to be determined.

$$0 = -\sum_{j=0}^{M} \Delta P_j \cdot \sum_{i=1}^{n_g+1} A_i \cdot e^{p_i(t_{\omega\min} - \tau_j)} \cdot U(t - \tau_j)$$

$$\Delta\omega_{\min} = \Delta\omega(t_{\omega\min}) \tag{2.17}$$

The instants of crossing the frequency threshold of the $(r+1)$th UFLS, $t_{\text{thrshld},r+1}$, stage can be computed by solving Equation 2.15 as

$$\Delta\omega_{\text{thrshld},r+1} = \sum_{j=0}^{r} \Delta P_j \cdot \sum_{i=1}^{n_g+1} \frac{A_i}{p_i} \cdot \left(1 - e^{p_i(t_{\text{thrshld},r+1} - \tau_j)}\right) \cdot U(t - \tau_j) \tag{2.18}$$

Equation 2.18 has a solution only if $\Delta\omega_{\min}$ is lower than $\Delta\omega_{\text{thrshld},r+1}$ and only after instants $t_{\text{thrshld},1}$ to $t_{\text{thrshld},r}$ are already known. These conditions imply that $t_{\text{thrshld},r+1}$ can only be computed in a sequential order starting from $r=0$; the procedure stops if $|\Delta\omega_{\min}| > |\Delta\omega_{\text{thrshld},r+1}|$. Note that Equations 2.17 and 2.18 are nonlinear and piecewise continuous functions of one unknown variable, which can be solved by a well-known algorithm such as the trust-region dogleg algorithm.

2.1.2.3 Prediction

It is finally possible to predict the response of the power system in terms of frequency. Prediction is done by means of machine learning techniques such as artificial neural networks (ANN). These techniques require a learning phase to adjust the relevant parameters before they can actually predict the system's response.

In Kottick and Or (1996), an SFD model with a single equivalent generating unit has been used to adjust the parameters of ANNs. During the learning phase, a learning set of 180 recorded disturbances has been used to train an ANN by a back-propagation learning algorithm to predict the minimum frequency deviation. This ANN uses 47 input variables corresponding to the actual and the available power of each generating unit as well as the recorded power shortage (p_{dist}). The output variable is the minimum frequency deviation. The absolute average prediction error is 0.044 Hz, with a 95% confidence interval of 0.128 Hz (i.e., 95% of all cases show an error smaller than 0.128 Hz). A second ANN is used to predict which UFLS stages actually act given a certain power shortage. This ANN uses three input variables: total power generation, loading of the forced outaged generating unit, and the minimum frequency deviation, which is the outcome of the first ANN. The second ANN has an average prediction error of 2.6% (7 stages have been wrongly predicted over 269 actually used stages for 40 analyzed power shortages).

In Mitchell et al. (2000), an ANN has been proposed to predict the initial ROCOF, the minimum and maximum frequency deviations, and the steady-state frequency deviation. The input variables are the actual power generation, the available power generation, the active load–generation level prior to the disturbance, the amount of load being shed, and the percentage of exponential type loads being shed. The ANN has been trained by a back-propagation algorithm with 14,046 simulated disturbances. When the ANN prediction was compared for two disturbances with corresponding simulated responses, a maximum prediction error of 0.06 Hz was obtained.

2.1.3 Sensitivity Studies

In this subsection, the impact of the equivalent inertia, of the generator gains and time constants, of generator output limitations, and of the load-damping factor and the UFLS scheme on frequency dynamics is evaluated. For this purpose, the SFD model of Figure 2.8 is taken to represent two generating units, with each generating unit represented by a first-order model. An initial power imbalance of 0.15 pu is applied. The load-damping factor is neglected except for the analysis of its impact. Likewise, generator output limitations are also neglected except for the analysis of their impact. Table 2.1 summarizes the assumed SFD model parameters.

The power system response to the disturbance is simulated in the time domain without loss of generality by means of Heun's integration method. The integration step should be defined as a fraction of the smallest time constant of the model formed by the two generators. An appropriate choice for the integration step is 0.1 s.

2.1.3.1 Inertia

By inspecting the nonlinear multigenerator SFD model of Figure 2.8 and by applying the final value theorem, the well-known expression for the steady-state frequency deviation can be obtained:

$$\Delta\omega_{ss} = -\frac{\Delta p_d}{\sum_{i=1}^{n_g} k_i + D} \tag{2.19}$$

TABLE 2.1

SFD Model Parameters Without Generator Output Limitations

Generator	H (s)	K	a_1 (s)	b_1 (s)	M_{base}(MW)	Δp_{min}	Δp_{max}
1	3	25	3	0	130	–	–
2	5	25	5	0	70	–	–

where:

n_g is the number of generators
D is the load-damping factor
k_i is the gain of the ith generator expressed according to the system base S_{base}

In a similar way, applying the initial value theorem to the derivative of frequency yields the initial value of the ROCOF (Anderson and Mirheydar 1990):

$$\Delta\dot{\omega}_0 = -\frac{\Delta p_d}{2H} = -\frac{\Delta p_d}{2\sum_{i=1}^{n_g}\frac{H_i \cdot M_{base,i}}{S_{base}}} \tag{2.20}$$

where:

H_i is the inertia of the ith generator
$M_{base,i}$ is its base rating

Figure 2.13 shows the impact of the inertia on frequency. According to Equation 2.20 and Figure 2.13, the most pronounced effect of increasing H_1 (which increases in turn the equivalent inertia) is to reduce the initial ROCOF and to delay and reduce the minimum frequency deviation. The equivalent

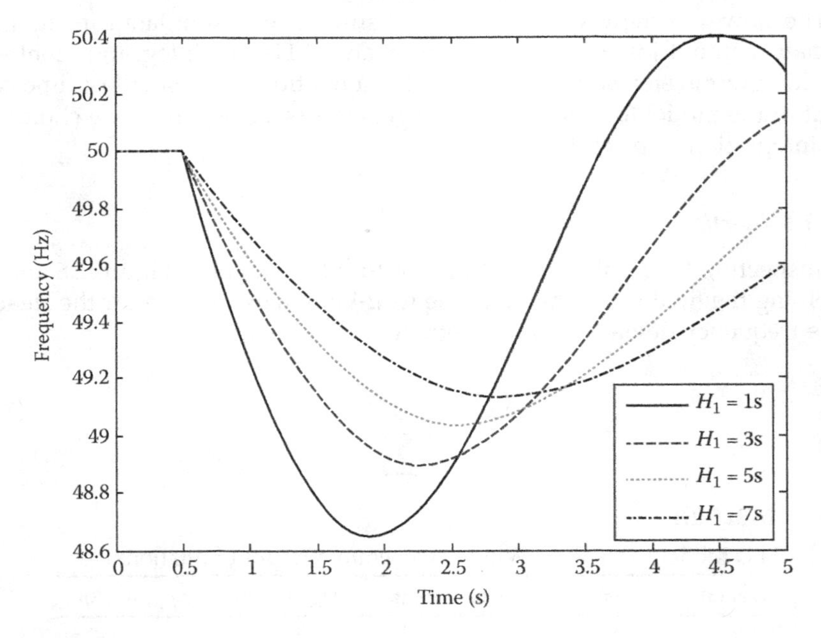

FIGURE 2.13
The impact of varying the inertia H_1 on frequency.

inertia does not affect the final steady-state value of frequency as stated in Equation 2.19. Higher inertia values result in a slower drop in frequency, which is logical, and also in a slower recovery. Since the response is slower for higher inertias, the governor has more time to respond, and therefore limits the minimum frequency deviation to smaller values.

2.1.3.2 Generator Gain and Time Constants

To show the effect of the generator gain, different values of K_1 are considered, corresponding to governor droops of 0.04, 0.05, 0.06, and 0.07 pu. In fact, actual observed system responses have sometimes shown the net system regulation to differ from the value of 0.04 or 0.05 pu (Anderson and Mirheydar 1990). Figure 2.14 shows the resulting impact on frequency.

As seen in Figure 2.14 and as stated in Equation 2.20, the generator gain has absolutely no effect on the initial ROCOF. Even if all governors are at the extreme valve-closing end of the individual backlash limits, such as following a gradual load decrease, a sudden power imbalance would require a rapid change to a valve or gate open condition. However, this cannot occur instantaneously. Higher generator gains shorten the recovery time and reduce the steady-state frequency deviation. In addition, the minimum frequency deviation is also reduced with increasing generator gains.

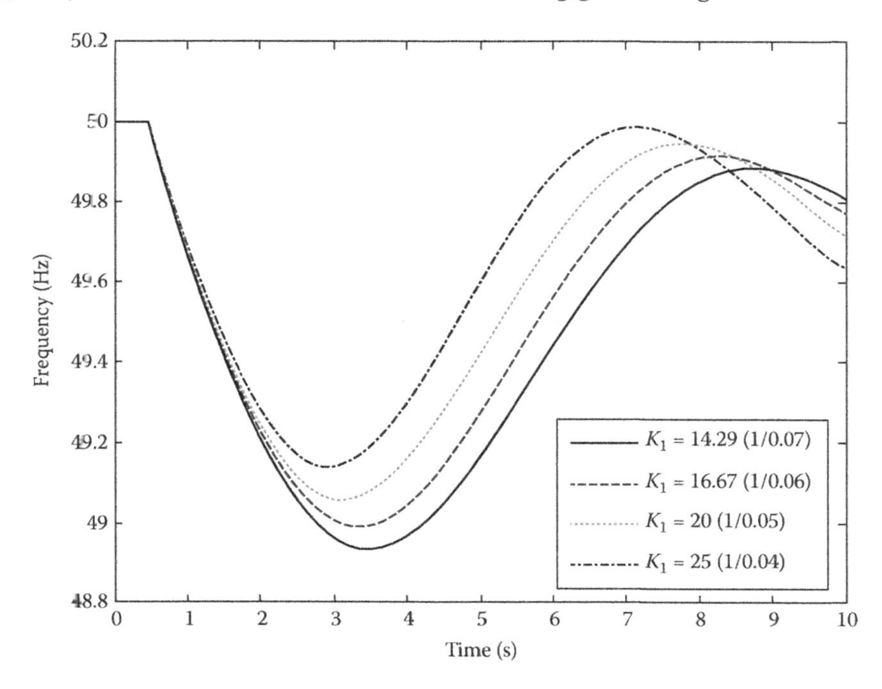

FIGURE 2.14
The impact of varying the generator gain K_1 on frequency.

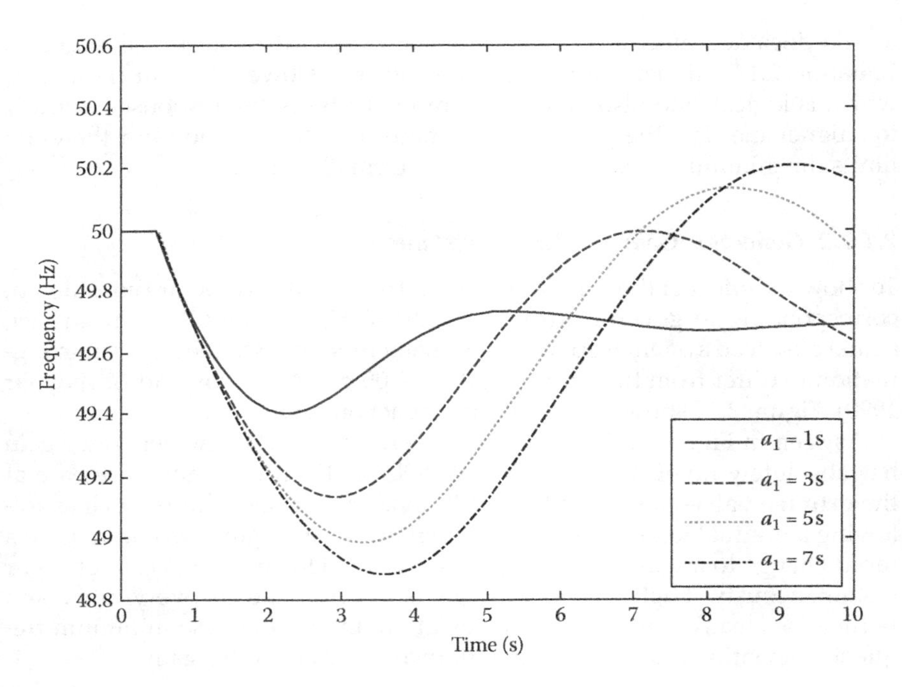

FIGURE 2.15
The impact of generator time constant a_1 on frequency.

Figures 2.15 and 2.16 depict the impact of the generator time constants a_1 and b_1 for different values. The time constants a_1 of 1, 3, 5, and 7 s are typical for gas, diesel, and steam turbines, respectively. The major effect of the generator time constant a_1 is to produce a lag in the response of the frequency following its initial dip. This lag also increases the minimum frequency deviation and delays the peak values of this maximum in proportion to the size of the time constant. In other words, this parameter has an effect on the response and the total time of exposure to low frequencies. Higher values also cause a larger frequency overshoot, as seen in Figure 2.15. Similar conclusions can be drawn for a second-order model.

The major effect of the generator time constant b_1 is to accelerate the response of the turbine-governor system and to decrease both the frequency deviation and the peak values of its oscillation in proportion to the size of the time constant. In other words, this parameter has an effect on the response and the total time of exposure to low frequencies. The accelerated response with higher values of b_1 is because this parameter represents the sensitivity of the model to the rate of change of its input, that is, the frequency. It is interesting that for b_1 larger than a_1, the oscillatory behavior of the response disappears.

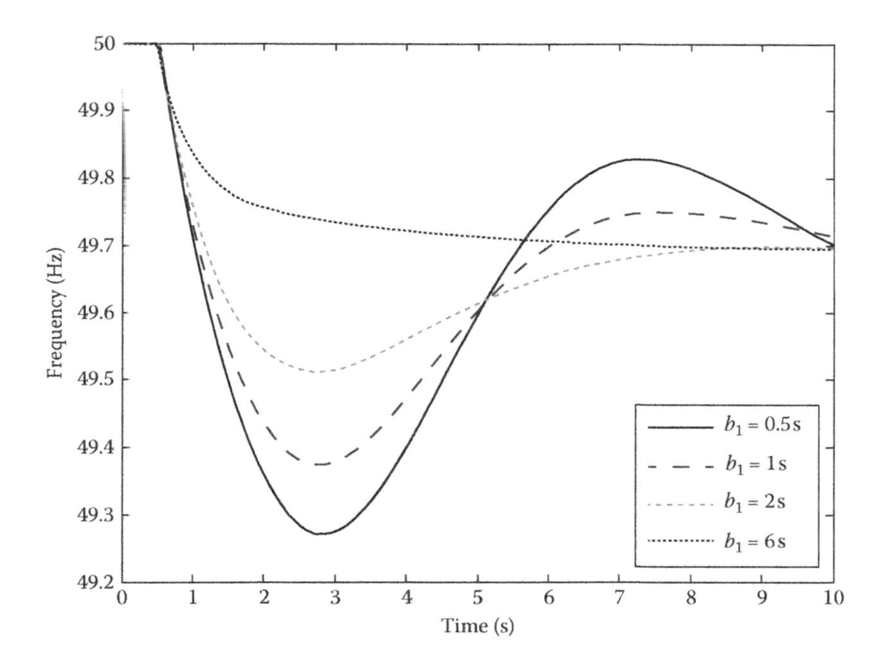

FIGURE 2.16
The impact of generator time constant b_1 on frequency.

2.1.3.3 Generator Output Limitations

Previous SFD models omitted generator output limitations for the analysis of short-term frequency dynamics. This assumption is, however, not valid for small isolated power systems, in which large disturbances can occur. Usually, spinning reserve is limited and covers for the outage of the largest generating unit. Figure 2.17 shows the impact of finite reserve on frequency. In fact, total generator output limitation Δp_{max} has been set to 0.17 and 0.1 pu, which is, respectively, slightly above and clearly below the amount of lost real power. Furthermore, for the case of 0.17 pu, both uniformly and nonuniformly distributed finite spinning reserve have been considered.

Obviously, the generator output limitation slowed down the frequency recovery. In addition, in the case of nonuniformly distributed reserve, frequency reduced further and gained its minimum value later than in the case of uniformly distributed spinning reserve. Thus, in multigenerator systems, generator output limitations could also retard the overall primary frequency control action if spinning reserve is nonuniformly distributed. Finally, if the amount of spinning reserve is lower than the amount of lost power, frequency decay cannot be arrested, and the system becomes frequency unstable. This is also illustrated in Figure 2.17. Thus, generator output limitations significantly

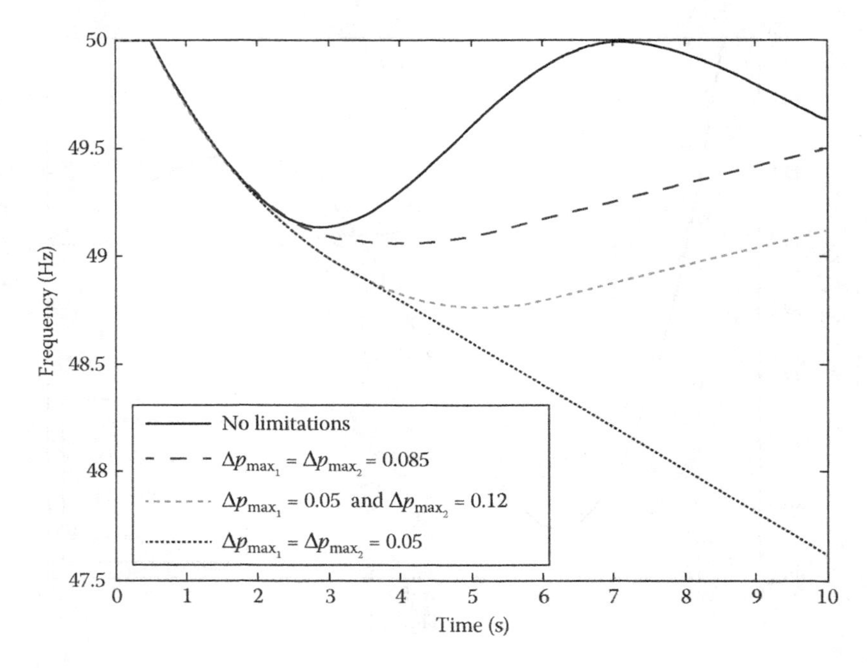

FIGURE 2.17
The impact of generator output limitations on frequency.

affect short-term frequency dynamics and frequency stability and must be taken into account in the power system model of small isolated power systems.

2.1.3.4 Load-Damping Factor

The effect of the load-damping factor D on frequency can be illustrated by plotting the system response for various values of this parameter. Figure 2.18 compares the impact on frequency for three different values. The Union for the Co-ordination of Transmission of Electricity (UCTE), for example, recommends a load-damping factor of 1 (UCTE 2004).

Comparing Figure 2.14, which shows the variation of K, and Figure 2.18 and taking into account Equation 2.19 shows that the effect of varying D and K is much the same, but K plays a much more important role than D. Usually, K amounts to 25 pu, whereas D ranges between 0.5 and 1 pu. Thus, even though the load has a frequency-dependent component, it is not nearly as important as the generator gain with respect to the impact on short-term frequency dynamics.

2.1.3.5 UFLS Scheme

UFLS schemes protect power systems against underfrequency conditions; therefore, the impact of a UFLS scheme on short-term frequency dynamics

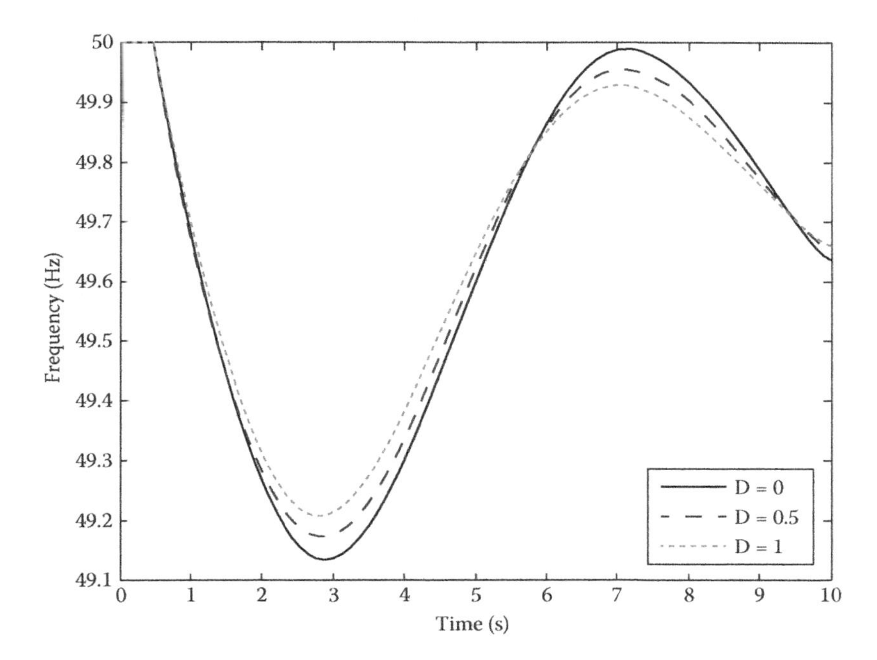

FIGURE 2.18
The impact of load-damping factor on frequency.

is evaluated. Again, the SFD model of Figure 2.8 is intended to represent two generating units, with each generating unit represented by a first-order model, as displayed in Table 2.2. A power loss of 0.15 pu is applied. The maximum generator output limitation is lower than the amount of lost real power.

Figure 2.17 has shown that for the SFD model defined by Table 2.2, the system becomes frequency unstable, since there is not enough spinning reserve available to cover the loss of 0.15 pu. Clearly, such an underfrequency condition would lead to a system blackout due to underfrequency turbine tripping. As a consequence, a simple single-stage UFLS scheme is implemented. Table 2.3 displays the parameters of the UFLS scheme. The value of the step size has been chosen such that the step size and the amount of available spinning reserve are at least equal to the amount of lost power.

TABLE 2.2
SFD Model Parameters With Generator Output Limitations

Generator	*H* (s)	*K*	a_1 (s)	b_1 (s)	M_{base}(MW)	Δp_{min}	Δp_{max}
1	3	25	3	0	130	–	0.05
2	5	25	5	0	70	–	0.05

TABLE 2.3

UFLS Scheme Parameters

Stage	Threshold (Hz)	Intentional Delay (s)	Delay (s)	Step Size (pu)
1	49	0	0	0.08

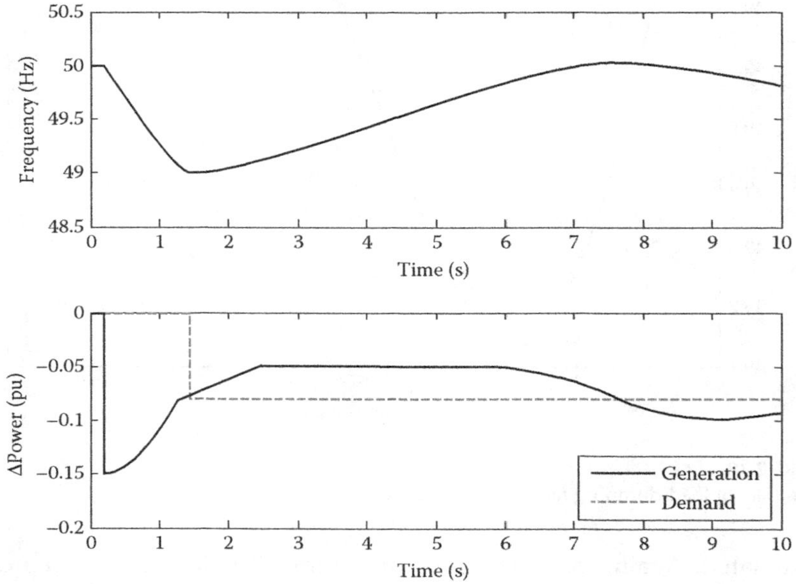

FIGURE 2.19
The impact of a UFLS scheme on frequency and real power.

Figure 2.19 shows the impact of the UFLS scheme on both the frequency and the power imbalance. The amount of 0.08 pu is shed 0.43 s after the contingency occurred. Clearly, the system remains stable. In fact, the frequency is a long way from the allowable minimum frequency of 47.5 Hz.

Figure 2.20 depicts the effect of the UFLS scheme in the ω–$d\omega/dt$ phase-plane. The initial ROCOF is −1.014 Hz, which corresponds with the theoretical amount determined by Equation 2.20. Furthermore, the UFLS action at 49 Hz is also clearly observable due to the abrupt change in the ROCOF.

2.2 Impact of Renewable Energy Sources

Currently, decoupled power generation (DPG) essentially suffers from two drawbacks: its limited primary frequency control capacity and its negligible

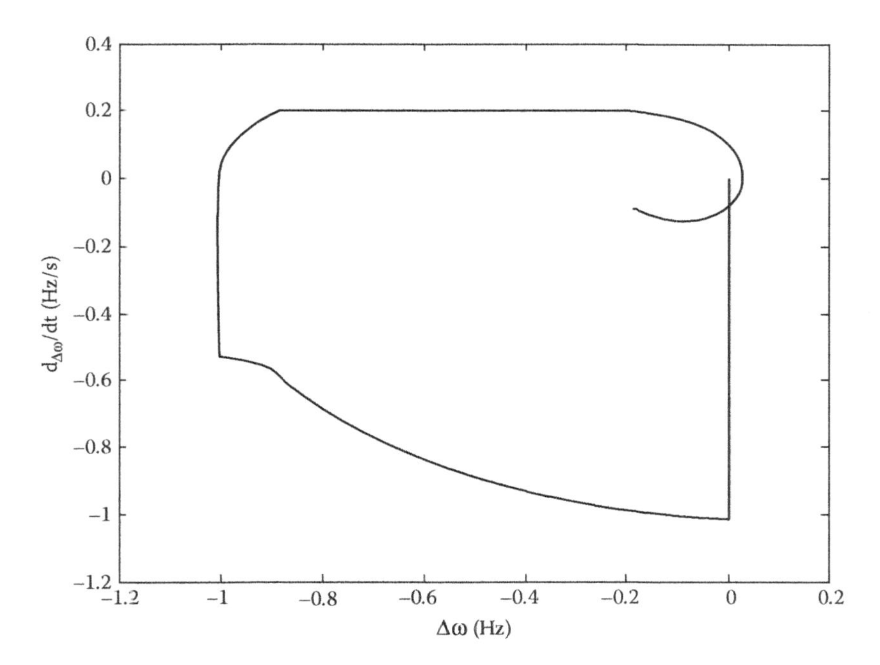

FIGURE 2.20
Phase-plane representation.

inertial response. Both the lack of primary frequency control and the lack of inertial response have a negative impact on frequency stability. Substituting conventional generation by DPG further weakens the system's ability to withstand large disturbances. In fact, the initial ROCOF increases with decreasing inertia. Furthermore, the availability of sufficient spinning reserve is critical for system stability. Thus, the smaller the inertia and the smaller the available spinning reserve, the larger the frequency deviations.

The differences between the characteristics of conventional and DPG and the provision of an increasing amount of installed capacity of DPG require contemplating scenarios with different levels of DPG penetration for the analysis of short-term frequency dynamics. Due to its specific characteristics, DPG can be represented as a negative load omitting any dynamics. In other words, and in terms of the nonlinear multigenerator SFD model of Figure 2.8, DPG is modeled like a generating unit but with zero inertia H_i and zero gain K_i.

To analyze the impact of DPG penetration on short-term frequency dynamics, consider the power system shown in Figure 2.21, consisting of a CG and, for example, a wind farm represented by an equivalent generator (DPG) (Rouco et al. 2008). This system can be represented by an SFD model, in which the DPG is represented as a generator without inertia and with zero gain. Table 2.4 displays the parameters of the SFD model.

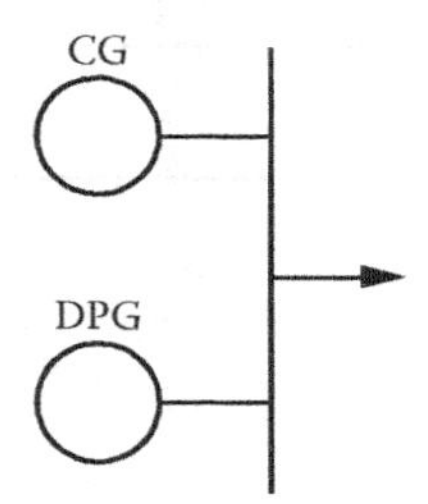

FIGURE 2.21
Simple power system with decoupled power generation.

TABLE 2.4
SFD Model Parameters With DPG

Generator	H (s)	K	a_1 (s)	b_1 (s)	M_{base}	Δp_{min}	Δp_{max}
CG	5	25	3	0	100	–	–
DPG	–	–	–	–	100	–	–

According to Equation 2.5, the equivalent inertia of the simple power system is 2.5 s. Figure 2.22 shows the impact of the presence of DPG on short-term frequency dynamics and compares the responses with and without DPG (see Table 2.1). Clearly, the initial ROCOF increases due to the reduction of equivalent inertia. Furthermore, primary frequency control capacity is equally affected, since the gain is reduced, according to Equation 2.7, to 12.5 pu. Thus, scenarios with DPG penetration need to be taken into account for the analysis of short-term frequency dynamics and for the design of UFLS schemes in particular. The proposed SFD model allows DPG to be included by simply adjusting its parameters.

Appendix 2A: Turbine-Governor System Reduction and Tuning

Figure 2A.1 shows the block diagram of a generic second-order model of a turbine-governor system. Governor reference is constant, since there is no secondary control, and therefore, frequency deviation is directly fed, with positive sign, into the turbine-governor second-order block. This model also includes the generator power-output limitations $\Delta p_{i,min}$ and $\Delta p_{i,max}$. The parameter K_i represents the gain of the turbine-governor system, which is usually the inverse of the governor droop and is not tunable. The parameters $a_{1,i}$ and $a_{2,i}$ and $b_{1,i}$ and $b_{2,i}$ correspond to the dominant poles and possible zeros of the turbine-governor system. The tunable parameters of the second-order model, that is, $a_{1,i}$, $a_{2,i}$, $b_{1,i}$, and $b_{2,i}$, are adjusted such that the response

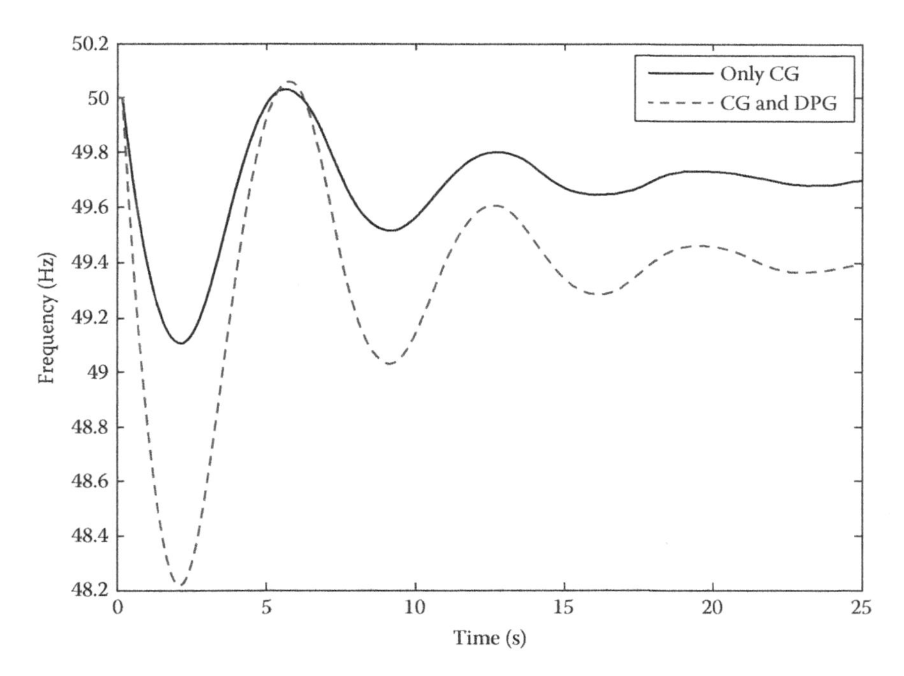

FIGURE 2.22
The impact of decoupled power generation on frequency.

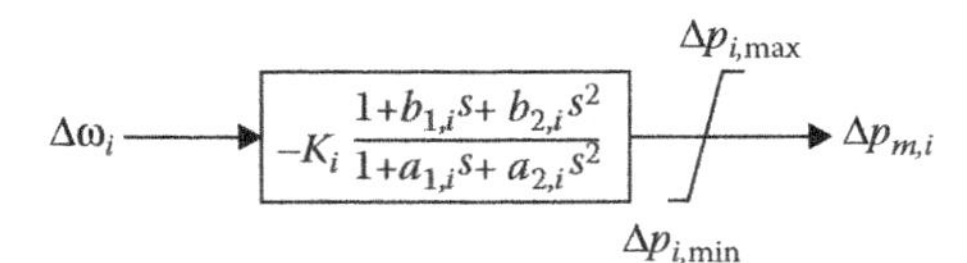

FIGURE 2A.1
Generic second-order model approximation of the turbine-governor system.

of the model resembles as much as possible the response of the complete model of the turbine-governor system. Optionally, these parameters can also be obtained from field tests; that is, the response of the simplified model is adjusted to resemble as much as possible the actual test responses.

Tuning of $a_{1,i}$, $a_{2,i}$, $b_{1,i}$, and $b_{2,i}$, consists in minimizing the difference between the simplified model response and the response of the detailed model of the turbine-governor system, which is basically a parameter estimation problem. Well-known algorithms exist for parameter estimation problems (Bates and Watts 1988; Ljung 1987). The estimation process used here is similar to the one proposed in Criado et al. (1994) and Rouco et al. (1999), where model responses are given by time-domain responses, and the model parameters are determined by means of a nonlinear estimation algorithm.

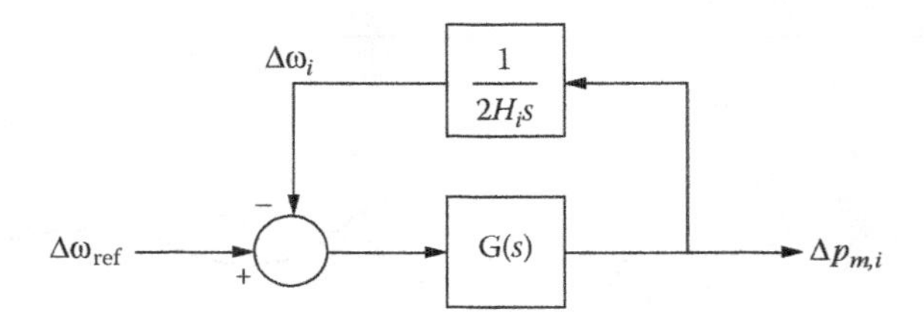

FIGURE 2A.2
Block diagram of the interaction between turbine-governor system and rotor.

The present parameter estimation problem essentially depends on four factors: the accuracy of the detailed model, the time interval chosen to simulate the simplified and the detailed model, the model structure (open loop or closed loop), and the selection of the simplified model type. The parameter estimation procedure is based on either an OL or a CL structure. An OL structure only considers the turbine-governor system $G(s)$, whereas a CL structure also contemplates the feedback loop depending on the rotor dynamics, as illustrated in Figure 2A.2.

The choice of the model structure normally depends on the response for which the simplified model should be adjusted. Field tests, for example, can be carried out with the generating unit either online or off-line, depending on the signal to be measured and on the test applied (Hannet and Khan 1993; Pereira et al. 2003). Speed deviations due to changes at the governor reference can only be measured with the generating unit off-line (Berube and Hajagos 1999), resulting in a CL parameter estimation. By contrast, online tests can be carried out if active power deviation due to a step at the governor reference is measured as in Tor et al. (2004), requiring an OL structure. If parameters of detailed turbine-governor systems are available, both OL and CL structures can be used.

In Egido et al. (2010), parameters of several first-order models have been estimated by applying a step input to the first-order model and the linearized detailed model and minimizing the difference of the OL outputs. A step input seems appropriate if online measurements to a change at the governor reference are used. Another possibility consists of applying a ramp input, since the response of the turbine-governor system and the system frequency could be represented by a ramp during the first few instants after a system-wide disturbance (Rouco et al. 2008). In fact, the impact of primary control on the frequency is still negligible, and frequency falls with a constant rate. Based on the OL responses to a ramp input, various adjustments have then been realized with different simulation times ranging from 1 to 4 s. This time range usually comprises the points in time when minimum frequency occurs. These points in time are of particular interest for the analysis of short-term frequency dynamics, and especially for the design and analysis of UFLS schemes. As an example, Table 2A.1 displays the parameters of the first-order model approximations of an

TABLE 2A.1

Comparison of Parameter Turnings with Different Simulation Times

Model	1s		3s		5s	
	K	a_1 (s)	*K*	a_1 (s)	*K*	a_1 (s)
IEEEG1	16.66	11.00	16.66	6.90	16.66	6.10
DEGOV1	25.00	0.51	25.00	0.75	25.00	0.92

IEEEG1 turbine-governor system using three different simulation time settings (1, 3, and 5 s). Figure 2A.3a compares the CL responses of the four first-order models with the CL response of the linearized complete IEEEG1 model. Clearly, with increasing simulation time, the accuracy of the first-order model increases as well. This is because the considered IEEEG1 turbine-governor system is a rather slow system. For faster turbine-governor systems, shorter simulation time settings lead to more accurate low-order models, as in the case of the DEGOV1 model and its first-order model shown in Figure 2A.3b and Table 2A.1. In fact, due to the extended simulation time, more weight has been given to posttransient data, distorting the curve-fitting problem.

Along with the tuning of simplified models based on OL responses to step or ramp inputs, CL responses to step inputs could also be used to adjust the simplified models. The feedback is formed by the interaction of the turbine-governor system and the rotor dynamics, as illustrated in Figure 2.2. With this model structure, the tuning process is still dependent on the simulation time, but to a much lesser extent. For example, too long a simulation time led to misleading results in the case of the OL adjusted DEGOV1 model, because adjustments between the detailed and simplified models toward the end of the simulation time outweighed the initial differences. Usually, simulation time settings around 3–5 s lead to good results and allow the dynamics of the turbine-governor system to be represented. A good estimation for the simulation time is given by twice the point in time of the maximum of the output of the complete turbine-governor system. In addition, the CL structure delivers a more appropriate input in terms of frequency deviation to the turbine-governor system than the ramp input. This could be important, since the input should resemble as much as possible those inputs for which the model should be most accurate (Levine 1999). Figure 2A.3 compares the CL responses of the simplified models adjusted by means of a CL structure with the responses of the linearized turbine-governor systems. In addition, the results are compared with those obtained when using an OL structure. For both the DEGOV1 and the GASTWD turbine-governor system, the CL-based parameter estimations seem to be more accurate.

Finally, the parameters of the five different first- and second-order model types are estimated by using both OL and CL structures. The most accurate simplified models are then selected and compared. The most accurate model type is the one that minimizes the parameter estimation problem, that is, the one that minimizes the difference between the responses of the simplified

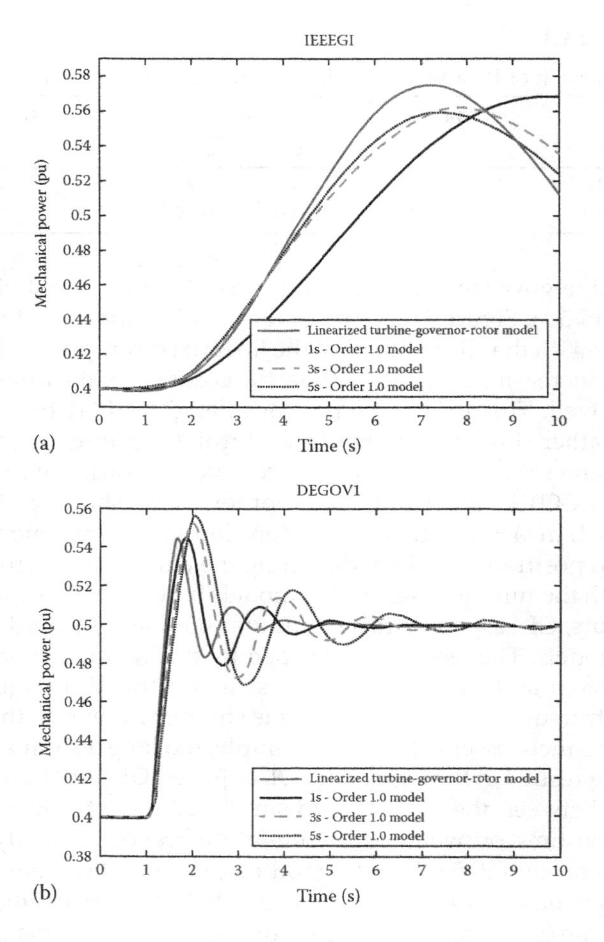

FIGURE 2A.3
Comparison of CL responses of (a) the linearized IEEG1 model and simplified models and (b) the linearized DEGOV1 model and simplified models adjusted by using different simulation times to a step of 0.1 pu.

and the original turbine-governor system model. Figure 2A.4 compares the responses of the most accurate simplified models, adjusted by using both OL and CL structures, with the responses of the linearized models of DEGOV1, GASTWD, and GAST2A turbine-governor systems.

For the turbine-governor systems shown, the results of the tuning process using either an OL or a CL model structure differ, and in general, the adjustment based on a CL model structure seems to be more accurate. Different turbine-governor systems require different models. For example, the GASTWD model of Figure 2A.4b seems to be best represented by a

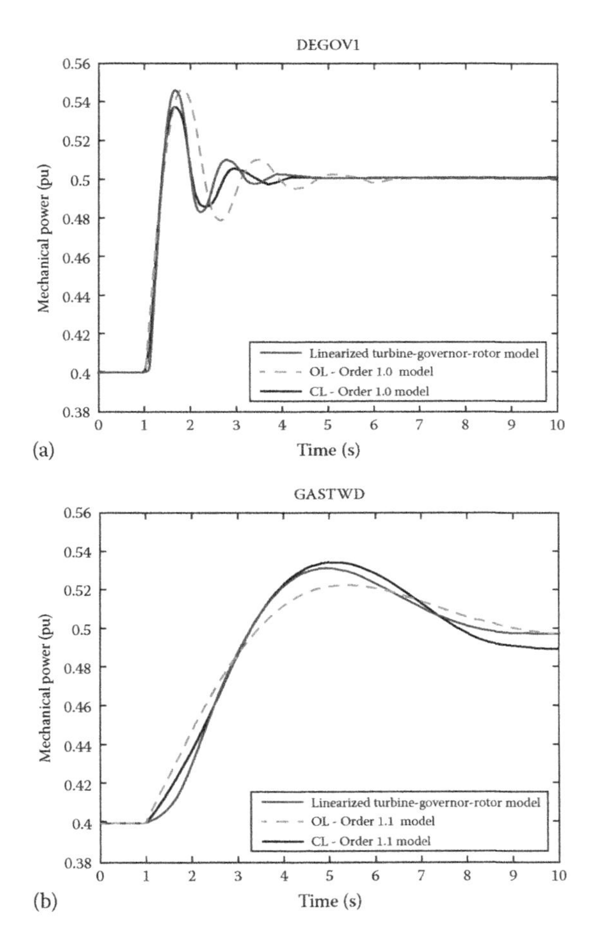

FIGURE 2A.4
Comparison of CL responses of (a) the linearized IEEEG1 model and most accurate simplified models and (b) the linearized DEGOV1 model and most accurate simplified models adjusted by using OL and CL model structure to a step of 0.1 pu.

second-order model with one zero (CL—Order 2.1), whereas the DEGOV1 model of Figure 2A.4a can be modeled by a second-order model without zeros (CL—Order 2.0). The GAST2A model also seems to be best represented by a second-order model without zeros (CL—Order 2.0). It is interesting that in the case of the DEGOV1 model, the OL structure yields to a second-order model with one zero, whereas the CL structure gives rise to a second-order model without zeros. By any measure, it seems that the simplified model represents the complete turbine-governor system model quite accurately.

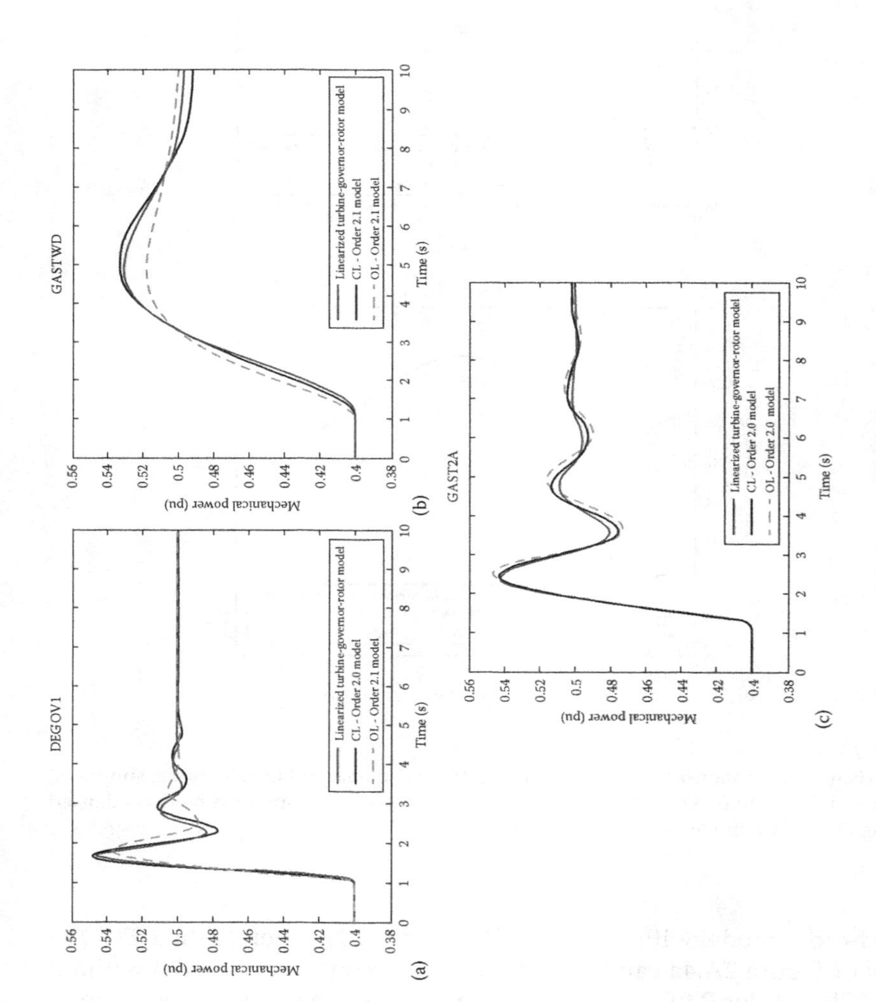

FIGURE 2A.5
Comparison of the responses of (a) the simplified and the linearized DEGOV1 model, (b) the simplified and the linearized GASTWD model, and (c) the simplified and the linearized GAST2A model with a step at the reference of the turbine-governor system.

References

Anderson, P. M. 1999. *Power System Protection*. Piscataway, NJ: IEEE Press.

Anderson, P. M., and M. Mirheydar. 1990. A low-order system frequency response model. *IEEE Transactions on Power Systems* 5 (3):720.

Bates, D. M., and D. G. Watts. 1988. *Nonlinear Regression and Its Applications*. New York: Wiley.

Berube, G. R., and L. M. Hajagos. 1999. Modeling based on field tests of turbine/governor systems. Power Engineering Society 1999 Winter Meeting, IEEE.

Chan, M. L., R. D. Dunlop, and F. Schweppe. 1972. Dynamic equivalents for average system frequency behavior following major disturbances. *IEEE Transactions on Power Apparatus and Systems* 91 (4):1637.

Crevier, D., and F. C. Schweppe. 1975. The use of Laplace transforms in the simulation of power system frequency transients. *IEEE Transactions on Power Apparatus and Systems* 94 (2):236.

Criado, R., M. Gutiérrez, J. Soto, R. Suarez, F. L. Pagola, L. Rouco, J. L. Zamora, and A. Zazo. 1994. Identification of excitation system models for power system stability studies. CIGRÉ Session 1994, Paris. Paper No. 38–304.

Denis Lee Hau, Aik. 2006. A general-order system frequency response model incorporating load shedding: Analytic modeling and applications. *IEEE Transactions on Power Systems* 21 (2):709.

Egido, I., F. Fernández, P. Centeno, and L. Rouco. 2010. Maximum frequency deviation calculation in small isolated power systems. *IEEE Transactions on Power Systems* 24 (4):1731–1738.

Elgerd, O. I. 1982. *Electric Energy Systems Theory: An Introduction*. New York: McGraw-Hill.

Germond, A. J., and R. Podmore. 1978. Dynamic aggregation of generating unit models. *IEEE Transactions on Power Apparatus and Systems* 97 (4):1060.

Hannet, L. N., and A. Khan. 1993. Combustion turbine dynamic model validation from tests. *IEEE Transactions on Power Systems* 8 (1):152.

Hoffman, J. D. 2001. *Numerical Methods for Engineers and Scientists*, 2nd Edition. New York: Marcel Dekker.

Horne, J., D. Flynn, and T. Littler. 2004. Frequency stability issues for islanded power systems. In *Power Systems Conference and Exposition, 2004*. IEEE PES. New York, NY.

IEEE. 1973. Dynamic models for steam and hydro turbines in power system studies. *IEEE Transactions on Power Apparatus and Systems* 92 (6):1904.

IEEE. 1991. Dynamic models for fossil fueled steam units in power system studies. *IEEE Transactions on Power Systems* 6 (2):753.

IEEE/CIGRE, Joint Task Force on Stability Terms and Definitions. 2004. Definition and classification of power system stability. *IEEE Transactions on Power Systems* 19 (2):1387–1401.

Kendall, A., W. Han, and D. Stewart. 2009. *Numerical Solution of Ordinary Differential Equations*. Chichester, UK: Wiley.

Kottick, D., and O. Or. 1996. Neural-networks for predicting the operation of an under-frequency load shedding system. *IEEE Transactions on Power Systems* 11 (3):1350.

Levine, W. S., ed. 1999. *The Control Handbook (Volume 1)*. Boca Raton, FL: CRC Press.

Ljung, L. 1987. *System Identification: Theory for the User*. Englewood Cliffs, NJ: Prentice-Hall.

Machowski, J., J. W. Bialek, and J. R. Bumby. 1997. *Power System Dynamics and Stability*. Chichester, UK: Wiley.

Mitchell, M. A., J. A. P. Lopes, J. N. Fidalgo, and J. D. McCalley. 2000. Using a neural network to predict the dynamic frequency response of a power system to an under-frequency load shedding scenario. *Power Engineering Society Summer Meeting, 2000*. IEEE.

New, W. C. 1977. *Load Shedding, Load Restoration and Generator Protection Using Solid-State and Electromechanical Underfrequency Relays*. Philadelphia, PA: General Electric Company.

Pereira, L., J. Undrill, D. Kosterev, D. Davies, and S. Patterson. 2003. A new thermal governor modeling approach in the WECC. *IEEE Transactions on Power Systems* 18 (2) 819–829.

Rouco, L., J. L. Zamora, A. Zazo, M. A. Sanz, and F. L. Pagola. 1999. A comprehensive tool for identification of excitation and speed-governing systems for power system stability studies. In *13th Power Systems Computation Conference (PSCC '99)*, Trondheim, Norway.

Rouco, L., J. L. Zamora, I. Egido, and F. Fernández. 2008. Impact of wind power generators on the frequency stability of synchronous generators. *Water and Energy Abstracts* 19 (2):46.

Siemens. 2005. PSS/E 30.2 Program Application Guide (Volume II). Schenectady, NY, USA.

Stanton, K. N. 1972. Dynamic energy balance studies for simulation of power-frequency transients. *IEEE Transactions on Power Apparatus and Systems* 91 (1):110.

Tor, O. B., U. Karaagac, and E. Benlier. 2004. Step-response tests of a unit at Ataturk hydro power plant and investigation of the simple representation of unit control system. *IEEE PES, 36th North American Power Symposium*, University of Idaho, Moscow.

UCTE. 2004. *Operation Handbook Policy 1: Load-Frequency Control and Performance*.

3

Frequency Protection

This chapter presents and describes frequency protection schemes for island power systems. Further, it provides insight into the design of conventional and advanced frequency protection schemes and their applications to island power systems.

The magnitude of frequency excursion mainly depends on the magnitude of the power imbalance and on the way the power system behaves following the disturbance. To avoid damage to generation and load-side equipment, it is crucial to operate the power system within an established, acceptable frequency range or, at least, to prevent longer frequency excursions out of this established range. Underfrequency operation is usually more problematic with regard to possible equipment damage than overfrequency operation due to the limited amount of spinning reserves.

Frequency protection consists of two fundamental protection systems of different scope: a system-wide underfrequency load-shedding (UFLS) scheme and underfrequency protection of individual generation and load equipment. The former can be considered as a higher-level frequency protection of the generation and load equipment protected by the latter (IEEE 2007). UFLS schemes are system protection schemes (in contrast to equipment protection, which tries to minimize equipment damage). Actually, the primary objective of UFLS schemes is to arrest frequency decay in due time, ideally preventing the action of generation-side underfrequency protection, which would finally lead to a cascading power outage. This implies that UFLS and underfrequency equipment protections need to be coordinated.

In theory, the power imbalance can be counterbalanced either by readjusting power generation by means of primary frequency control or by readjusting load demand. However, for large disturbances, primary frequency control action is usually not fast enough to restore the power balance and to confine frequency excursions to acceptable values. Furthermore, the available spinning reserve could be insufficient to cover the lost power generation. However, some efforts have been made to reduce load shedding during the dispatch phase by redispatching committed generation to reduce potential load shedding (Padron et al. 2016). Alternatively, frequency decay can be altered by shedding specific load blocks, reducing the initial power imbalance caused, for example, by a generation outage. If the system survives as a result of successful load shedding, fewer customers are affected and system restoration is accomplished within a much shorter time (in the order of minutes). Hence, UFLS schemes not only protect generation and load-side equipment, but also improve power system reliability and security and minimize the socioeconomic impact on utilities and customers.

The rest of the chapter is dedicated to the review, design, and application of UFLS schemes to island power systems. First, current practices and classes of UFLS are reviewed. Then, the design of a conventional UFLS scheme and its application to island power systems are presented. Conventional UFLS schemes are designed by means of off-line simulation studies, rendering them inherently inadaptable. In a second step, advanced UFLS schemes are presented and applied. These schemes are adaptive in nature by adjusting their response as a function of the power imbalance and even the system's state prior to the disturbance.

3.1 Review of Frequency Protection

UFLS schemes can be divided into manual and automatic UFLS schemes. Manual UFLS actions are generally applied to frequency restoration problems and are rarely involved in the confinement of frequency excursions. By contrast, automatic UFLS schemes are principally used to arrest frequency decay and secondarily to bring frequency back to an acceptable steady-state value if necessary. To date, automatic UFLS schemes are mostly conventional UFLS schemes (NERC 2008; IEEE 2007), which measure frequency and optionally the rate of change of frequency (ROCOF) by means of type 81 relays and shed a predefined amount of load in the case of frequency and/or ROCOF fall below a certain threshold. Advanced automatic UFLS schemes, overcoming the major disadvantages of conventional UFLS schemes, have been discussed in the literature since the early 1990s, but to date they have hardly been used in island power systems.

3.1.1 Current Practices

UFLS practices vary from power system to power system, depending on the characteristics of the system and its operation criteria imposed by the system operator. For example, the European Network of Transmission System Operators for Electricity (ENTSO-e) states that the required amount of reserve amounts to the size of the synchronous zone's reference incident (e.g., 3000 MW for continental Europe). A disturbance equivalent in size to the reference incident should not activate load shedding. Generating units are allowed to disconnect if frequency falls to 47.5 Hz (UCTE 2004).

Table 3.1 summarizes the operation criteria of several interconnected and island power systems. A state of the art of grid codes of selected island and interconnected power systems can be found in Merino et al. (2014).

TABLE 3.1

Operation Criteria in Terms of Most Demanding Frequency Operation Range and of Primary Reserve Requirements

System	Frequency Range	Primary Reserve Requirement
Continental Europe (ENTSO-e 2013)	47.5–48.5 Hz during >30 min	Reference incident of 3000 MW
Nordel (Nordel 2007; ENTSO-e 2013)	47.5–48.5 Hz during 30 min	Normally largest unit
NGET (NGET 2009)	47–47.5 Hz during 20 s	Loss of generation or power infeed of external interconnections
Ireland (EirGrid & SONI) (EirGrid 2009; Ela et al. 2010)	47–47.5 Hz during 20 s	Largest unit or power infeed plus contribution to RES
Iceland	47.5–48.5 Hz during 30 min	70 MW
Cyprus (Merino et al. 2014)	47.3 Hz for 5 s	60–70 MW[a]
EDF-SEI	46 Hz for 0.4 s	Largest unit
HECO (HECO 2008; Matsuura 2009)	56 Hz for 6 s	Largest unit in Ohau
HELCO and MECO	56 Hz for 6 s	RES increase[b]
SEIE (MITyC 2006)	47.5 Hz for 3 s	Largest unit or large RES decrease or largest demand increase

[a] Trade-off between load-shedding and reserve.
[b] Load shedding is part of reserve.

Usually, reserve requirements are such that the loss of the largest generating unit is covered. Larger, interconnected systems such as the European continental system consider a reference incident, for example, the loss of a complete generation station. By contrast, the reserve requirement for the Spanish island systems, for instance, explicitly states that reserves must cover at least 1.5 times the power generation of a gas turbine of two-plus-one combined-cycle power plants (i.e., loss of about the power of the gas turbine and its contribution to the steam turbine). Some systems also specifically include expected variations of nondispatchable renewable energy sources. Finally, certain island systems include load shedding within their reserves to reduce costs, partially shifting load shedding from a system protection to a system operation level. In fact, Cheng and Shine (2003) consider costs related to both load shedding and spinning reserve provision for the design of the UFLS scheme.

Table 3.1 might indicate that larger systems are more demanding with regard to frequency ranges, since longer underfrequency operations are reported. Note, however, that larger systems only specify underfrequency operation down to 47.5 Hz, in general between 48.5 and 47.5 Hz, and they usually allow generating units to disconnect for frequencies below 47.5 Hz.

Smaller systems, by contrast, specify underfrequency operations down to 46 Hz, although with operation times in the order of seconds. The specified frequency ranges apply in general to thermal generating units, which are more sensitive to frequency deviations. Diesel engines, for example, can withstand transitory frequency levels as low as 43 Hz. Further, not all of these requirements for generating units might be satisfied during real disturbances, because grid codes are not retroactive, and older generating units are excluded from new requirements. Moreover, generating units have to comply with these requirements only if equipment safety is not jeopardized, and finally, the action of other protection devices (such as overcurrent relays) could trip the generating unit before critical frequency values are reached.

Operation criteria, and especially underfrequency operation requirements for generating units, define UFLS practices, since turbine underfrequency protection and the UFLS scheme need to be coordinated. Table 3.2 compares UFLS practices according to the grid codes of selected system operators. Others can also be found in Transpower (2009). Settings attributed to the National Grid Electricity Transmission (NGET) coincide with the settings of the NGET power system and do not include settings of the Scottish Power Transmission (SPT) and Scottish Hydro-Electric Transmission (SHETL) systems. Further, EirGrid settings are those of the Electricity Supply Board (ESB) (IEEE 2007). Finally, it should be kept in mind that this is only a sample of possible settings. UCTE recommendations have also been included.

It can be inferred from Table 3.2 that load under relief is usually divided into three to six discrete blocks with frequency thresholds in the range of 48–49 Hz in the case of larger systems and with frequency thresholds between 47 and 49 Hz for smaller power systems. Further, these discrete blocks correspond to step sizes of 5%–15% of system demand. In any case, settings are determined by the system operator. A distinction between smaller and larger systems can also be made with respect to anticipated actions. In fact, smaller systems only shed interruptible loads as an anticipated measure, whereas larger systems also use anticipated measures such as disconnection of pumped storage units or connection of quick-start power plants. NGET and EirGrid grid codes plan to manually disconnect demand and to reconnect load shed by the UFLS scheme if load shed by the UFLS is not restored within a reasonable period of time. The purpose of such action is to ensure that a subsequent fall in frequency will again be contained by the operation of the UFLS scheme. Finally, Red Eléctrica de España (REE) grid code proposes the use of ROCOF steps due to the small size of the isolated power systems and their inherent sensitivity to large disturbances. Although not presented in Table 3.2, ROCOF relays are also employed among others by the Argentinean system operator Compañia Administradora del Mercado Mayorista Eléctrico (CAMMESA) (Cammesa 2008) and the operator of the central Chilean system Cuenta oficial del Centro de Despacho Económico de Carga del Sistema Interconectado Central (CDEC-SIC) (CDEC-SIC 2008).

TABLE 3.2

Underfrequency (UF) Load-Shedding Practices (Partially Taken from IEEE [2007])

Entity	UFLS Scheme								
	Anticipated Actions	Step 1	Step 2	Step 3	Step 4	Step 5	Step 6	Step 7	Step 8
UCTE	Quick-start turbines, disconnection pumping units	49 Hz 10%–20%	48.7 Hz 10%–15%	48.4 Hz 10%–15%					
Nordel	Quick-start turbines, HVDC emergency action	48.8 Hz 6%–10%	48.6 Hz 6%–10%	48.4 Hz 6%–10%	48.2 Hz 6%–10%	48 Hz 6%–10%			
NGET[a]	Quick-start turbines, disconnection pumping units, interruptible loads	48.8 Hz 5%	48.7 Hz 15%	48.6 Hz 15%	48.5 Hz 7.5%	48.4 Hz 7.5%	48.2 Hz 7.5%	48 Hz 5%	47.8 Hz 5%
EirGrid[b]	Disconnection pumping units, interruptible loads	48.5 Hz 12%	48.4 Hz 12%	48.3 Hz 12%	48.2 Hz 12%				
HECO	Interruptible loads	58.5 Hz 5%	58 Hz 10%	57.7 Hz 10%	57.4 Hz 10%	57.2 Hz 10%	57.0 Hz 5%–10%		
REE	Interruptible loads	UF and ROCOF: Frequency thresholds within 50–47 Hz							

[a] NGET settings only.
[b] ESB settings.

3.1.2 Classes of Automatic UFLS

Two different categories of automatic UFLS schemes can be broadly devised: conventional and advanced automatic UFLS schemes. Conventional schemes can be further differentiated into static and semi-adaptive schemes. Static schemes use only underfrequency relays, whereas semi-adaptive schemes also employ ROCOF relays (Mohd Zin et al. 2004; Delfino et al. 2001). Even though static UFLS schemes are very quick and simple to implement, they present disadvantages such as over- or undershedding in the presence of contingencies that these schemes were not designed for. The measurement of the ROCOF enables semi-adaptive UFLS schemes to distinguish between smaller and larger disturbances (thus the term *semi-adaptive*), reducing but not avoiding the risk of overshedding. The ROCOF threshold may be implemented in a separate step or in combination with an existing step of the static UFLS scheme. In addition, they improve the frequency response because they shed loads earlier in the case of larger disturbances. However, ROCOF measurement is delicate, since ROCOF is distorted by intermachine oscillations (Concordia et al. 1995; Novosel et al. 1996), making the measurement slow and unreliable and requiring, for example, integrated (Li and Jin 2006) or average ROCOF measurements. Finally, off-line contingency simulations are carried out to determine the relay parameters and the amount to be shed of both static and semi-adaptive UFLS schemes.

By contrast, advanced UFLS schemes also use additional information and measurements such as power generation and voltage to compute the amount of load that needs to be shed. The amount of load to be shed is highly dependent on the actual power imbalance, which can be estimated by measuring the initial ROCOF succeeding the disturbance. Advanced UFLS schemes are thus highly adaptive in the sense that they adjust their response as a function of the actual power system conditions. Advanced UFLS schemes comprise adaptive (also known as dynamic) and centralized UFLS schemes. Adaptive schemes in principle work like their conventional counterparts, but instead of shedding a predefined amount of load at each stage, they determine the quantity of load that needs to be curtailed at each stage. The structure of centralized schemes is completely different from that of adaptive and conventional UFLS schemes. Unlike these schemes, centralized UFLS schemes lack stages, and they determine and shed load as a function of specific system variables such as frequency, ROCOF, voltage, and power generation.

3.2 Design and Application of Conventional Frequency Protection Schemes

Several methods have been reported in the literature to design conventional UFLS schemes, but there exists no generally accepted method for the design of UFLS schemes (Thalassinakis and Dialynas 2004). UFLS schemes should be simple, quick, and effective to avoid possible blackouts (Delfino et al. 2001; Concordia et al. 1995). Power utilities adopt different approaches to this problem, which are mainly based on their experience and the robustness of the system (Thalassinakis and Dialynas 2004; Concordia et al. 1995) and usually follow typical design criteria on the number of steps, step size, settings of the frequency thresholds, and so on (Concordia et al. 1995; Anderson and Mirheydar 1992; Delfino et al. 2001; IEEE 1975). A systematic method for the design of UFLS schemes for island power systems has been proposed in Sigrist et al. (2012a).

Principally, two design approaches can be distinguished to date: experimental and optimal designs. Experimental designs employ trial–error procedures or choose the best scheme of a set of candidate schemes. In Lokay and Burtnyk (1968), a trial–error procedure is outlined to adjust the UFLS scheme parameters. An improved and automated version of this trial–error procedure is described in Jones and Kirkland (1988). In Concordia et al. (1995), the design of UFLS schemes is based on a screening process of a set of manually selected candidate UFLS schemes. Monte-Carlo simulations have also been applied to the problem of UFLS scheme design (Thalassinakis and Dialynas 2004; Shrestha and Lee 2005). Another screening process has been applied in Cheng and Shine (2003), in which the selected scheme is the one with the lowest total cost, including costs due to load shedding and available spinning reserve. Trial–error or screening processes, however, do not guarantee that a minimum amount of load is shed.

Optimal designs, by contrast, make use of optimization techniques to adjust the parameter settings of the UFLS schemes. Deterministic algorithms, such as steepest descent and quasi-Newton methods, have been applied to solve the optimization problem (Jenkins 1983; Halevi and Kottick 1993; Denis Lee Hau 2006). In its most generic formulation, decision variables of the optimization problem correspond to the frequency and ROCOF thresholds, the intentional time delays, and the step sizes of underfrequency and ROCOF stages. Different underlying power system models (single-generator system frequency dynamics [SFD] models or multigenerator SFD models) have been used, and different UFLS scheme parameters have been considered as decision variables. In Ceja-Gomez et al. (2012), the optimization problem associated with the UFLS scheme design has been formulated as a mixed-integer linear programming problem. However, a substantial drawback of deterministic algorithms lies in the fact that they may get caught in local minima. The optimal solution is therefore highly dependent on the initial guess of the decision variables, as indicated in Denis Lee Hau (2006). Due to the problem's nonlinear (generator models) and discontinuous (UFLS scheme) nature, heuristic algorithms seem to be more adequate (Denis Lee Hau 2006; Lopes et al. 1999). Mitchell et al. (2000), Mitchell (2000), and Martínez et al. (1993) optimally adjusted frequency thresholds and the step size by means of a genetic algorithm. Hong and Wei (2010) determined the step size, the number of steps, and the intentional delay by minimizing the amount of shed load and maximizing the lowest swing frequency by dint of a hierarchical genetic algorithm. Extensions to the prior work by considering fuzzy loads or by contemplating multiple wind and photovoltaic (PV) generation scenarios and generating unit outages with their appropriate probabilities have been presented in Hong and Chen (2012) and Hong et al. (2013). Probabilistic approaches to quantify load or active power imbalance variations have also been considered in Sigrist et al. (2012b) and Bogovic et al. (2015) and included in the design problem (Sigrist and Rouco 2014). Finally, the objective function, which aims at minimizing the amount of shed load, becomes discontinuous and shows a steplike shape if step sizes are not considered as decision variables (which is usually not possible in small island power systems), impeding the use of gradient-based methods.

Whatever the design approach, the adjustment of the parameters of the UFLS scheme depends very much on the selected operating and contingency scenarios. An operating and contingency scenario is defined as a particular disturbance for a particular system operating condition. A robust and efficient design must then tackle both the selection of operating and contingency scenarios and the adjustment of the parameters.

3.2.1 Disturbance Selection

To design a robust UFLS scheme, different operating and contingency scenarios need to be considered. In general, the tuning is such that the UFLS scheme protects the power system against the maximum disturbance (Lokay and Burtnyk 1968; Anderson and Mirheydar 1992; Thompson and Fox 1994). However, a design taking into account only the maximum disturbance may cause overshedding for smaller, less critical disturbances. On the contrary, if the scheme has been designed for smaller disturbances, it may not shed sufficient load to avoid a system collapse in the case of a large disturbance. Thus, it is crucial to select credible, appropriate operating and contingency scenarios (Lokay and Burtnyk 1968; Concordia et al. 1995).

In general, many contingency scenarios under many power system conditions can be considered (line tripping, generation outages, etc.). In large interconnected power systems, an underfrequency condition is usually caused by system separation into islands because of line tripping. By contrast, in small isolated power systems, outages of generating units lead to pronounced frequency deviations, since any individual generating unit infeed presents a substantial portion of the total demand; therefore, only generator outages are usually considered as contingency scenarios for isolated power systems (Concordia et al. 1995). However, it is computationally not very efficient or even impossible to design a UFLS scheme for all possible generator outages and for all power system conditions.

Traditionally, after a major disturbance, settings of protection devices and control actions are revised and possibly readjusted to enable the system to withstand the same disturbance next time (Jung et al. 2002). The principal drawback of a design based on a worst-case scenario is the possibly poor performance of the UFLS scheme for less severe disturbances. In Denis Lee Hau (2006), four different contingency scenarios of distinct severity have been considered. However, severity alone is not sufficient, since other factors such as available spinning reserve, system inertia, or generation mix, all dependent on the operating point, influence the system response as well. This can be resolved by selecting an appropriate set of possible operating and contingency scenarios, since the power system may behave in the same way for different operating and contingency scenarios. A common practice is therefore to determine the system conditions corresponding to different load–demand levels (e.g., minimum and maximum) and to design the UFLS scheme taking into account the outages of the largest, the smallest, and finally a medium-size generator for each of these system conditions (Concordia et al. 1995; Thompson and Fox 1994). Nevertheless, this kind of scenario selection does not necessarily guarantee a selection of representative scenarios, that is, those scenarios that best represent the patterns in frequency dynamics due to power imbalances.

In Sigrist et al. (2010), a method based on data mining is proposed to identify the representative operating and contingency scenarios, that is, to find those scenarios that best represent all other possible scenarios, reducing the computational cost and increasing the robustness of the UFLS scheme design. Figure 3.1 illustrates the method of finding representative operating and contingency scenarios. Two representative operating and contingency scenarios (bold lines) describe in this example two clearly distinct groups of power system responses given in terms of frequency (dashed lines). These representative scenarios are real disturbances given by the outage of a particular generating unit for a particular operating scenario. In fact, this method is based on the assumption that the power system behaves in a similar way for different disturbances of different operating conditions. It should therefore be possible to find different patterns in the behavior of the power system.

An operating and contingency scenario is defined as the outage of a single generator G_i or multiple simultaneous outages of generators $[G_i, G_{i+1}, \ldots]$ for

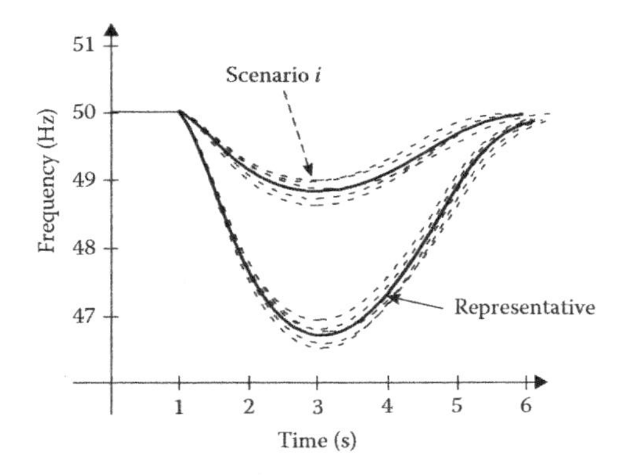

FIGURE 3.1
Illustration of representative operating and contingency scenarios.

a particular system operating condition. For convenience, a system operating condition will be henceforth described by its corresponding generation dispatch scenario. Thus, representative operating and contingency scenarios are those single or multiple outages of a particular system operating condition that represent all other possible operating and contingency scenarios. In other words, representative scenarios are those scenarios that best describe the patterns in frequency dynamics due to power imbalances.

The nonlinearities introduced by the generator output limitations require a suitable simulation tool to analyze operating and contingency scenarios. Simulations are carried out using the nonlinear multigenerator SFD model presented in Chapter 2, Section 2.1.1.4. The operating and contingency scenarios are then given by the corresponding responses of the system frequency to the considered outages:

$$\Omega = \left[\omega_1(t) \quad \ldots \quad \omega_j(t) \quad \ldots \quad \omega_M(t) \right]$$

where:

$\omega_j(t)$ is the response of the system frequency to the jth operating and contingency scenario
M is the number of scenarios

The simulation time and the integration step are fixed to ensure that all vectors $\omega_j(t)$ are of the same length. This is important, because clustering algorithms employ distance measures (e.g., Euclidean distance), which require vector inputs of equal length.

However, it is not necessary to simulate the SFD model over a longer period, since the time window of interest only contains the first few seconds after a contingency. It is during these first seconds that the UFLS scheme has to act. Thus, the necessary time window can be reduced, but differences due to slower or faster turbine-governor response or insufficient spinning reserve should still be perceptible (Sigrist et al. 2010). A solution is to set the simulation time equal to the maximum instant of minimum frequency of all operating and contingency scenarios, or

$$t_{\text{sim}} = \max\left(t_{\omega\,\text{min},j}\right), \quad j = 1,\ldots,M \tag{3.1}$$

where $t_{\omega\text{min},j}$ is the instant of minimum frequency of scenario j. In practice, the instant of minimum frequency is unknown, but it can be estimated (Sigrist et al. 2010).

The identification of representative operating and contingency scenarios can be realized by means of several data mining techniques. Data mining techniques emerged in the early 1990s and allow relevant information to be processed and extracted from large databases. Cluster analysis forms part of data mining. Cluster analysis refers to the partitioning of a data set into clusters, so that the data in each subset ideally share some common trait and differ from the data in other subsets. Different methods based on advanced statistical and modeling techniques, such as K-Means, Fuzzy C-Means, and Kohonen Self Organizing Map (KSOM), are commonly employed.

3.2.2 Adjustment of the Parameters of a Frequency Protection Scheme

The efficiency of the UFLS scheme depends on the tuning of its parameters, which is proposed to be carried out as a function of the selected representative operating and contingency scenarios. The parameters are tuned such that the UFLS scheme fulfills some specified performance criteria. The performance criteria mainly concern frequency levels and the amount of curtailed load and are strongly related to the power system of interest. Typical criteria for an UFLS scheme of a small isolated power system include

- The minimum quantity of load should be shed (Anderson 1999).
- The frequency deviation should be as small as possible. The earlier the load is shed, the lower the frequency deviation (Anderson and Mirheydar 1992; Lukic et al. 1998).
- The frequency should not stay below a minimum allowable value for more than a certain amount of time. Typically, the frequency should not stay below 47.5 Hz for more than 3 s (IEEE 1987; REE 2005).
- The frequency overshoot should be less than a maximum allowable value (e.g., 51.5–52 Hz).

Additional criteria sometimes reported are

- The posttransient steady-state frequency value should be higher than the predefined value (e.g., 49.5 Hz) (Thompson and Fox 1994).
- The UFLS scheme should respect the priority of the load (Blackburn and Damin 2006).
- Load shedding should take place where the disturbance occurs (Prasetijo et al. 1994).
- No load should be shed once the frequency has passed its minimum value, since frequency is recovering (Sigrist et al. 2012a).

The problem of UFLS tuning can be formulated as an optimization problem as shown in Equation 3.2.

$$\begin{aligned} &\underset{x}{\text{minimize}}\, f(x) \quad \text{s.t.} \\ &lb \le x \le ub \\ &g(x) \le 0 \end{aligned} \tag{3.2}$$

The main objective is to minimize the amount of shed load for the selected representative operating and contingency scenarios while satisfying constraints on system stability and load shedding performance; therefore the objective function $f(x)$ aims at minimizing the amount of shed load, whereas stability and desired load-shedding performance are guaranteed by imposing appropriate inequality constraint functions $g(x)$; lb and ub are the lower and upper bounds of the decision variables $\mathbf{x}$. In its most generic formulation, decision variables of the optimization problem correspond to the frequency and ROCOF thresholds, the intentional time delays, and the step sizes of underfrequency and ROCOF stages:

$$x = \begin{bmatrix} \omega_{uf} & \omega_{\text{rocof}} & d\omega/dt & t_{\text{int},uf} & t_{\text{int,rocof}} & p_{\text{step},uf} & p_{\text{step},uf} \end{bmatrix}^T$$

Although the step size of each UFLS stage ($p_{\text{step},uf}$ or $p_{\text{step,rocof}}$, respectively) has usually been considered as a decision variable in the literature, the actual implementation of the step size is, after all, rather difficult for small island systems (Sigrist et al. 2012a). Nevertheless, even if the step size is not a decision variable for small island power systems, the step size of a stage can be varied, since it is possible to combine adjacent stages without compromising the priority constraint. Another possibility could consist in relaxing the priority of stages, which allows stages to be rearranged, similarly to a somehow rough step size optimization. By contrast, in the case of larger systems, it is possible to adjust the step size, since feeder blocks are usually a small fraction of the total load under relief and thus, can be readily grouped to sum up to the desired step size. In either case, it is supposed that the implementable step size coincides with actual step size, thus neglecting step size variations, which have been analyzed in detail in Sigrist et al. (2012b).

Equation 3.3 formulates a possible objective function complying with the principal objective:

$$f_1(x) = \sum_{j=1}^{M} \alpha_{f,j} p_{\text{shd},j}(x) \tag{3.3}$$

where:

$\alpha_{f,j}$ is a weighting factor
$p_{\text{shd},j}$ is the amount of shed load in pu (i.e. per unit) for the jth contingency
M is the total number of contingencies

The weighting factor $\alpha_{f,j}$ could, for instance, be a function of the amount of lost power in the jth scenario or the number of total operating and contingency scenarios associated with the jth representative scenario (Sigrist 2010). Figure 3.2 shows the shape of the objective function f_1 in case of a very simple UFLS scheme with only one stage and where only the frequency threshold is

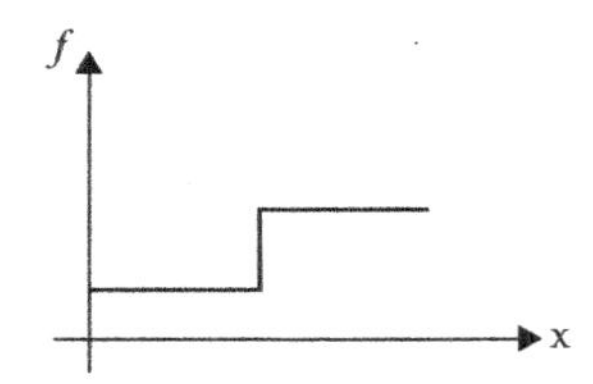

FIGURE 3.2
Shape of the objective function f_1.

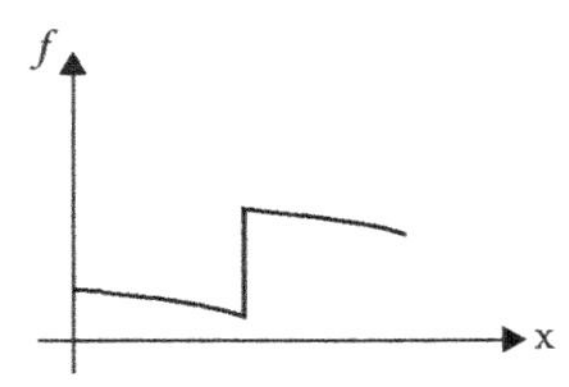

FIGURE 3.3
Shape of the objective function f_2.

optimized. Clearly, the objective function f_1 is discontinuous. Furthermore, the definition of gradients of this objective function would be useless.

Another possible and more generic formulation of the objective function is given by Equation 3.4. The composite objective function f_2 adds to the objective function f_1 a term related to the minimum frequency of each contingency j:

$$f_2(x) = \sum_{j=1}^{M} \alpha_{f,j} p_{\text{shd},j}(x) + \sum_{j=1}^{M} \beta_{f,j} \Delta\omega_{\min,j}(x) \tag{3.4}$$

where $\alpha_{f,j}$ and $\beta_{f,j}$ are weighting factors. Again, the weighting factors could be inversely proportional to the amount of lost power generation in pu. Another possibility is to choose both $\alpha_{f,j}$ and $\beta_{f,j}$ equal to unity (Denis Lee Hau 2006). In this case, the minimum frequency–related term simply smooths the objective function. Figure 3.3 shows the shape of the objective function f_2 in the case of a very simple UFLS scheme with only one stage and where only frequency threshold is optimized. The objective function f_2 is still discontinuous, but the effect of the term related to the minimum frequency appears clearly.

Finally, two other possible formulations of the objective function are given by Equations 3.5 (Halevi and Kottick 1993) and 3.6. Again, $\alpha_{f,j}$, $\beta_{f,j}$, $\gamma_{f,j}$, and $\delta_{f,j}$ are weighting factors of the composite objective functions f_3 and f_4.

$$f_3(x) = \sum_{j=1}^{M} \alpha_{f,j} p_{\text{shd},j}(x) + \sum_{j=1}^{M} \gamma_{f,j} \int \frac{1}{2} \Delta\omega_j(x,t)^2 \, dt \tag{3.5}$$

$$f_4(x) = \sum_{j=1}^{M} \beta_{f,j} \Delta\omega_{\min,j}(x) + \sum_{j=1}^{M} \delta_{f,j} \Delta\omega_{ss,j}(x) \tag{3.6}$$

The difference between the objective functions f_2 and f_3 is that f_3 attempts to minimize, apart from the amount of shed load, the overall frequency deviation and not only the term related to the minimum frequency. The objective function f_4 is rather exotic, since it does not include the amount of shed load, but instead, the postcontingency steady-state frequency deviation.

Constraints can be divided into two principal categories. The first category contains the constraints imposed by the power system, whereas the second covers the constraints with respect to the performance of the UFLS scheme. The constraints of the first category are (1) the minimum and (2) the maximum allowable frequency values. In general, the constraint of minimum allowable frequency is accompanied by a maximum time duration during which the frequency can stay below the minimum allowable frequency. Typical values for small island power systems are 47.5 Hz during a maximum 3 s (REE 2005). The maximum allowable frequency varies between 51.5 and 52 Hz. Equations 3.7 and 3.8 implement these two constraints for the jth operating and contingency scenarios.

$$g_{1,j}(\mathbf{x}) = t_{\omega \le \omega_{\min,\text{allowable}},j}(\mathbf{x}) - t_{\omega_{\min,\text{allowable}}} \tag{3.7}$$

$$g_{2,j}(\mathbf{x}) = \omega_{\max,j}(\mathbf{x}) - \omega_{\max,\text{allowable}} \tag{3.8}$$

The constraint g_1 requires that the time frequency below $\omega_{\min,\text{allowable}}$ is smaller than $t_{\omega\min,\text{allowable}}$. The constraint g_2 requires that the maximum frequency, $\omega_{\max}$, is always smaller than or equal to $\omega_{\max,\text{allowable}}$.

The constraints with respect to the performance of the UFLS scheme are (3) the UFLS scheme does not act once the frequency has passed its minimum value, $t_{\omega\min}$, since frequency is returning toward its nominal value (instant of shedding); (4) the amount of shed load, p_{shd}, is smaller than or at most equal to the amount of lost real power, p_{loss} (amount of shed load); and (5) the UFLS scheme respects the priority of loads (priority).

Functions $g_3(\cdot)$, $g_4(\cdot)$, and $g_5(\cdot)$ in Equations 3.9 through 3.11 implement the constraints associated with the second category. Constraint $g_3(\cdot)$ restricts the last instant of shedding $t_{\text{shd},j}$ of the jth operating and contingency scenario to the instant of minimum frequency $t_{\omega\min,j}$. Constraint $g_4(\cdot)$ limits the amount of shed load $p_{\text{shd},j}$ to the amount of lost generation $p_{\text{loss},j}$. Function $g_5(\cdot)$ imposes a minimum difference of $\sigma\omega$ Hz between the frequency thresholds ω_{thrshld} of two consecutive stages, controlling the criterion on priority of the loads indirectly but only partially.* If $is_{\text{shd},j,i}$ denotes that the ith UFLS step has been activated due to the jth operating and contingency scenario, Equation 3.12 guarantees for all activated steps I that the priority has been respected as long as $g_6(\cdot)$ is lower than or equal to zero. If at least one equation $g_{6,j,k}$ is positive, the priority of the (k-1)th step has been violated.

$$g_{3,j}(\mathbf{x}) = t_{\text{shd},j}(\mathbf{x}) - t_{\omega\text{min},j}(\mathbf{x}) \tag{3.9}$$

$$g_{4,j}(\mathbf{x}) = p_{\text{shd},j}(\mathbf{x}) - p_{\text{loss},j} \tag{3.10}$$

$$g_5(\mathbf{x}) = \left[\begin{array}{ccccc|ccccc|ccc|c|c|c|ccc} -1 & 1 & \cdots & & & 0 & & \cdots & & 0 & 0 & \cdots & 0 & \cdots & \cdots & \cdots & 0 & \cdots & 0 \\ 0 & 0 & \vdots & & & 0 & & \vdots & & 0 & & & & \vdots & \vdots & \vdots & & \vdots & \\ 0 & 0 & 0 & -1 & 1 & 0 & & \cdots & & 0 & & \vdots & & \vdots & \vdots & \vdots & & \vdots & \\ 0 & & \cdots & & & -1 & 1 & \cdots & & & & \vdots & & \vdots & \vdots & \vdots & & \vdots & \\ 0 & & \vdots & & & 0 & 0 & \vdots & & & & & & & & & & & \\ 0 & & \cdots & & & 0 & 0 & 0 & -1 & 1 & 0 & \cdots & 0 & \cdots & \cdots & \cdots & 0 & \cdots & 0 \end{array}\right] \mathbf{x} - \begin{bmatrix} \sigma_\omega \\ \vdots \\ \vdots \\ \sigma_\omega \end{bmatrix} \tag{3.11}$$

$$g_{6,j,i}(\mathbf{x}) = is_{\text{shd},j,i}(\mathbf{x}) - is_{\text{shd},j,i-1}(\mathbf{x}) \qquad \forall i \in I \tag{3.12}$$

The optimization problem is solved by a simulated annealing algorithm (Sigrist et al. 2012a). It has been shown in Sigrist et al. (2012a) that different heuristic algorithms, particularly simulated annealing and genetic algorithms, lead to similar results.

3.2.3 Application to Island Power Systems

The method for a robust and efficient design of UFLS schemes of island power systems is applied to two island power systems with different features. The smaller island system, La Palma, has a peak demand of about

* Priority is guaranteed for cases where the intentional time delay is very small, and frequency thresholds are used to differentiate the UFLS stages. In these cases, there exists no overlapping of UFLS stages. However, time delays are not always very short, and thus, the (i + 1)th stage may act without the ith stage acting.

35 MW compared with the 530 MW peak demand of the larger island system, Gran Canaria. First, representative operating and contingency scenarios are selected by means of clustering techniques. Subsequently, the UFLS scheme parameters are optimally adjusted to minimize the total amount of shed load of the representative operating and contingency scenarios.

The representative operating and contingency scenarios are determined by means of the K-Means algorithm. In the case of the island system of La Palma, only N-1 outages have been considered, since they include generation losses of over 50% of the total demand. Multiple outages (N-x) have been contemplated for the Gran Canaria island power system. It has been found by applying the K-Means algorithm iteratively that four clusters are sufficient to represent all possible operating and contingency scenarios. Figure 3.4 shows the quadratic quantization error as a function of the number of clusters for both island power systems, where four clusters describe the knee point.

Figure 3.5a shows the representative operating and contingency scenarios of the La Palma power system determined by the K-Means algorithm and compares them with the operating and contingency scenarios given by the common practice, that is, the operating and contingency scenarios corresponding to the outages of the largest and smallest generating units for the maximum and minimum load–demand level. Note that during the process of selecting operating and contingency scenarios, the UFLS scheme is not involved and the worst operating and contingency scenarios are in both cases nearly the same. According to Figure 3.5a, the clustering-based method covers a wider range of possible system responses. In Figure 3.5b, the first and second principal components of the representative scenarios are superposed on the principal components of all operating and contingency scenarios and the principal components of common practice. For example, the first three operating and contingency scenarios of common practice seem to be very close, whereas the clustering-based scenarios are rather widespread and thus better cover the variance within all the possible operating and contingency scenarios. Finally, Figure 3.6 compares the representative operating and contingency scenarios of the La Palma power system for different clustering algorithms. The differences are rather small, and during the first few seconds, the operating and contingency scenarios determined by different clustering techniques are practically the same.

Once the operating and contingency scenarios for which the UFLS scheme will be designed have been selected, the existing UFLS scheme can now be redesigned using the proposed method for the design of a robust and efficient UFLS scheme. The decision variables x are the frequency thresholds ω_{uf} and ω_{rocof}, the ROCOF thresholds dω/dt, and the intentional time delays t_{int} of all maximally required steps. Step sizes of the underfrequency and ROCOF stages have not been considered as a decision variable, and therefore, the

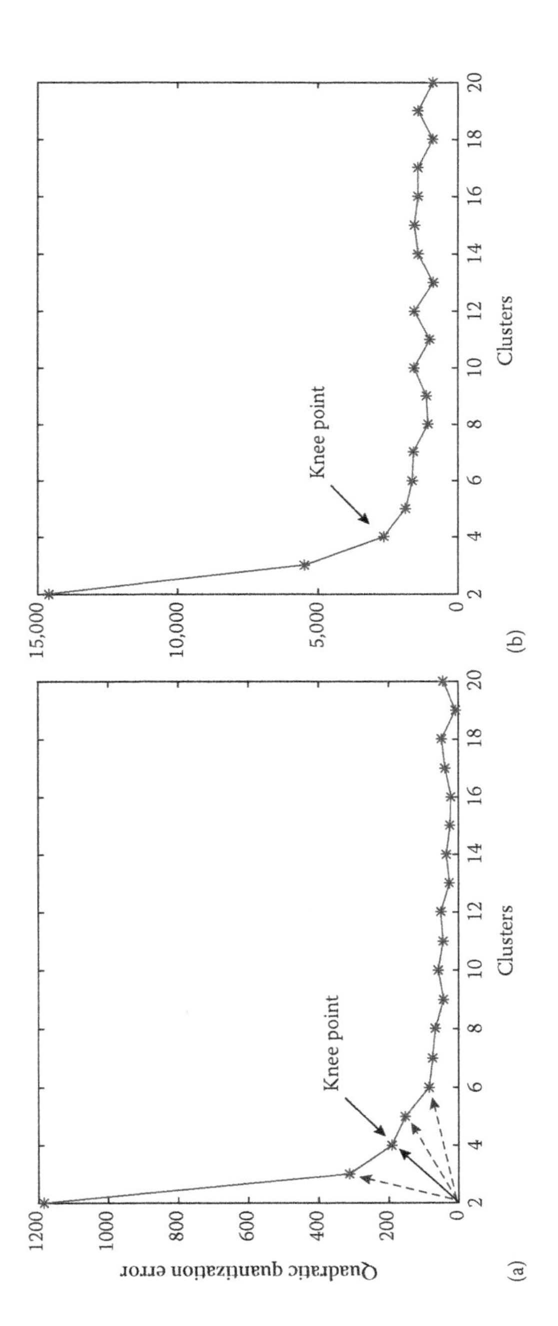

FIGURE 3.4
Quadratic quantization error (the sum of the squared distance of each cluster from its associated data points) as a function of the number of clusters: (a) for the smaller island and (b) for the larger island.

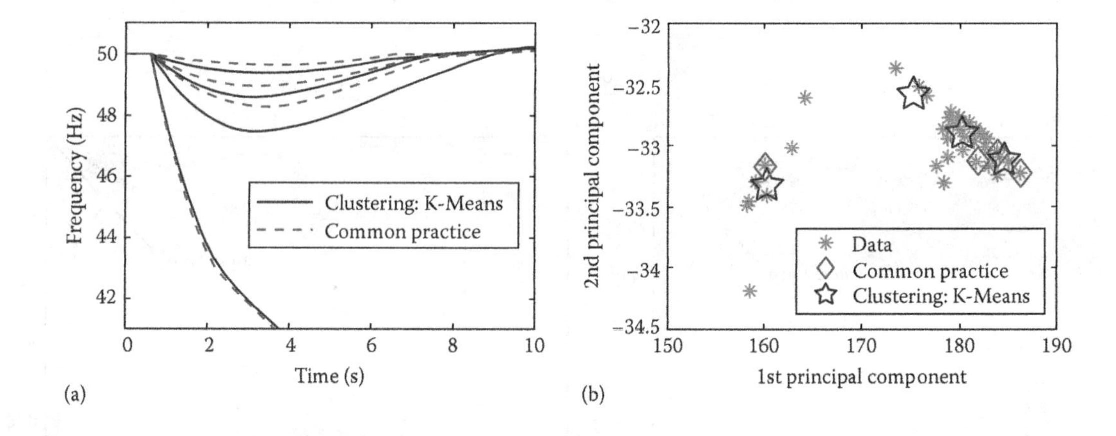

FIGURE 3.5
Comparison of operating and contingency scenarios of the La Palma power system determined by the common practice and by the K-Means clustering algorithm: (a) time domain and (b) principal component analysis.

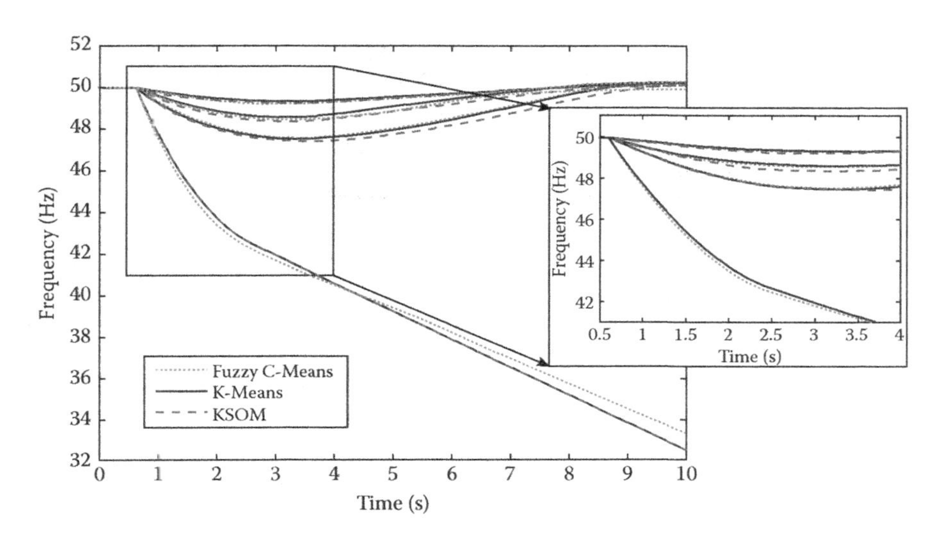

FIGURE 3.6
Comparison of the performance of different clustering techniques in the case of the smaller island.

step sizes of the existing UFLS scheme have been used. The UFLS scheme parameters are bounded as follows:

- $47\text{ Hz} \leq \omega_{uf} \leq 49\text{ Hz}$
- $49.3\text{ Hz} \leq \omega_{\text{rocof}} \leq 49.8\text{ Hz}$
- $-1.5\text{ Hz/s} \geq \mathrm{d}\omega/\mathrm{d}t \geq -2.5\text{ Hz/s}$
- $0\text{ s} \leq t_{\text{irt},uf} \leq 0.5\text{ s}$

The number of maximally required steps k is such that their accumulated step sizes equal at least the largest power imbalance within the selected operating and contingency scenarios:

$$\sum_{i=1}^{k} p_{\text{step},i} \geq \max_{j}\left(p_{\text{lost},j}\right) \tag{3.13}$$

The optimization constraints are the minimum and maximum allowable frequencies, the instant of shedding, the amount of shed load, and the priority. In particular, and taking into account Equation 3.7, minimum allowable frequencies $\omega_{\text{min,allowable}}$ of 48 and 47 Hz with corresponding $t_{\omega\text{min,allowable}}$ of 2 and 0 s, respectively, have been imposed. The maximum allowable frequency $\omega_{\text{max,allowable}}$ is set to 52 Hz. Note that the constraints imposed on minimum allowable frequency are more restrictive than those required by system operators (e.g., frequency < 47.5 Hz for maximum 3 s). This leads to a more conservative scheme and therefore gives a certain safety margin. Further, more restrictive constraints might also benefit UFLS scheme performance in situations where assumed step sizes do not correspond with real step sizes due to feeder-load variation, feeder outages, or breaker failures (Sigrist et al. 2012b; Sigrist and Rouco 2014).

Table 3.3 shows the systematically designed UFLS scheme of the smaller island system. The number of stages adjusted depends on the maximum amount of lost power generation in the contemplated operating and contingency scenarios according to Equation 3.13. Since the worst operating and contingency scenario implies a loss of about 50% of power generation (see Table 3.4), the system can be stabilized by using the first six stages. The remaining steps (7–10) could be used as backup steps in the case of step-size variations or nonresponding turbine-governor systems. Backup steps support the UFLS scheme during such events and guarantee system integrity, and minimizing the amount of shed load is no longer paramount when tuning their settings. In comparison with the existing UFLS scheme, frequency thresholds ω_{uf} and intentional time delays t_{int} of the underfrequency relays have been lowered, whereas frequency thresholds ω_{rocof} and $dt/\omega t$ of the ROCOF relays have been increased. The latter allows a faster UFLS scheme intervention in the case of severe disturbances.

Figure 3.7a and Table 3.4 compare the two UFLS schemes. Figure 3.7a shows the system responses in terms of frequency to the four representative operating and contingency scenarios. It can be inferred that the minimum and maximum frequency deviations clearly exceed the minimum and maximum allowable frequency deviations in the case of the existing UFLS scheme. Frequency overshoot also indicates overshedding. In the case of the systematically designed UFLS scheme, frequency is confined within the allowable frequency range. Table 3.4 indicates whether the UFLS scheme satisfies the constraints on its performance (state) and in addition, provides a cause in the case of constraint violation. It can be inferred that the existing UFLS scheme sheds too much load for the first three contingencies (cause a). Furthermore, load is shed after the frequency has passed its minimum (cause b). Overshedding is also the reason for the large frequency overshoot and the resulting frequency instability, since generators cannot lower their output to the new demand level. Finally, it can be seen from Table 3.4 that the systematically designed UFLS scheme sheds less load than the existing UFLS scheme and that no constraint on its performance is violated. Thus, the existing UFLS scheme has been successfully optimized with respect to the considered representative operating and contingency scenarios.

To come full circle, both the existing and the systematically designed UFLS scheme of the La Palma island system are applied to all possible N-1 operating and contingency scenarios (i.e., all possible single generation outages for all operating scenarios). Figure 3.7b and c compare their time-domain responses in terms of frequency. The first two rows of Table 3.5 compare the performance of both UFLS schemes. According to Table 3.5, the

TABLE 3.3

Existing (Existing) and Systematically Designed (Systematic) UFLS Scheme of the La Palma Island System

	ω_{uf} (Hz)		t_{int} (s)			
Step	**Existing**	**Systematic**	**Existing**	**Systematic**	t_{opn} (s)[a]	p_{step} (%)[b]
1	48.81	48.52	0.6	0.07	0.2	7.1
2	48.81	48.04	0.9	0.14	0.2	0.6
3	48.66	47.89	1.3	0.27	0.2	14.5
4	48.66	47.75	1.8	0.36	0.2	3.6
5	48.66	47.60	2.3	0.06	0.2	7.3
6	48.00	47.35	1.2	0.13	0.2	13.6
7	48.00	–	1.7	–	0.2	12.5
8	47.00	–	1.8	–	0.2	4.2
9	47.00	–	2.1	–	0.2	15.1
10	47.00	–	2.4	–	0.2	20.5

	ω_{rocof} (Hz)		$d\omega/dt$ (Hz/s)		t_{int} (s)			
Step	**Existing**	**Systematic**	**Existing**	**Systematic**	**Existing**	**Systematic**	t_{opn} (s)[a]	p_{step} (%)[b]
1	49.50	49.78	−1.8	−2.2	0.1	0.16	0.2	7.1
2	49.50	49.78	−1.8	−2.2	0.1	0.16	0.2	0.6
3	49.50	49.78	−1.8	−2.2	0.1	0.16	0.2	14.5
4	49.50	49.78	−1.8	−2.2	0.1	0.16	0.2	3.6

[a] t_{opn} includes measurement and opening delays.

[b] p_{step} is given as a percentage of the demand.

TABLE 3.4

Comparison of the Performance of the Existing and the Systematically Designed UFLS Scheme of the La Palma Island System

			p_{shed} (MW) [%][a]		State (Error Type[b])	
Scenario	p_{dem} (MW)	p_{loss} (MW) (%)[a]	Existing	Systematic	Existing	Systematic
1	32.15	6.7 [20.8]	8.29 [25.8]	2.28 [7.1]	Error (a,b)	Correct
2	18.29	8.96 [49]	10.83 [59.2]	8.54 [46.7]	Error (a,b)	Correct
3	29.98	4.74 [15.8]	2.31 [7.7]	0	Error (a)	Correct
4	30.74	2.35 [7.6]	0	0	Correct	Correct

[a] As a percentage of demand.
[b] (a) overshedding, (b) late sheddin.

amount of shed load has been remarkably reduced after implementing the systematically designed UFLS scheme. In addition, fewer errors (e.g., overshedding, low frequency) occurred thanks to the systematically designed UFLS scheme. Total frequency deviations ($\Sigma\Delta\omega min$) have also been slightly reduced. Table 3.5 also displays the results for what happens when (1) operating and contingency scenarios are selected by means of common practice instead of representative operating and contingency scenarios, (2) the minimum frequency constraint is relaxed (from 48 Hz–2 s to 47.5 Hz–3 s), (3) the priority constraint is relaxed, and (4) step sizes are also optimized.

Clearly, the selection of operating and contingency scenarios affects the robustness (common practice yields a larger total amount of shed load). By comparing the results obtained by relaxing the priority constraint and the inclusion of the step sizes as decision variables, it is interesting to see that relaxing priorities yields a somewhat rough step size optimization by reordering the UFLS stages. Relaxing minimum frequency constraints leads to an even lower total amount of shed load.

Whereas the La Palma island system basically exhibits an underdamped system response in terms of frequency, larger systems, such as the Gran Canaria power system, show a rather overdamped system response. Table 3.6 shows the systematically designed UFLS scheme of the Gran Canaria island system. It is worthwhile mentioning that the upper and lower bounds of the ROCOF thresholds have been modified with respect to the design applied to the La Palma island system: particularly, -0.5 Hz/s $\geq d\omega/dt \geq -1.5$ Hz/s.

Larger island systems exhibit smaller ROCOF than the smaller island systems, mainly due to the larger inertia and because individual generating units produce smaller fractions of the total demand.

Figure 3.9a compares the existing and the systematically designed UFLS scheme in terms of system frequency responses to the four representative operating and contingency scenarios. Frequency overshoots occur in the case

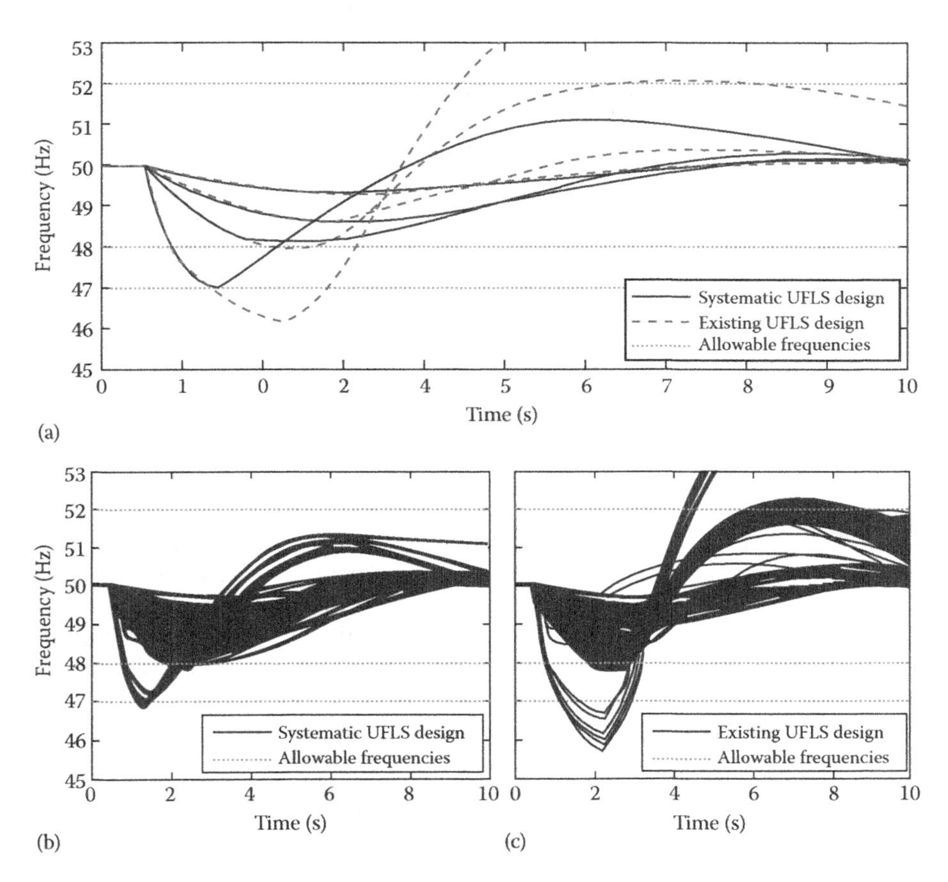

FIGURE 3.7
Time-domain responses in terms of frequency for the La Palma island system: (a) comparison of the impact of the existing and the systematically designed UFLS on the representative operating and contingency scenarios, (b) the impact of the systematically designed UFLS scheme on all possible operating and contingency scenarios, and (c) the impact of the existing UFLS scheme on all possible operating and contingency scenarios.

of the existing scheme, indicating overshedding. In contrast, minimum frequencies are lower in the case of the systematically designed UFLS scheme, but clearly above the imposed constraints. It can, however, be seen that frequency stalls below or close to 49 Hz, requiring an additional constraint on the minimum frequency operation: 49 Hz–5 s. For power systems exhibiting an underdamped system response type, restrictions on steady-state frequency values become important. With this new constraint, frequency recovers to values at or above 49 Hz. Table 3.7 shows that the amount of load has been effectively reduced.

Finally, and to come full circle, both the existing and the systematically designed UFLS scheme of the Gran Canaria island system are applied to

TABLE 3.5

Comparison of Different UFLS Scheme Designs for the La Palma Island System

Case	UFLS Performance		
	$\Sigma\, \Delta\omega_{min}$ (Hz)	Total Shed (MW)	No. Error States
Existing	225.43	528.68	164
Base case (systematic)	213.2	231.64	6
Common practice	228.58	364.93	81
47.5 Hz–3 s	246.07	100.49	6
No priority	217.23	195.94	9
Complete optimization	220.25	167.05	10

all possible N-x operating and contingency scenarios (i.e., all possible single and multiple generation outages for all operating scenarios). Figure 3.9b and c compare their time-domain responses in terms of frequency, highlighting again the problem of low-frequency stalling even with the UFLS design imposing a constraint of 49 Hz–5 s. Finally, Table 3.8 compares the performance of the existing and the systematically designed UFLS scheme. According to Table 3.8, the amount of shed load has been remarkably reduced after implementing the systematically designed UFLS scheme. Note that the additional constraints increased load shedding by 27%.

3.3 Design and Application of Advanced Frequency Protection Schemes

Advanced UFLS overcome the major drawbacks of conventional UFLS schemes that result in their lack of adaptiveness to the actual active power imbalance and power system state. Advanced UFLS schemes can be further divided into adaptive and centralized UFLS schemes. Adaptive UFLS schemes differ from centralized UFLS schemes by maintaining the structure of conventional schemes (i.e., steps) and by rendering at least one UFLS scheme parameter adaptive.

3.3.1 A Review of Adaptive Frequency Protection Schemes

Adaptive UFLS schemes can be broadly categorized into schemes that adapt the step size to the actual power imbalance and other schemes that adapt and enrich, with other triggering signals, relay parameters such as frequency thresholds and/or intentional time delay settings.

TABLE 3.6

Existing (Existing) and Systematically Designed (Systematic) UFLS Scheme of the Gran Canaria Island System

Step	ω_{uf} (Hz)		t_{int} (s)		t_{opn} (s)[a]	p_{step} (%)[b]
	Existing	Systematic	Existing	Systematic		
1	49.00	48.66	0.1	0.08	0.2	3.43
2	48.92	48.28	0.15	0.08	0.2	5.27
3	48.85	48.17	0.2	0.26	0.2	5.03
4	48.79	48.05	0.3	0.04	0.2	5.68
5	48.72	47.92	0.4	0	0.2	4.17
6	48.66	–	0.5	–	0.2	9.95
7	48.60	–	0.6	–	0.2	5.46
8	48.55	–	0.7	–	0.2	3.05
9	48.50	–	0.8	–	0.2	4.07
10	48.35	–	0.9	–	0.2	5.89
11	48		1			10.34

Step	ω_{rocof} (Hz)		$d\omega/dt$ (Hz/s)		t_{int} (s)		t_{opn} (s)[a]	p_{step} (%)[b]
	Existing	Systematic	Existing	Systematic	Existing	Systematic		
1	49.30	49.48	−0.8	−1.4	0.15	0.08	0.2	9.97

[a] t_{opn} includes measurement and opening delays.

[b] pstep is given as a percentage of the demand (see footnote table 3.3)

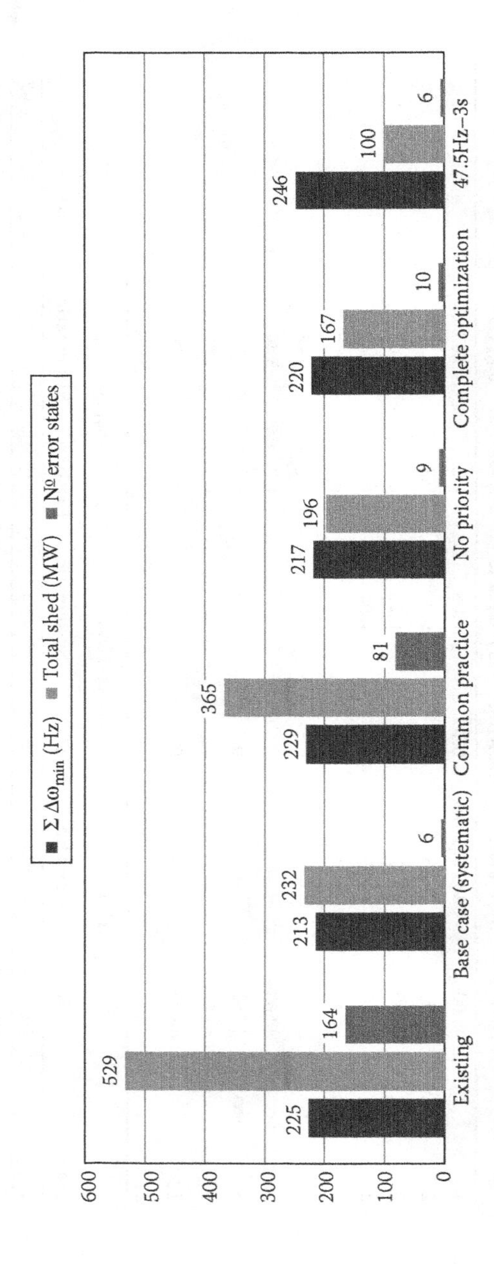

FIGURE 3.8
Comparison of different UFLS scheme designs.

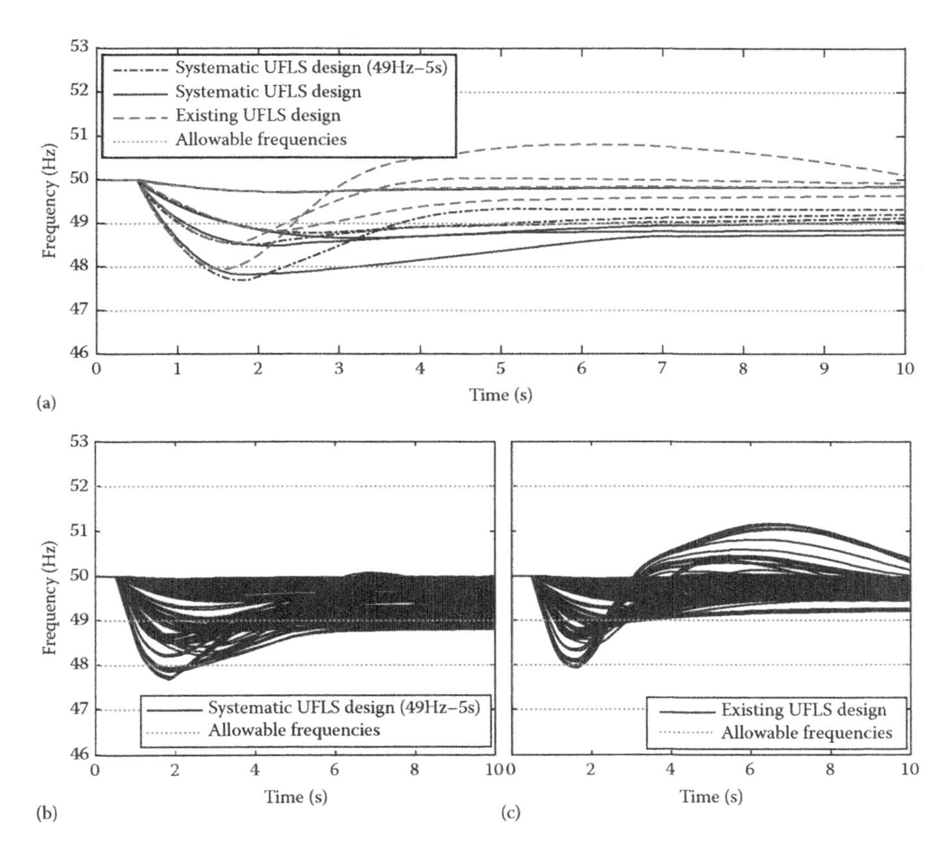

FIGURE 3.9
Time-domain responses in terms of frequency for the Gran Canaria island system: (a) comparison of the impact of the existing and the systematically designed UFLS on the representative operating and contingency scenarios, (b) the impact of the systematically designed UFLS scheme (49 Hz–5 s) on all possible operating and contingency scenarios, and (c) the impact of the existing UFLS scheme on all possible operating and contingency scenarios.

Adaptive UFLS schemes of the first category use the ROCOF as an additional input signal to compute the load to be shed and to accelerate the load-shedding process (Anderson and Mirheydar 1992; Thompson and Fox 1994). In fact, the initial ROCOF is proportional to the amount of active power imbalance. Adaptivity then refers to the fact that the amount of load to be shed is computed as a function of the power imbalance. Equation 2.6 relates the ROCOF at the center of inertia (COI) to the active power imbalance Δp given on the basis of the system rating S_{base}:

$$2H\Delta\dot{\omega} = \Delta p \tag{3.14}$$

where the equivalent inertia H has been defined as

TABLE 3.7

Comparison of the Performance of the Existing and the Systematically Designed UFLS Scheme of the Gran Canaria Island System

			p_{shed} (MW) [%][a]		State (Error Type[b])	
Scenario	p_{dem} (MW)	p_{loss} (MW) [%][a]	Existing	Systematic	Existing	Systematic
1	536.21	269.47 [50.3]	278.92 [52]	126.5 [23.6]	Error (a,b)	Correct
2	517.42	140.48 [27.2]	71.05 [13.7]	0	Error (b)	Correct
3	527.65	49.92 [9.5]	0	0	Error (b)	Correct
4	536.21	204.97 [38.2]	179.85 [33.5]	71.9 [13.4]	Correct	Correct

[a] As a percentage of demand.
[b] (a) overshedding, (b) late sheddin.

TABLE 3.8

Comparison of Different UFLS Scheme Designs for the Gran Canaria Island System

	UFLS Performance		
Case	Total Shed (MW)	$\Sigma\,\Delta\omega_{min}$ (Hz)	No. of Error States
Existing	8304.93	168.05	56
Systematic	3346.24	176.57	6
Systematic (49 Hz–5 s)	4594.15	176.02	22

$$H = \sum_{i=1}^{n} \frac{H_i \cdot M_{base,i}}{S_{base}}$$

By measuring the initial ROCOF at the COI and by knowing the equivalent inertia H, one can estimate the initial active power imbalance. Whereas, for example, Anderson and Mirheydar (1992) and Terzija (2006) assume that H is known beforehand or that it varies only slightly with each generating unit outage, Thompson and Fox (1994) consider the variation in H (and in the available reserve) after a generating unit outage by assuming that this variation is proportional to the amount of lost power generation with respect to the total demand. Further, the estimation of the amount of active power imbalance is based on ROCOF measurements, which is delicate, since ROCOF measurements are highly distorted by local dynamics. Although neither frequency nor ROCOF is uniform, the difference between individual frequencies and the frequency at COI is less pronounced than the corresponding differences for the ROCOF. Figure 3.10 illustrates the response in terms of individual

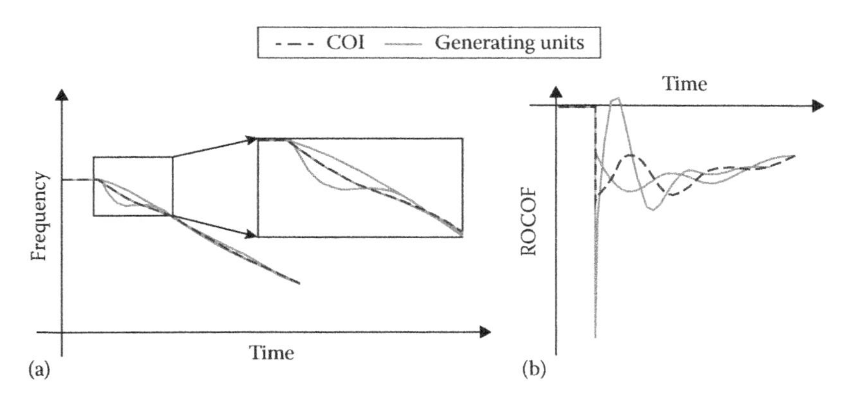

FIGURE 3.10
Illustration of individual frequencies and ROCOF, frequency and ROCOF of the center of inertia, and their inherent oscillations after a disturbance.

frequencies and ROCOF by comparing them with the frequency and ROCOF at the COI.

The amount of load to be shed can be deduced from the estimated active power imbalance in different ways. Terzija and Koglin (2002) and Terzija (2006) directly equalize the amount of load to be shed to the estimated active power imbalance. Anderson and Mirheydar (1992), by contrast, define a critical active power imbalance, and the amount of load to be shed is equal to the difference between the actual and the critical active power imbalance. Load shedding is triggered whenever the actual active power imbalance is larger than the critical one. Equation 2.11 expresses the frequency deviation in terms of the active power imbalance for a linear SFD model:

$$\Delta\omega(t) = \frac{R\Delta p}{DR+1}\left(1+\alpha e^{-\varsigma\omega_n t}\sin(\omega_r t+\phi)\right) \tag{3.15}$$

The critical active power imbalance Δp is the one that leads to a frequency deviation just equal to the minimum allowable frequency $\omega_{\text{min,allowable}}$:

$$\begin{aligned}\Delta p_{\text{crit}} &= \Delta p\big|_{\Delta\omega\geq\Delta\omega_{\text{min,allowable}}} \\ &= \frac{\Delta\omega_{\text{min,allowable}}(DR+1)}{R\left(1+\alpha e^{-\frac{\varsigma\omega_n}{\omega_r}\tan^{-1}\left(\frac{\omega_r T_R}{\varsigma\omega_n T_R-1}\right)}\sin\left(\tan^{-1}\left(\frac{\omega_r T_R}{\varsigma\omega_n T_R-1}\right)\right)+\phi\right)}\end{aligned} \tag{3.16}$$

Due to linearity, the amount of load to be shed p_{shed} is then expressed in Equation 3.17:

$$\Delta p_{\text{shed}} = \Delta p - \Delta p_{\text{crit}} \tag{3.17}$$

This amount would be exactly sufficient if load shedding were triggered immediately after the active power imbalance, but load shedding usually starts at a frequency of around 49–49.5 Hz. Anderson and Mirheydar (1992) multiply the estimated amount of load to be shed in Equation 3.17 by a factor of 1.05 to take into account the delay in initiating load shedding. In Huang and Huang (2001), pumped-storage units have been included in the adaptive load-shedding process, activated once the estimated power imbalance exceeds a predefined value based on the steady-state frequency value. Finally, estimation of the initial active power imbalance does not consider any load dynamics. Nonetheless, a generating unit outage also leads to a voltage drop, reducing the active power demand too. Rudez and Mihalic (2011a) contemplate voltage dependency of loads when estimating the initial active power imbalance.

Once the active power imbalance has been estimated, the corresponding amount of load to be shed needs to be distributed over the steps of the adaptive UFLS schemes. In this way, not all the estimated power deficit is shed, since the governor starts acting and certain steps do not enter the load shedding. Usually, the first step sheds half the amount of load, whereas the other half is distributed over the remaining steps (Anderson and Mirheydar 1992; Lukic et al. 1998). In Terzija and Koglin (2002) and Terzija (2006), the estimated active power imbalance is distributed over several steps, both uniformly and nonuniformly. Another way is to shed one-third in the first step and two-thirds in the second step for less critical disturbances, whereas for severe disturbances, the total amount of active power imbalance is shed (Parniani and Nasri 2006). In Rudez and Mihalic (2011a), not only is the estimated active power imbalance distributed over the steps, but each step size is additionally adapted online by taking into account primary frequency control contribution. Similarly, Ketabi and Fini (2015) distribute the estimated power imbalance over four predefined steps and also take into account primary frequency control action, canceling UFLS action when governor power is larger than the UFLS step. Parniani and Nasri (2006) further take into account the location of the disturbance by distributing the estimated amount of shed load on the buses according to their inertia and/or the electrical distance from the disturbance. Actually, load shedding in larger power systems should take into account the disturbance location (Terzija 2006). In fact, due to the larger extension of the power system, frequency decay is not immediate everywhere throughout the system. Thus, shedding in a location remote from the disturbance might not alter frequency decay. Ghaleh et al. (2011) and Prasetijo et al. (1994), describing a centralized scheme, propose to measure voltages and distribute the amount of load to be shed over load buses according to the measured voltage dips.

Several works address adaptiveness by adjusting relay parameter settings adaptively, improving the UFLS performance. For example, in Elkateb and

Dias (1993a,b), the relay timer settings have been made dependent on the ROCOF, speeding up UFLS scheme action. To protect against consecutive disturbances, the relay settings of the steps that did not actuate during the first disturbance can be adjusted to improve the performance for a possible consecutive disturbance (Chuvychin et al. 1996). In You et al. (2002), an adaptive UFLS scheme is superimposed on a conventional scheme. This adaptive scheme acts whenever ROCOF is larger than a predefined critical value, which is periodically updated. Lin et al. (2008) propose a proportional-integral-derivative (PID) controller to adapt the intentional time delays of load-shedding steps. A distinction is made between rough-adjusting and fine-adjusting steps. Whether or not load needs to be shed, or even to be reconnected (fine-adjusting steps), depends on some ROCOF thresholds. In Ning et al. (2011), the first underfrequency threshold setting is made dependent on the ROCOF. The larger the ROCOF, the higher this setting. Further, the step size is distributed according to the ROCOF of each area. Ghaleh et al. (2011) enrich the frequency relay operation by analyzing the phase-plane described by frequency and the integral of the instantaneous voltage dips. The integral of the voltage dip not only indicates the location of the disturbance but also differentiates between the duration of low-voltage conditions. Load-shedding steps are then defined in terms of frequency and the integral of the voltage dips. Saffarian and Sanaye-Pasand (2011) extend the approach of Ghaleh et al. (2011). In addition, load shedding is accelerated for more severe disturbances by increasing the frequency threshold setting, below which load shedding is initiated, according to the ROCOF. In Abedini et al. (2014), underfrequency threshold settings are adaptively adjusted as a function of the rate of change of voltage (ROCOV) and the ROCOF. ROCOV indicates where disturbance occurs—a larger ROCOV leads to an increase in the underfrequency threshold setting in the region where the disturbance occurs with regard to the initial setting. Additionally, different ROCOF threshold settings are put in place to "adapt" the number of relays to trip for a given ROCOF (corresponding to a rough and static approximation of the active power imbalance estimation). In Hoseinzadeh et al. (2015a), frequency thresholds of relays at a given load bus are made a function of the voltage dip suffered, adapting the settings online during a disturbance. The time delay is inherently taken into account by putting consecutive frequency thresholds sufficiently apart as a function of the actual ROCOF. Underfrequency thresholds are only adapted once frequency and ROCOF fall below given thresholds. Hoseinzadeh et al. (2016) add an anti-stalling scheme to the previous work and limit the curtailable loads to active and reactive power-absorbing loads.

3.3.2 A Review of Centralized Frequency Protection Schemes

Centralized schemes continuously measure critical system variables such as frequency and real power generation, and in the case of a contingency, act according to the actual power system condition. A centralized system initiates the load-shedding process depending on the received measurements, either obtained locally or by means of supervisory control and data acquisition (SCADA), wide area measurement systems (WAMS), and so on, and they typically abandon the structure of conventional UFLS schemes, divided into several steps. Centralized schemes require advanced technologies and rely on the availability of sufficiently fast communication equipment. For example, a programmable logic controller (PLC) can send specified load-shedding signals to the relays of interest (Rodríguez et al. 2005; Shokooh et al. 2005). Basically, event-based and/or response-based centralized UFLS schemes have been proposed, where the former initiate load shedding according to a certain event recorded by the breaker status of monitored elements, whereas the latter initiate load shedding on the basis of the system response, for example, in terms of ROCOF, and by estimating the active power imbalance as adaptive schemes do. The majority of the proposed centralized schemes are response-based schemes. Note that conventional underfrequency relays still might be used as backup elements (Mitchell et al. 2000).

Early applications of centralized load-shedding schemes shed a specified amount of load immediately after the detection of a contingency (Nirenberg et al. 1992). Nowadays, a substantial effort is made to minimize the amount of load to be shed according to the actual power system condition (De Tuglie et al. 2000; Bonian et al. 2005). Constraints with regard to the steady-state frequency or post-shedding ROCOF threshold have also been incorporated (Larsson and Rehtanz 2002; Rodríguez et al. 2005). Further studies take into account the difference between power loss and system reserve or between the postfault generation capability and the maximum amount of power that can be imported through interconnections (Apostolov et al. 1994; Shokooh et al. 2005). Recent studies primarily focus on the distribution of the amount of load to be shed (in general estimated according to Anderson and Mirheydar [1992] or Terzija [2006]) among the available load buses of the power system.

Hsu et al. (2005) trained the data set of an artificial neural network (ANN) model by performing the stability analysis of a power system for various fault scenarios. When a fault occurs, the proposed ANN controller will determine the minimum amount of load to be shed quickly according to the input data captured by the SCADA system in real time (i.e., total load demand, total generation, and ROCOF). In Sanaye-Pasand and Davarpanah (2006), the configuration and status of the power system are periodically monitored and in each period, the optimal amount of load to be shed is determined for every possible generating unit outage and stored in a look-up table. Unlike a response-based scheme, this scheme is event based; that is, the breakers of the generating units are monitored. Similarly, Dong et al. (2008) optimally tuned the UFLS scheme for a given generation scenario and disturbance online. Seyedi and Sanaye-Pasand (2009b) included both response-based and event-based UFLS schemes. The former estimates the active power imbalance by using ROCOF at the COI, as in Anderson and Mirheydar (1992), and the latter estimates the power mismatch on the basis of the circuit breaker status of the monitored elements (generating units, tie lines, etc.). Total load is shed in one step and distributed among the buses with largest voltage dips and additionally with the lowest voltage-reactive power (VQ) margins. In Girgis and Mathure (2010), the active power imbalance is repetitively estimated in terms of ROCOF. At each repetition (step), load shedding is distributed according to the ROCOV as a function of the active power deficit (the steeper the ROCOV at a bus, the more load is shed). Mahat et al. (2010) propose a load-shedding scheme for islanded distribution systems in which loads are ranked according to willingness to pay. A look-up table determines the amount of load to be shed as a function of the measured ROCOF. In Rudez and Mihalic (2011b), the prediction of the minimum frequency by using the second derivative of frequency is presented for load shedding. The second derivative of frequency represents the rate of change of governor action. It is assumed that the second derivative of frequency has exponential-like behavior, such that a generic exponential function can be formulated and identified by measurements. The amount of load to be shed is a linear function of the predicted minimum frequency. Several consecutive load-shedding actions might be necessary to keep the minimum frequency above a given value. Seethalekshmi et al. (2011) compute the amount of load to be shed, like Anderson and Mirheydar (1992). If the amount of load to be shed is above a critical value, the distribution among load buses is done according to the initial voltage dips experienced by load buses due to the disturbance (i.e., the buses with the largest voltage dips share the estimated amount of load to be shed). Otherwise, the distribution among load buses is done according to the voltage stability risk index (if this index is violated). In Tang et al. (2013), load shedding is triggered not only by frequency and ROCOF measurements but also by the determination of the minimum eigenvalue of the Jacobian matrix (a negative eigenvalue stands for voltage collapse). Further, the distribution of the amount of load to be shed (estimated as in Rudez and Mihalic [2011a]) is determined by means of indices depending on active and reactive power flow tracing and frequency deviation and VQ sensitivity. In Abdelwahid et al. (2014), a hardware-in-the-loop implementation of a centralized UFLS scheme is presented. The total amount of load to be shed, estimated according to Anderson and Mirheydar (1992), is equal to the amount of active power imbalance and is shed in one single step. Load buses are selected as in Seethalekshmi et al. (2011). Adamiak et al. (2014) describe and demonstrate a load-shedding scheme based on the estimated load–generation balance of an industrial power complex by measuring power generation and power import. Load is shed if the load–generation balance is positive. Reddy et al. (2014) estimate the amount of load to be shed according to Anderson and Mirheydar (1992) but distribute the amount of load to be shed according to the sensitivity of frequency of the COI with respect to a change of load. More sensitive load buses are selected first to shed load. Manson et al. (2014) describe the application of an inertia-tracking and load-compensation load-shedding scheme to a facility. Inertia is tracked by identifying which generator and major load equipment are attached to which island section. Loads are tracked thanks to proper time stamping of the frequency and ROCOF measurements at load buses (i.e., time-synchronized phasor measurements). The amount of load to be shed is estimated by estimating the active power imbalance. In Gu et al. (2014), a multiagent-based UFLS scheme is proposed. In the first stage, each agent (e.g., distributed generator) monitors its ROCOF, predicts its power imbalance, and communicates its power imbalance to the remaining agents, finally reaching a consensus on the total power imbalance. In the second step, the amount of load to be shed is distributed among the available loads according to their costs, expressed as a function of their priority. Prasad et al. (2014) present an ANN-based load shedding scheme, triggered when the active power imbalance is larger than a critical value defined as in Anderson and Mirheydar (1992). The distribution of the amount of load to be shed among the load buses is done in terms of the distance from the disturbance, measured by the electrical distance (related to the inverse of the submatrix dQ/dV of the Jacobian). Shekari et al. (2016) compute the minimum amount of load to be shed in the case of a disturbance, quite similarly to Anderson and Mirheydar (1992) and Denis Lee Hau (2006) but by redefining the equivalent parameters of a linear SFD for multimachine systems. The power imbalance is obtained as in the case of event-based UFLS schemes by measuring a tripping signal of a generating unit. The distribution of the amount of load to be shed is carried out using a power tracing criterion and a voltage stability criterion, considering both ROCOV and voltage dip magnitude. Tofis et al. (2015) propose a centralized load shedding based on local frequency measurement only. The active power deficit is estimated by continuously estimating the behavior of the swing equation. A control law based on Lyapunov is derived to shed load and to bound estimation errors. In Rudez and Mihalic (2015), a WAMS-based load-shedding scheme is presented that predicts frequency behavior with simple extrapolation functions (linear during the inertial response interval, polynomial during the governor response interval). The amount of load to be shed follows linear laws of predicted minimum frequencies and/or time to reach the minimum frequency. Finally, Anderson and Mirheydar (1992), Rudez and Mihalic (2011a), and Sigrist et al. (2013) all estimate the active power imbalance as a function of the ROCOF, and the amount of load to be shed is then deduced from the active power imbalance. Initial ROCOF could, however, be used not only to estimate the amount of active power imbalance, but also to estimate the amount of load to be shed directly (Sigrist 2015). Whereas Seyedi and Sanaye-Pasand (2009a) defined a steplike relation between the amount of load to be shed and ROCOF, Sigrist (2015) shows that the relation between the ROCOF of a disturbance and the associated optimum amount of load to be shed follows a simple linear or quadratic law, as illustrated in Figure 3.11.

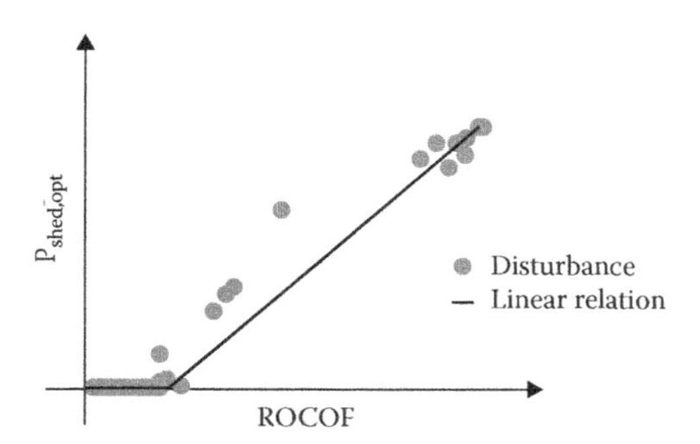

FIGURE 3.11
Illustration of the approximate linear relationship between ROCOF and the optimum amount of load to be shed.

3.3.3 Proposals for Centralized UFLS Schemes for Island Power Systems

Since the major problem is to avoid frequency instability, it would be of interest to know a frequency stability boundary, which could be incorporated into a centralized UFLS scheme for island power systems. A boundary curve to detect generation deficiency has been proposed in Chuvychin et al. (1996), without being able to determine whether a system remains stable after a disturbance. Seyedi and Sanaye-Pasand (2009a) employ ω–dω/dt phase-plane representations to define frequency and ROCOF thresholds (semi-adaptive scheme), or ROCOF thresholds and intentional time delay. The frequency stability boundary can be meaningfully presented in a phase-plane. The formulation of the frequency stability boundary is based on a simplified version of the SFD model described in Chapter 2, where generating units are represented by a first-order system. The boundary is finally described by a function $f(\Delta\omega)$, drawing an ellipse the phase-plane (Sigrist et al. 2013):

$$\Delta\dot{\omega} = f(\Delta\omega) \le \sqrt{\Delta\dot{\omega}_{max}^2\left(1 - \frac{\Delta\omega^2}{\Delta\omega_{min}^2}\right)} \tag{3.18}$$

where

$$\Delta\omega_{min} = -\frac{\Delta p}{\hat{K}}\sqrt{\frac{\hat{K}}{2H}}, \quad \Delta\dot{\omega}_{max} = -\frac{\Delta p}{2H} \tag{3.19}$$

The main idea of this centralized scheme is that as long as system trajectories are within the frequency stability boundary after a disturbance, no load-shedding actions need be taken. By contrast, if the system trajectory is outside the frequency stability boundary, load should be shed such that the system trajectory falls within the frequency stability boundary. Moreover, if the system trajectory is initially within the stability boundary, but its directionality (i.e., its tendency in the ω–dω/dt phase-plane) points to the boundary, load should be shed to correct the directionality. The measured directionality is given by

$$g_{ms} = \left\| \frac{\Delta(\Delta\dot{\omega}_{ms})}{\Delta(\Delta\omega_{ms})} \right\|$$

where:

$\Delta\omega_{ms}$ is the measured frequency
$\Delta\dot{\omega}_{ms}$ is the ROCOF

The theoretical directionality for a given frequency is computed as follows:

$$g = \left\| \left. \frac{\partial f(\Delta\omega)}{\partial\Delta\omega} \right|_{\Delta\omega=\Delta\omega_{ms}} \right\|$$

Figure 3.12 summarizes the main idea of the proposed centralized UFLS scheme. Note that the stability boundary $f(\Delta\omega)$ computes a ROCOF, which is negative as long as frequency has not reached its minimum. In addition, a higher minimum allowable frequency value results in a more conservative, but safer, stability boundary.

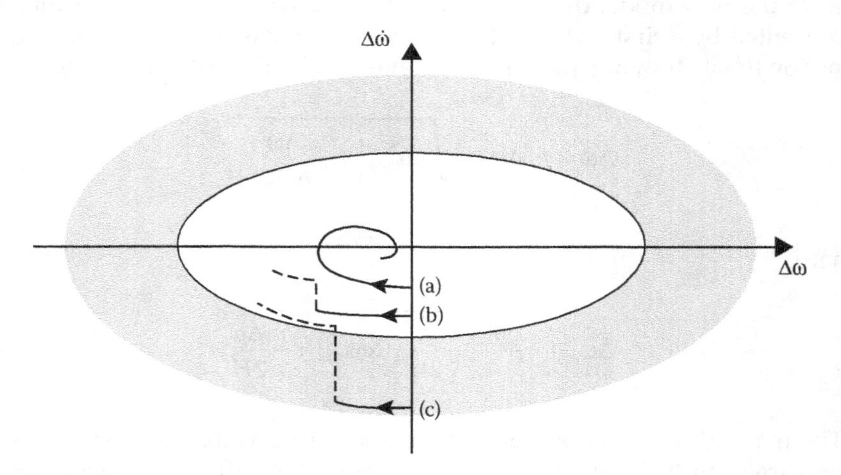

FIGURE 3.12
Principles of the centralized UFLS scheme: (a) no load shedding is needed, (b) load shedding is needed to correct the directionality, and (c) load shedding is needed to bring back the trajectory within the stability boundary.

The operating rules of the centralized UFLS scheme are then

1. If $\Delta\dot{\omega}_{ms} < \Delta\dot{\omega}_{\max}$, then the load to be shed is given by Equation 3.20:

$$p_{\text{shed}} = 2H\left(\Delta\dot{\omega}_{\max} - \Delta\dot{\omega}_{ms}\right) \tag{3.20}$$

   Equation 3.20 implies that if the disturbance is larger than the critical loss Δp, an amount of load is shed equivalent to the difference between the actual loss and the critical loss. The method to compute the amount of load to be shed is similar to the one proposed in Anderson and Mirheydar (1992), except that the definition of the critical loss is simpler, and it also takes into account the existence of several generating units. For case (c) in Figure 3.12, the amount of load to be shed is determined by Equation 3.20. Load shedding takes place once the load to be shed has been computed.

2. If $\Delta\dot{\omega}_{ms} \leq 0$ and as long as $g_{ms} \leq g$, then the load to be shed is estimated according to

$$p_{\text{shed}} = 2H\left(f\left(\Delta\omega^*\right) + \Delta\dot{\omega}_{ms}\right) \tag{3.21}$$

   where

$$\Delta\omega^* = -g_{ms}\Delta\omega_{\min} \Big/ \sqrt{\frac{\Delta\dot{\omega}_{\max}^2}{\Delta\omega_{\min}^2} + g_{ms}^2} \tag{3.22}$$

   Equation 3.22 computes the equivalent frequency for which the theoretic directionality coincides with the measured directionality. This equivalent frequency has an associated equivalent ROCOF given by $f(\Delta\omega^*)$. Equations 3.21 and 3.22 imply then that load is shed such that the measured directionality g_{ms} is corrected to the theoretical directionality g according to the stability boundary. For case (b) in Figure 3.12, the amount of load to be shed is determined by Equations 3.21 and 3.22. In this way, this operating rule effectively makes use of the available spinning reserve. Load shedding takes place once the load to be shed has been estimated. Note that Rule 2 needs to be coordinated with Rule 1 in a timely way to avoid unnecessary load shedding.

3. If $\Delta\dot{\omega}_{ms} \geq \Delta\dot{\omega}_{\max}$ and $g_{ms} > g$, then no load-shedding actions need be taken.

   In an island power system, the impact of the extension of the power system on frequency decay is not paramount, and congruently, the distribution of the estimated amount of load to be shed among the load buses is a minor problem. In general, the feeders and the corresponding relays to shed load are selected by simple iterative methods if the number of feeders is limited (Larsson and Rehtanz 2002) or by optimization algorithms if the number of feeders is large (Rodríguez et al. 2005; Apostolov et al. 1994).

3.3.4 Application to Island Power Systems

In this section, the proposed centralized scheme making use of the frequency stability boundary is applied to the island power system of La Palma and the island power system of El Hierro.

The centralized UFLS scheme needs three main parameters: the equivalent inertia H, $\Delta\omega_{min}$, and $\Delta\dot{\omega}_{max}$. These parameters are usually known for a given generation dispatch scenario, but they do not necessarily coincide with those parameters used in the formulation of the frequency stability boundary, since the identity of the tripped generating unit is in general unknown. Thus, the three parameters for the proposed centralized UFLS scheme are analyzed first.

Figure 3.13 shows the evolution of the equivalent inertia and $\Delta\omega_{min}$ and $\Delta\dot{\omega}_{max}$ for all possible operating and contingency scenarios of the La Palma island system. Figure 3.13a compares the actual equivalent inertia after a generator tripping with the equivalent inertia defined for the corresponding generation dispatch scenario (i.e., with all generators connected). It can be seen that the equivalent inertia defined for each generation dispatch scenario coincides with the mean value of the actual equivalent inertias and that the committed error in using the equivalent inertia defined for each dispatch scenario is smaller than 20%. This error mostly affects the estimation of the amount of lost generation.

Figure 3.13b displays the two parameters $\Delta\dot{\omega}_{max}$ and $\Delta\omega_{min}$. $\Delta\omega_{min}$ does not vary, since it indicates the minimum allowable frequency, set here to −0.06 pu (47 Hz), and it is therefore constant for all generator trippings. However, $\Delta\dot{\omega}_{max}$ varies, since the parameters of the connected generating units vary, but this variation is quite small, and a mean value can be used.

Knowing the main parameters of the proposed centralized UFLS scheme, the UFLS scheme is now used to protect the island power system of La Palma against all single generating unit outages of a particular generation dispatch scenario. Figure 3.14a shows the responses to the five generator trippings in the ω–$d\omega/dt$ phase-plane using the centralized UFLS scheme. It can be seen that the system responses are within the boundary curve after load shedding takes place. Actually, for the tripping of generator G17, a first block of load is shed, since the amount of lost generation is larger than the critical loss, and two further blocks of load are shed afterward, since the trajectory's directionality points to the frequency stability boundary. Note that the simulation time has been set to 10 s, and therefore, ROCOF is not zero at the end. Figure 3.14b shows the system response to the consecutive outages of two generators (G17 and G12) in the ω–$d\omega/dt$ phase-plane. Generator G17 trips first, which is also

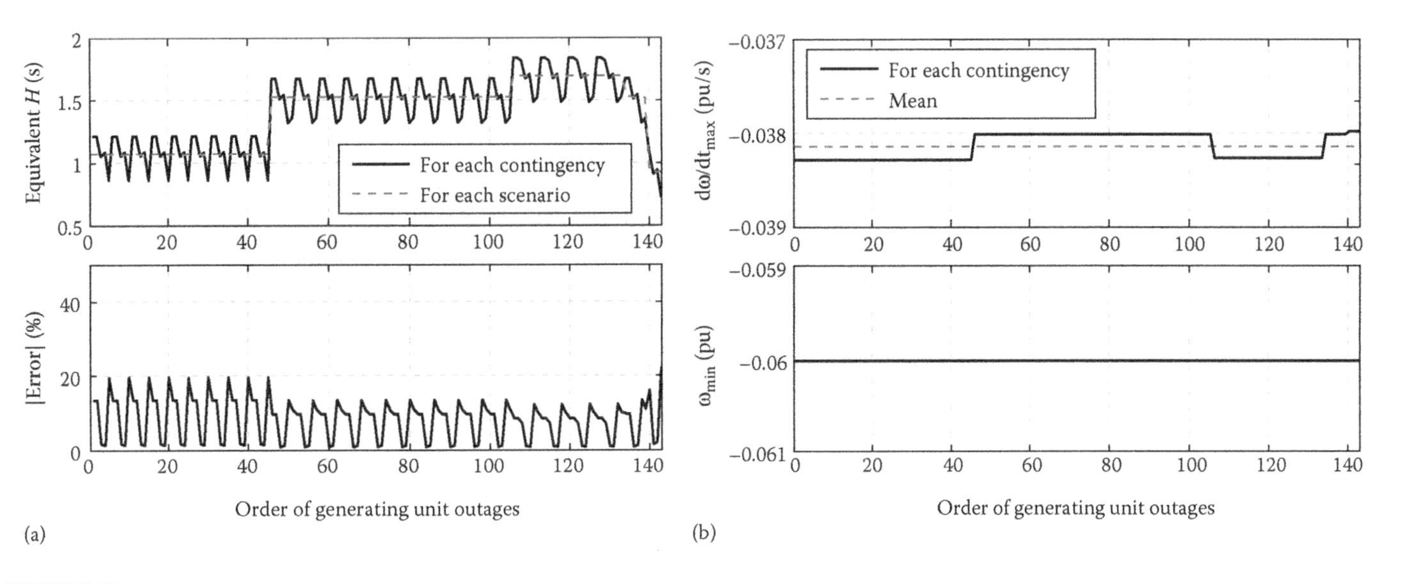

FIGURE 3.13
Evolution of the (a) equivalent inertia and (b) $\Delta\omega_{min}$ and $\Delta\dot{\omega}_{max}$ along the considered operating and contingency scenarios of the island system of La Palma.

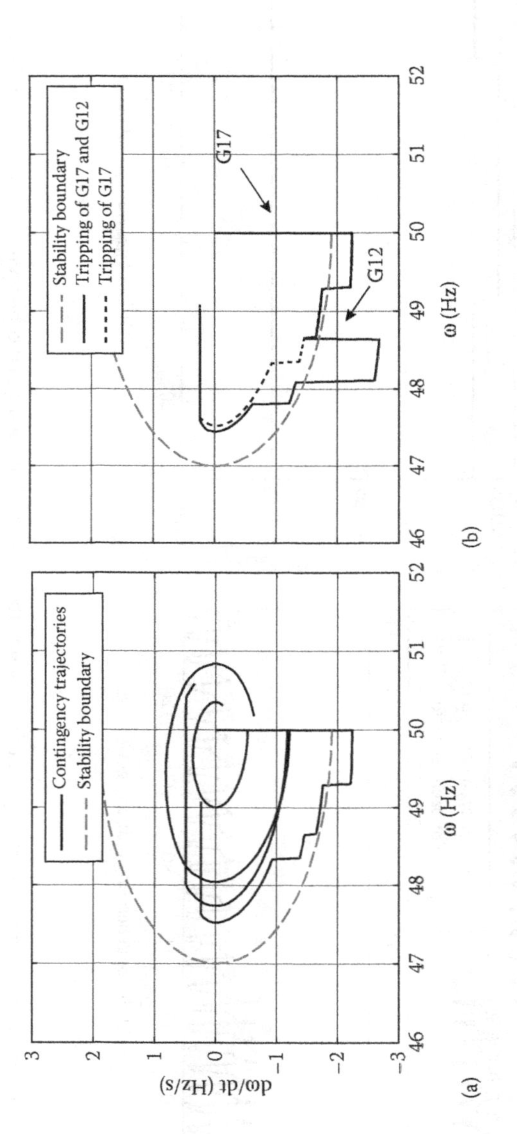

FIGURE 3.14
System responses in the ω–dω/dt phase-plane to (a) the five considered single generator trippings and (b) the consecutive tripping of two generators.

TABLE 3.9

Comparison of the Performance of the Existing, an Optimal, and the Centralized UFLS Scheme

	Existing			Optimal			Centralized		
Generator	ω_{min} (Hz)	ω_{ss} (Hz)	p_{shed} (MW)	ω_{min} (Hz)	ω_{ss} (Hz)	p_{shed} (MW)	ω_{min} (Hz)	ω_{ss} (Hz)	p_{shed} (MW)
G17	48.56	49.78	5.93	47.3	49.39	3.13	47.52	49.53	3.71
G16	48.07	50.1	5.93	47.73	49.54	0	47.73	49.54	0
G15	48.21	50.01	5.1	48.04	49.64	0	48.04	49.64	0
G13	49	49.8	0	49	49.8	0	49	49.8	0
G12	49	49.8	0	49	49.8	0	49	49.8	0

indicated in Figure 3.14b by the dashed line. Note that due to spinning reserve activation, G12 increases its generation from 2.55 MW to nearly 4 MW, its maximum power limit. Again, after the consecutive tripping of generator G12, a block of load is shed to bring back the trajectory within the frequency stability boundary, and a further block of load is shed afterward, since the trajectory's directionality appoints the frequency stability boundary.

Table 3.9 gives some additional information on the amount of shed load (p_{shed}), the minimum frequency deviation (ω_{min}), and the quasi-steady-state frequency (ω_{fin}). In addition, the centralized scheme is compared with the existing conventional UFLS scheme and with an optimal conventional scheme, which sheds the minimum amount of load to prevent the frequency from falling below the minimum allowable frequency. Note that the optimal scheme is the best possible scheme for each contingency, but it is necessary to tune this optimal scheme for each contingency separately and in advance. By contrast, the proposed centralized scheme is able to deal with different operating and contingency scenarios without readjusting its main parameters $\Delta\omega_{min}$ and $\Delta\dot{\omega}_{max}$. As shown in Table 3.9, the existing scheme is clearly outperformed, and it is interesting to see that the centralized scheme is nearly as efficient as the optimal one. Quasi-steady-state frequency values are acceptable for all three schemes.

Both the existing and the centralized UFLS scheme are applied to all possible operating and contingency scenarios. Figure 3.15 shows the system responses in terms of frequency to all operating and contingency scenarios for the centralized and for the existing scheme, respectively. It can be seen that the system responses in the case of the centralized scheme are confined within the allowable frequencies, whereas in the case of the existing scheme, large frequency overshoots exist (in some cases the system turns unstable, since generators cannot lower their output below the technical limit). It can also be seen that the centralized scheme allows larger deviations within the allowable frequencies to make use of the existing spinning reserve. The total amount of shed load has been reduced from 393.8 to 40.2 MW with the centralized UFLS scheme.

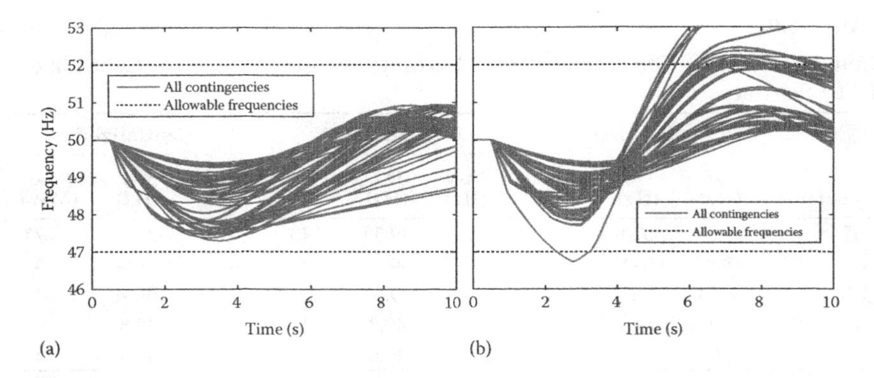

FIGURE 3.15
System responses of the La Palma island power system in terms of frequency to all possible operating and contingency scenarios by applying (a) the centralized UFLS scheme and (b) the existing conventional UFLS scheme.

TABLE 3.10
Comparison of the Performance of the Existing, an Optimal, and the Centralized UFLS Scheme

	Existing			Optimal			Centralized		
Generator	ω_{min} (Hz)	ω_{ss} (Hz)	p_{shed} (MW)	ω_{min} (Hz)	ω_{ss} (Hz)	p_{shed} (MW)	ω_{min} (Hz)	ω_{ss} (Hz)	p_{shed} (MW)
G09	49.53	49.88	0	49.53	49.88	0	49.53	49.88	0
G08	47.74	49.87	1.15	47.74	49.4	0	47.74	49.4	0
G07	48.09	49.67	0.38	48.09	49.55	0	48.09	49.55	0
G06	49	49.73	0	49	49.73	0	49	49.73	0
G05	48.52	49.73	0.38	48.52	49.58	0	48.52	49.58	0
G02	49.34	49.82	0	49.34	49.82	0	49.34	49.82	0

In the case of the El Hierro island system, the centralized UFLS scheme is first applied to a single generation dispatch scenario and then to all possible operating and contingency scenarios. Table 3.10 shows for a particular generation dispatch scenario the amount of shed load, the minimum frequency deviation, and the quasi-steady-state frequency, and it compares the centralized scheme with the existing conventional UFLS scheme and with an optimal conventional scheme. For both the optimal and the centralized scheme, no load shedding occurs, since the frequency does not fall below the minimum allowable frequency of 47 Hz (therefore, in both cases p_{shed} is equal to 0 MW, and minimum frequencies are the same). By contrast, the existing scheme sheds load, and in addition, it sheds load after the frequency has attained its minimum value (this is indicated by the fact that all three schemes show the same minimum frequencies). The main reason for this is the large intentional time delays of the existing UFLS scheme. Quasi-steady-state frequency values are acceptable for all three schemes.

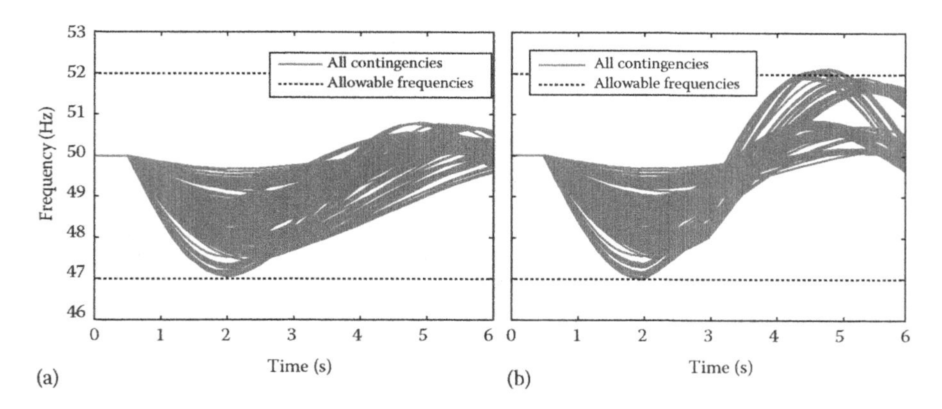

FIGURE 3.16
System responses of the El Hierro island power system in terms of frequency to all possible operating and contingency scenarios by applying (a) the centralized UFLS scheme and (b) the existing conventional UFLS scheme.

Figure 3.16 shows the system responses in terms of frequency to all possible operating and contingency scenarios for the centralized and for the existing scheme of the El Hierro island power system, respectively. It can be seen that the system responses in the case of the centralized scheme are confined within the allowable frequencies as for the La Palma power system. The system response in the case of the existing UFLS scheme also remains within the allowable frequencies, but some high-frequency overshoots can be detected, smaller than those of the La Palma island system and indicators of overshedding. It can also be seen that both schemes allow for frequency deviations within the allowable frequencies to make use of the existing spinning reserve. The total amount of shed load has been reduced from 38.5 to 3.4 MW with the centralized UFLS scheme.

References

Abdelwahid, S., A. Babiker, A. Eltom, and G. Kobet. 2014. Hardware implementation of an automatic adaptive centralized underfrequency load shedding scheme. *IEEE Transactions on Power Delivery* 29 (6):2664.

Abedini, M., M. Sanaye-Pasand, and S. Azizi. 2014. Adaptive load shedding scheme to preserve the power system stability following large disturbances. *IET Generation, Transmission & Distribution* 8 (12):2124.

Adamiak, M., M. J. Schiefen, G. Schauerman, and B. Cable. 2014. Design of a priority-based load shed scheme and operation tests. *IEEE Transactions on Industry Applications* 50 (1):182.

Anderson, P. M. 1999. *Power System Protection*. Piscataway, NJ: IEEE Press.

Anderson, P. M., and M. Mirheydar. 1992. An adaptive method for setting underfrequency load shedding relays. *IEEE Transactions on Power Systems* 7 (2):647.

Apostolov, A. P., D. Novosel, and D. G. Hart. 1994. Intelligent protection and control during power system disturbance. In *Proceedings of the American Power Conference,* Chicago, IL. pp. 1175–1181.

Blackburn, J. L., and T. J. Damin. 2006. *Protective Relaying. Principles and Applications.* Boca Raton, FL: CRC Press.

Bogovic, J., U. Rudez, and R. Mihalic. 2015. Probability-based approach for parametrisation of traditional underfrequency load-shedding schemes. *IET Generation, Transmission & Distribution* 9 (16):2625.

Bonian, S., X. Xiaorong, and H. Yingduo. 2005. *WAMS-Based Load Shedding for Systems Suffering Power Deficit, IEEE/PES Transmission and Distribution Conference and Exhibition: Asia and Pacific.* Dalian, China: IEEE.

Cammesa. 2008. Los procedimientos: Anexos PT 4: Anexos "A" a "H".

CDEC-SIC, Dirección de Operación. 2008. Estudio Esquemas de Desconexión Automáticos de Carga 2008–2009.

Ceja-Gomez, F., S. S. Qadri, and F. D. Galiana. 2012. Under-frequency load shedding via integer programming. *IEEE Transactions on Power Systems* 27 (3):1387.

Cheng, C. P., and C. Shine. 2003. Under-frequency load shedding scheme design based on cost-benefit analysis. In *6th International Conference on Advances in Power System Control, Operation and Management. Proceedings. APSCOM 2003,* 514. Stevenage, UK: IEE.

Chuvychin, V. N., N. S. Gurov, S. S. Venkata, and R. E. Brown. 1996. An adaptive approach to load shedding and spinning reserve control during underfrequency conditions. *IEEE Transactions on Power Systems* 11 (4):1805.

Concordia, C., L. H. Fink, and G. Poullikkas. 1995. Load shedding on an isolated system. *IEEE Transactions on Power Systems* 10 (3):1467.

De Tuglie, E., M. Dicorato, M. La Scala, and P. Scarpellini. 2000. A corrective control for angle and voltage stability enhancement on the transient time-scale. *IEEE Transactions on Power Systems* 15 (4):1345.

Delfino, B., S. Massucco, A. Morini, P. Scalera, and F. Silvestro. 2001. Implementation and comparison of different under frequency load-shedding schemes. *2001 Power Engineering Society Summer Meeting,* Piscataway, NJ.

Denis Lee Hau, A. 2006. A general-order system frequency response model incorporating load shedding: Analytic modeling and applications. *IEEE Transactions on Power Systems* 21 (2):709.

Dong, M., C. Lou, and C. Wong. 2008. Adaptive under-frequency load shedding. *Tsinghua Science and Technology* 13 (6):823.

EirGrid. 2009. *EirGrid Grid Code.* Version 3.4, 2009, Dublin, Ireland.

Ela, E., B. Kirby, E. Lannoye, M. Milligan, D. Flynn, B. Zavadil, and M. O'Malley. 2010. Evolution of operating reserve determination in wind power integration studies. *Power and Energy Society General Meeting, 2010 IEEE,* July 25–29, 2010.

Elkateb, M. M., and M. F. Dias. 1993a. New proposed adaptive frequency load shedding scheme for cogeneration plants. In *Fifth International Conference on Developments in Power System Protection,* London, UK: IEE.

Elkateb, M. M., and M. F. Dias. 1993b. New technique for adaptive-frequency load shedding suitable for industry with private generation. *IEE Proceedings C: Generation, Transmission and Distribution* 140 (5):411.

ENTSO-e. 2013. Appendix 1 of System Operation Agreement. Brussels, Belgium (available at: www.entsoe.eu)

Ghaleh, A. P., M. Sanaye-Pasand, and A. Saffarian. 2011. Power system stability enhancement using a new combinational load-shedding algorithm. *IET Generation, Transmission & Distribution* 5 (5):551.

Girgis, A. A., and S. Mathure. 2010. Application of active power sensitivity to frequency and voltage variations on load shedding. *Electric Power Systems Research* 80 (3):306.

Gu, W., W. Liu, J. Zhu, B. Zhao, Z. Wu, Z. Luo, and J. Yu. 2014. Adaptive decentralized under-frequency load shedding for islanded smart distribution networks. *IEEE Transactions on Sustainable Energy* 5 (3):886.

Halevi, Y., and D. Kottick. 1993. Optimization of load shedding system. *IEEE Transactions on Energy Conversion* 8 (2):207.

HECO. 2008. *Integrated Resource Plan*. Available at www.heco.com.

Hong, Y. Y., and P. H. Chen. 2012. Genetic-based underfrequency load shedding in a stand-alone power system considering fuzzy loads. *IEEE Transactions on Power Delivery* 27 (1):87.

Hong, Y. Y., M. C. Hsiao, Y. R. Chang, Y. D. Lee, and H. C. Huang. 2013. Multiscenario underfrequency load shedding in a microgrid consisting of intermittent renewables. *IEEE Transactions on Power Delivery* 28 (3):1610.

Hong, Y. Y., and S. F. Wei. 2010. Multiobjective underfrequency load shedding in an autonomous system using hierarchical genetic algorithms. *IEEE Transactions on Power Delivery* 25 (3):1355.

Hoseinzadeh, B., F. F. da Silva, and C. L. Bak. 2016. Decentralized coordination of load shedding and plant protection considering high share of RESs. *IEEE Transactions on Power Systems* PP (31):5, pp. 3607–3615.

Hoseinzadeh, B., F. M. Faria da Silva, and C. L. Bak. 2015b. Adaptive tuning of frequency thresholds using voltage drop data in decentralized load shedding. *IEEE Transactions on Power Systems* 30 (4):2055.

Hsu, C. T., M. S. Kang, and C. S. Chen. 2005. Design of adaptive load shedding by artificial neural networks. *IEE Proceedings: Generation, Transmission and Distribution* 152 (3):415.

Huang, S., and C. Huang. 2001. Adaptive approach to load shedding including pumped-storage units during underfrequency conditions. *IEE Proceedings: Generation, Transmission and Distribution* 148 (2):165.

IEEE. 1975. A status report on methods used for system preservation during underfrequency conditions. *IEEE Transactions on Power Apparatus and Systems* 94 (2):360.

IEEE. 1987. IEEE guide for abnormal frequency protection for power generating plants. *ANSI/IEEE Std C37.106–1987.*

IEEE. 2007. IEEE guide for the application of protective relays used for abnormal frequency load shedding and restoration. *IEEE Std C37.117–2007*:c1.

Jenkins, L. 1983. Optimal load shedding algorithm for power system emergency control. In *1983 Proceedings of the International Conference on Systems, Man and Cybernetics*. Bombay and New Delhi, India: IEEE.

Jones, J. R., and W. D. Kirkland. 1988. Computer algorithm for selection of frequency relays for load shedding. *Computer Applications in Power, IEEE* 1 (1):21.

Jung, J., L. Chen-Ching, S. L. Tanimoto, and V. Vittal. 2002. Adaptation in load shedding under vulnerable operating conditions. *IEEE Transactions on Power Systems* 17 (4):1199.

Ketabi, A., and M. H. Fini. 2015. An underfrequency load shedding scheme for hybrid and multiarea power systems. *IEEE Transactions on Smart Grid* 6 (1):82.

Larsson, M., and C. Rehtanz. 2002. Predictive frequency stability control based on wide-area phasor measurements. *2002 IEEE Power Engineering Society Summer Meeting*, Piscataway, NJ.

Li, Z., and Z. Jin. 2006. UFLS design by using f and integrating df/dt. In *2006 IEEE/PES Power Systems Conference and Exposition*, Atlanta, GA: IEEE.

Lin, X., H. Weng, Q. Zou, and P. Liu. 2008. The frequency closed-loop control strategy of islanded power systems. *IEEE Transactions on Power Systems* 23 (2):796.

Lokay, H. E., and V. Burtnyk. 1968. Application of underfrequency relays for automatic load shedding. *IEEE Transactions on Power Apparatus and Systems* PAS-87 (3):776.

Lopes, J. A. P., Wa Wong Chan, and L. M. Proenca. 1999. Genetic algorithms in the definition of optimal load shedding strategies. In *PowerTech Budapest 99*, Budapest, Hungary: IEEE.

Lukic, M., I. Kuzle, and S. Tesnjak. 1998. An adaptive approach to setting underfrequency load shedding relays for an isolated power system with private generation. In *MELECON '98. 9th Mediterranean Electrotechnical Conference. Proceedings*, Tel Aviv, Israel.

Mahat, P., Z. Chen, and B. Bak-Jensen. 2010. Underfrequency load shedding for an islanded distribution system with distributed generators. *IEEE Transactions on Power Delivery* 25 (2):911.

Manson, S., G. Zweigle, and V. Yedidi. 2014. Case study: An adaptive underfrequency load-shedding system. *IEEE Transactions on Industry Applications* 50 (3):1659.

Martínez, M., G. Winter, B. Martín, B. Galván, P. Cuesta, and J. Abderraman. 2003. La optimización mediante algoritmos genéticos del proceso de deslastre y reposición de carga. *Vector Plus*. (21): pp. 13–26.

Matsuura, M. 2009. Island breezes. *Power and Energy Magazine, IEEE* 7 (6):59.

Merino, J., P. Mendoza-Araya, and C. Veganzones. 2014. State of the art and future trends in grid codes applicable to isolated electrical systems. *Energies* 7 (12):7936.

Mitchell, M. A. 2000. Optimization of under-frequency load shedding strategies through the use of a neural network and a genetic algorithm. Master Thesis, INESC, Universidade do Porto.

Mitchell, M. A., J. A. P. Lopes, J. N. Fidalgo, and J. D. McCalley. 2000. Using a neural network to predict the dynamic frequency response of a power system to an under-frequency load shedding scenario. *Power Engineering Society Summer Meeting, 2000*. IEEE. Seattle, WA.

MITyC, Ministerio de Industria, Turismo y Comercio. 2006. "RESOLUCIÓN de 28 de abril de 2006, de la Secretaría General de Energía, por la que se aprueba un conjunto de procedimientos de carácter técnico e instrumental necesarios para realizar la adecuada gestión técnica de los sistemas eléctricos insulares y extrapeninsulares. ANEXO." In *Boletín oficial del estado*.

Mohd Zin, A. A., H. Mohd Hafiz, and M. S. Aziz. 2004. A review of under-frequency load shedding scheme on TNB system. In *National Power and Energy Conference. PECon 2004. Proceedings*. Kuala Lumpur, Malaysia: IEEE.

NERC. 2008. UFLS Regional Reliability Standard Characteristics. Available at: www.nerc.com

NGET. 2009. The grid code. Warwick, UK. Available at: www2.nationalgrid.com

Ning, X., H. Jiang, and X. Zhu. 2011. An improved under frequency load shedding scheme in multi-machine power system based on the rate of change of frequency. In *2011 International Conference on Advanced Power System Automation and Protection (APAP)*, October 16–20, 2011. IEEE, Beijing, China.

Nirenberg, S. A., D. A. McInnis, and K. D. Sparks. 1992. Fast acting load shedding. *IEEE Transactions on Power Systems* 7 (2):873.

Nordel. 2007. *Nordic Grid Code*. Available at www.entsoe.eu

Novosel, D., K. T. Vu, D. Hart, and E. Udren. 1996. Practical protection and control strategies during large power-system disturbances. In *1996 IEEE Transmission and Distribution Conference Proceedings. Proceedings of the 1996 IEEE Power Engineering Society*, New York.

Padron, S., M. Hernandez, and A. Falcon. 2016. Reducing under-frequency load shedding in isolated power systems using neural networks. Gran Canaria: A case study. *IEEE Transactions on Power Systems* 31 (1):63.

Parniani, M., and A. Nasri. 2006. SCADA based under frequency load shedding integrated with rate of frequency decline. In *2006 IEEE Power Engineering Society General Meeting*, Montreal, Quebec, Canada: IEEE.

Prasad, M., K. N. Satish, Kuldeep, and R. Sodhi. 2014. A synchrophasor measurements based adaptive underfrequency load shedding scheme. *2014 IEEE, Innovative Smart Grid Technologies: Asia (ISGT Asia)*, May 20–23, pp. 424–428. IEEE. Kuala Lumpur, Malaysia.

Prasetijo, D., W. R. Lachs, and D. Sutanto. 1994. A new load shedding scheme for limiting underfrequency. *IEEE Transactions on Power Systems* 9 (3):1371.

Reddy, C. P., S. Chakrabarti, and S. C. Srivastava. 2014. A sensitivity-based method for under-frequency load-shedding. *IEEE Transactions on Power Systems* 29 (2):984.

REE. 2005. Criterios generales de protección de los sistemas eléctricos insulares y extrapeninsulares. REE.

Rodríguez, J., M. González, J. R. Diago, C. Domingo, I. Egido, P. Centeno, F. Fernández, and L. Rouco. 2005. Un esquema centralizado de deslastre de cargas para pequeños sistemas aislados. Ninth Spanish Portuguese Congress on Electrical Engineering (9CHLIE), Marbella, Spain, June 30–July 2.

Rudez, U., and R. Mihalic. 2011a. Monitoring the first frequency derivative to improve adaptive underfrequency load-shedding schemes. *IEEE Transactions on Power Systems* 26 (2):839.

Rudez, U., and R. Mihalic. 2011b. A novel approach to underfrequency load shedding. *Electric Power Systems Research* 81 (2):636.

Rudez, U., and R. Mihalic. 2016. WAMS based underfrequency load shedding with short-term frequency prediction. *IEEE Transactions on Power Delivery* 31(4):1912–1920.

Saffarian, A., and M. Sanaye-Pasand. 2011. Enhancement of power system stability using adaptive combinational load shedding methods. *IEEE Transactions on Power Systems* 26 (3):1010.

Sanaye-Pasand, M., and M. Davarpanah. 2006. A new adaptive multidimensional load shedding scheme using genetic algorithm. In *2005 Canadian Conference on Electrical and Computer Engineering*. Saskatoon, Saskatchewan, Canada: IEEE.

Seethalekshmi, K., S. N. Singh, and S. C. Srivastava. 2011. A synchrophasor assisted frequency and voltage stability based load shedding scheme for self-healing of power system. *IEEE Transactions on Smart Grid* 2 (2):221.

Seyedi, H., and M. Sanaye-Pasand. 2009a. Design of new load shedding special protection schemes for a double area power system. *American Journal of Applied Sciences* 6 (2):317–327.

Seyedi, H., and M. Sanaye-Pasand. 2009b. New centralised adaptive load-shedding algorithms to mitigate power system blackouts. *IET Generation, Transmission & Distribution* 3 (1):99.

Shekari, T., F. Aminifar, and M. Sanaye-Pasand. 2016. An analytical adaptive load shedding scheme against severe combinational disturbances. *IEEE Transactions on Power Systems* (31):5, pp. 4135–4143.

Shokooh, S., T. Khandelwal, F. Shokooh, J. Tastet, and J. J. Dai. 2005. Intelligent load shedding need for a fast and optimal solution. IEEE PCIC, pp. 1–10 Europe, Basel.

Shrestha, G. B., and K. C. Lee. 2005. Load shedding schedules considering probabilistic outages. In *Power Engineering Conference, 2005.* IPEC, 2005. The 7th International. Singapore.

Sigrist, L. 2010. Design of UFLS schemes of small isolated power systems. PhD, IIT, Pontificias Comillas.

Sigrist, L. 2015. A UFLS scheme for small isolated power systems using rate-of-change of frequency. *IEEE Transactions on Power Systems* 30 (4):2192.

Sigrist, L., I. Egido, and L. Rouco. 2012a. A method for the design of UFLS schemes of small isolated power systems. *IEEE Transactions on Power Systems* 27 (2):951.

Sigrist, L., I. Egido, and L. Rouco. 2012b. Performance analysis of UFLS schemes of small isolated power systems. *IEEE Transactions on Power Systems* 27 (3):1673.

Sigrist, L., I. Egido, and L. Rouco. 2013. Principles of a centralized UFLS scheme for small isolated power systems. *IEEE Transactions on Power Systems* 28 (2):1779.

Sigrist, L., I. Egido, E. F. Sanchez-Ubeda, and L. Rouco. 2010. Representative operating and contingency scenarios for the design of UFLS schemes. *IEEE Transactions on Power Systems* 25 (2):906.

Sigrist, L., and L. Rouco. 2014. Design of UFLS schemes taking into account load variation. In 2014 Power Systems Computation Conference (PSCC), August 18–22, 2014. Available at www.pscc-central.org.

Tang, J., J. Liu, F. Ponci, and A. Monti. 2013. Adaptive load shedding based on combined frequency and voltage stability assessment using synchrophasor measurements. *IEEE Transactions on Power Systems* 28 (2):2035.

Terzija, V. V. 2006. Adaptive underfrequency load shedding based on the magnitude of the disturbance estimation. *IEEE Transactions on Power Systems* 21 (3):1260.

Terzija, V. V., and H. J. Koglin. 2002. Adaptive underfrequency load shedding integrated with a frequency estimation numerical algorithm. *IEE Proceedings: Generation, Transmission and Distribution* 149 (6):713.

Thalassinakis, E. J., and E. N. Dialynas. 2004. A Monte-Carlo simulation method for setting the underfrequency load shedding relays and selecting the spinning reserve policy in autonomous power systems. *IEEE Transactions on Power Systems* 19 (4):2044.

Thompson, J. G., and B. Fox. 1994. Adaptive load shedding for isolated power systems. *IEE Proceedings-Generation, Transmission and Distribution* 141 (5):491.

Tofis, Y., Y. Yiasemi, and E. Kyriakides. 2015. A plug-and-play selective load shedding scheme for power systems. *IEEE Systems Journal* PP (99):1.

Transpower. 2009. Automatic under-frequency load shedding (AUFLS) technical report Appendix A: A collation of international policies for under-frequency load shedding. Available at www.systemoperator.co.nz.

UCTE. 2004. *Operation Handbook Policy 1: Load-Frequency Control and Performance.*

You, H., V. Vittal, and Z. Yang. 2002. Self-healing in power systems: An approach using islanding and rate of frequency decline based load shedding. *Power Engineering Review, IEEE* 22 (12):62.

4

Voltage Stability

This chapter describes modeling and computation issues for voltage stability studies in island power systems. It further contains the definition and mathematical background of the problem, a description of the most efficient techniques to compute the margin to voltage collapse, and sensitivity analysis techniques. This chapter also discusses critical topics for voltage stability in island power systems, such as the reactive power reserve and the impact of renewable energy resources.

4.1 Voltage Stability of Electric Power Systems

Voltage stability refers to the ability of a power system to maintain acceptable bus voltages under normal conditions and after being subjected to a disturbance (Kundur et al. 1994). Low voltages imply high currents to maintain the power demand. Since power systems are nonlinear, not only voltage profile decreases with demand growth, but also its slope; thus, the voltage fall will be faster as the demand increases. This situation usually leads to a cascade failure of the different system devices, followed by a system blackout.

Voltage collapse can occur over a wide variety of time frames, as Figure 4.1 depicts (IEEE Power System Relaying Committee 1996). In the transient region, voltages and angular stabilities are closely related in some cases, since low voltages may accelerate angular instability; likewise, the loss of synchronism may result in voltage depression. However, transient voltage collapse can also emerge without loss of angular stability. In the long-term region, angles of voltages are also affected, but only in the moments immediately before the complete blackout. In most cases, analyzing and preventing long-term voltage collapses during the planning of the operation also reduces the possibility of transient voltage collapse situations.

From an electrical point of view, there are different reasons why a voltage collapse can occur in a power system (van Cutsem 2007; Taylor et al. 1994). It is very common in highly loaded scenarios, where some branches in the network are supporting high power flows and large voltage drops that undergo voltage collapse. This can also happen if interarea energy transfers are overestimated or if a large generation–demand imbalance exists. However, in voltage collapse prevention, the most relevant variable is the

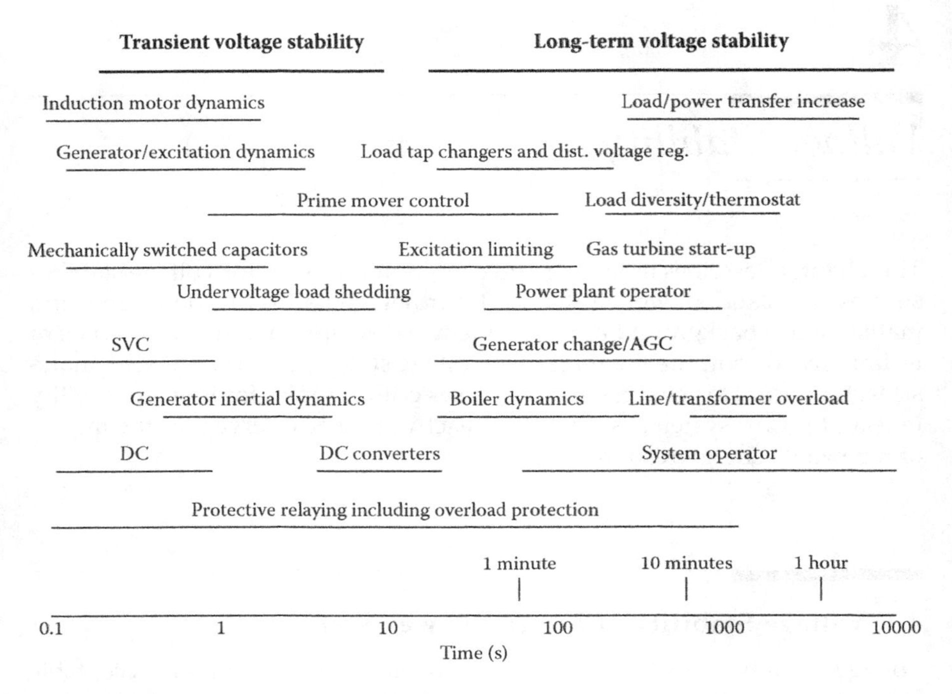

FIGURE 4.1
Time frames for voltage stability phenomena.

reactive power. A severe lack of reactive power margin is by far the quickest way to a voltage collapse situation. Improperly adjusted voltage control devices or saturation of reactive power limits in generation units puts voltage stability at risk. Also, the situation becomes worse in the case of one or more network devices being disconnected from the network under a contingency.

From a mathematical point of view, voltage collapse phenomena are due to the proximity of the system operating point to a saddle-node bifurcation in the power system equations (Dobson et al. 1992). Under secure operating conditions, voltages barely descend as the demand grows, but as the system gets close to a saddle-node bifurcation, this gradient tends to minus infinity till the voltage collapse point, where the system Jacobian becomes singular. Far from this point, the steady-state equations of the power systems have no solution. Therefore, the analysis of how far the system is from the voltage collapse point represents a key tool to keep the system safe from voltage instability situations. This distance is known as the *voltage collapse margin*.

4.1.1 Concept and Definition of Voltage Collapse Margin

Voltage stability is concerned with the ability of a power system to maintain acceptable bus voltages under normal conditions and after being subjected to a disturbance. Voltage stability is mainly a dynamic phenomenon that implies detailed models of each device of the power system: generators, transformers, loads, and so on. However, not only can voltage instability be detected using steady-state analysis, but also preventive control actions can be optimized to maintain the system far enough from the voltage collapse point.

From a steady-state point of view, voltage collapse is illustrated by the trajectories of bus voltages as system demand grows, usually called *nose curves*. When active power demand grows and so does the active power generation, voltages tend to decrease. This is because increasing the power flow implies more current through lines and transformers, and hence higher voltage drop. Nevertheless, this behavior becomes highly nonlinear as the load of the systems grows, so the voltage slope tends to be vertical, and the voltage fall emerges as critical. This boundary is known as the *voltage collapse point*, and the distance from the starting base case to the maximum load is considered as the margin to voltage collapse.

Figure 4.2 depicts the evolution of the active power demand of an area of the Spanish electric power system, and the voltage trajectory of the area pilot bus, in a long-term close-to-collapse scenario that occurred in 2001. At 16:40, the local demand reaches a local minimum of 4000 MW and starts to grow. As a consequence of an ill-conditioned system, the pilot bus voltage falls dramatically for several minutes from 400 to 350 kV. At 18:30, since the voltages are falling out of control, an emergency load shedding of almost 300 MW is applied to stabilize the system. Once the load has been shed, the system voltages change their tendency and recover reasonable values and hence, stability.

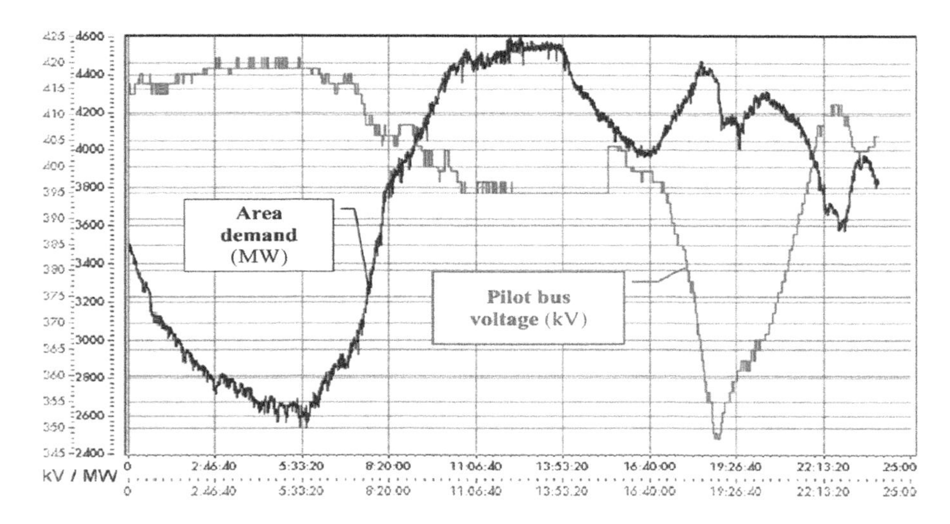

FIGURE 4.2
Time trajectories of area demand and pilot bus voltage in an actual scenario of the Spanish power system.

From an operational point of view, there are several factors that may lead the system to a voltage collapse: highly loaded scenarios, large generation–demand imbalance, inadequate reactive power reserve, or an incorrect adjustment of the voltage control devices. The situation nearly always gets worse when a contingency occurs and one or more network devices get disconnected. However, in all these cases, the system behavior is similar; thus, as the demand grows, the voltages will fall ever faster, till a maximum loading is reached. This maximum loading point is the voltage collapse point, where the gradient of bus voltages becomes infinite due to a saddle-node bifurcation; thus, the system becomes unsolvable for demand values greater than the critical value.

To illustrate the voltage collapse concept, consider a two-buses network, in which a generation unit and a load are interconnected by a 1.0 pu reactance. In addition, the generation unit voltage is set to 1.0 (van Cutsem 2007). The diagram of the network and the corresponding power flow equations are represented in Figure 4.3.

Depending on the active (p) and reactive (q) power demands, the system described in Figure 4.3 can present two solutions (feasible), one solution (critical), or no solution (unfeasible). Figure 4.4 describes this phenomenon

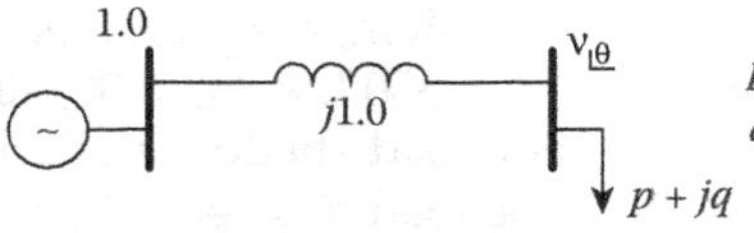

1.0

$j1.0$

$v_{\angle\theta}$

$p + jq$

$$p = -v \cdot \sin\theta$$
$$q = -v^2 + v \cdot \cos\theta$$

FIGURE 4.3
Two-buses network diagram and steady-state equations.

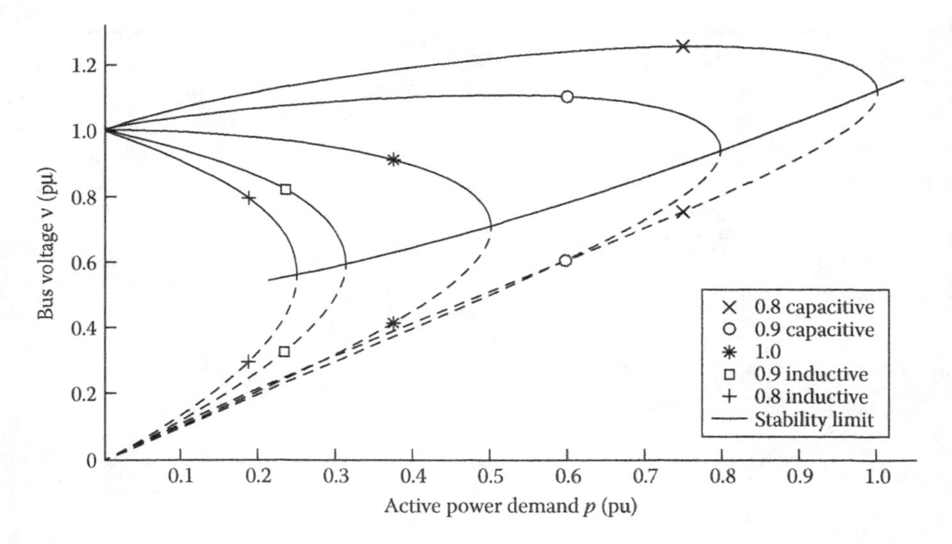

FIGURE 4.4
Voltage trajectories for demand growth, considering different load power factors.

by depicting different full voltage v trajectories as active power demand p increases for various values of the load power factor. These curves are also known as *nose curves*.

In Figure 4.4, each load power factor corresponds to a single manifold with a singularity in the power axis, the voltage collapse point. This singularity implies two different trajectories for the voltages considering the same demand evolution. The difference between them is that the upper trajectory is stable and the lower is unstable. Electric power systems are designed to work considering voltages close to 1.0 pu and negative gradients with respect to demand. Moreover, even though the system is working on the upper branch, some events may take the system to a new trajectory on an unstable manifold, undergoing a voltage collapse. These events include the outage of one or more network devices, such as generation units, lines, or transformers, or the reactive power saturation in one or more generation units (see Section 4.4), among others.

It can be figured out that the power flow equations system of the two-buses network will be feasible if the discriminant Δ is greater than zero, where the discriminant Δ is equal to

$$\Delta = \frac{1}{4} - p^2 - q \tag{4.1}$$

Thus, the feasibility boundary in the pq-space will correspond to a parabola, below which the power flow equations are feasible. Figure 4.5 depicts the feasibility boundary for the two-buses network and the geometrical interpretation of the different type of solutions in the θv-space.

Point (a) in Figure 4.5 represents a feasible point of work, since it is situated below the feasibility boundary. This point in the pq-space corresponds with two curves in the θv-space, which intersect at two points, that is, the stable (S) and the unstable (U) solutions of the steady-state equations. In addition, the gradient vectors $\mathbf{p}_x$ and $\mathbf{q}_x$ have different directions, which means that the Jacobian is not singular. On the contrary, point (c) in Figure 4.5 represents an unfeasible point of work, since it is situated outside the feasibility region. In this case, this point in the pq-space corresponds with two curves in the θv-space, which don't intersect at any point; that is, the steady-state equations system has no solution. Finally, point (b) is on the feasibility boundary; thus, it corresponds with the saddle-node bifurcation of the steady-state equations, which is a voltage collapse point. This point in the pq-space corresponds with two curves in the θv-space that are tangent to each other at one point; that is, the stable (S) and the unstable (U) solutions of the steady-state equations are the same. At this point, the Jacobian matrix of the steady-state equations becomes singular, which is in accordance with the fact that the gradient vectors in the θv-space $\mathbf{p}_x$ and $\mathbf{q}_x$ are parallel to each other.

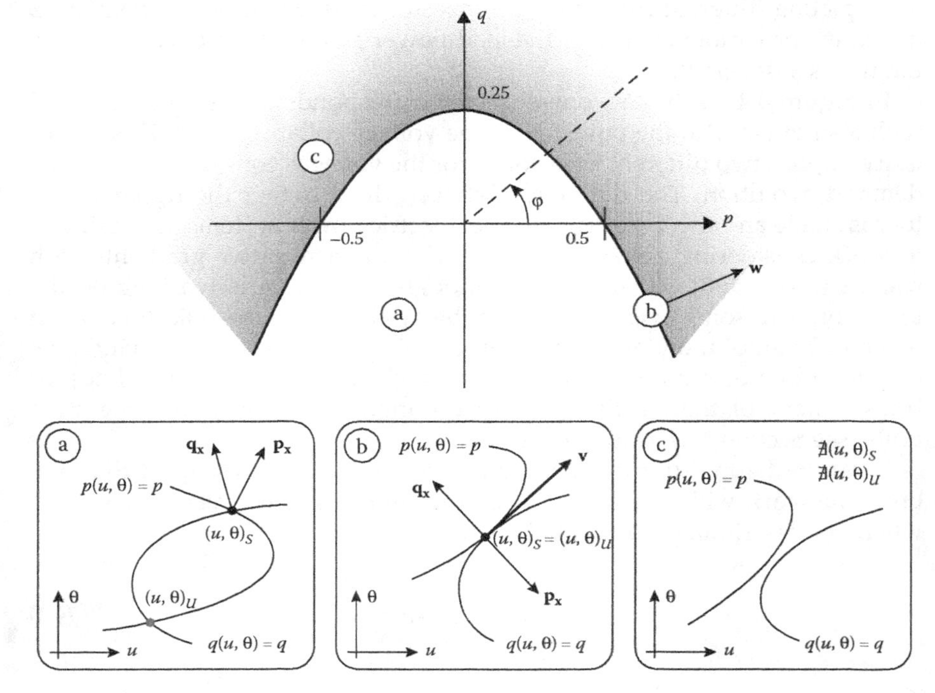

FIGURE 4.5
Feasibility analysis of the two-buses power system.

In point (b), the singular eigenvalue comes with its corresponding left and right eigenvectors, and the interpretation of both of them has been depicted in Figure 4.5. On the one hand, the right eigenvector appears in the θv-space (**v**) as the normal vector to the hyperplane formed by the gradients of all the steady-state equations. On the other hand, the left eigenvector appears in the *pq*-space (**w**) as the normal vector of the feasibility boundary. For this reason, the left eigenvector describes the optimal direction in the *pq*-space to move away from the feasibility boundary and move the system away from a voltage collapse.

Power flow steady-state equations represent the power balance between the net power produced $\mathbf{S}^0$ and the sum S of power injected into the network at each bus.

$$\mathbf{S}^0 = \mathbf{S}(\mathbf{V}, \theta) \tag{4.2}$$

In Equation 4.2, power S represents either active or reactive power. On the left-hand side of Equation 4.2, there is net power S^0 produced at the bus, that is, the difference between the generated and demanded power. On the right-hand side, there is the power injected into the network as a function of the state variables of the system, that is, the magnitude and the angle of bus voltages. In the case of actual networks, the voltage collapse concept is the same; that is, there exists a feasibility boundary in the bus power space that represents a saddle-node bifurcation. Therefore, any power dispatch inside the boundary represents a feasible power flow problem, and any power dispatch outside the boundary corresponds with an unfeasible power flow problem.

To evaluate the risk of a voltage collapse, it is not necessary to compute the whole feasibility boundary within the bus power space. The most widely used procedure to evaluate the voltage stability is to assume that the power dispatch of the system is going to evolve in a concrete direction. This evolution vector ΔS is parameterized using a factor λ; thus, the bus power dispatch becomes a linear function of λ. Rearranging Equation 4.2 and introducing the power dispatch variation vector:

$$\mathbf{S}^0 + \lambda \cdot \Delta\mathbf{S} - \mathbf{S}(\mathbf{V}, \theta) = 0 \tag{4.3}$$

System Equation 4.3 will be feasible or unfeasible depending on the parameter λ, as Figure 4.6 depicts.

In Figure 4.6a, the bus power dispatch evolves from the initial point of work $\mathbf{S}^0$ following the direction defined by $\Delta\mathbf{S}$. For each value of the load margin λ, a new feasible power dispatch is defined, till the critical load margin λ_C is reached. At this value of the load margin, the corresponding bus power dispatch is on the feasibility boundary, that is, the saddle-node bifurcation. Therefore, the resulting distance $\lambda_C \cdot \Delta\mathbf{S}$ from the base case to the feasibility boundary is considered to be the margin to voltage collapse.

In Figure 4.6b, the state variables, that is, the magnitude and angle of bus voltages, evolve from the initial point of work following a concrete path defined by the system equation in Equation 4.3. The end point of the state variable trajectories is the voltage collapse point, located at the critical load margin λ_C. At this point, the slope of the trajectories becomes infinite, since the voltage collapse point corresponds with a saddle-node bifurcation, and hence the Jacobian matrix of the system is singular.

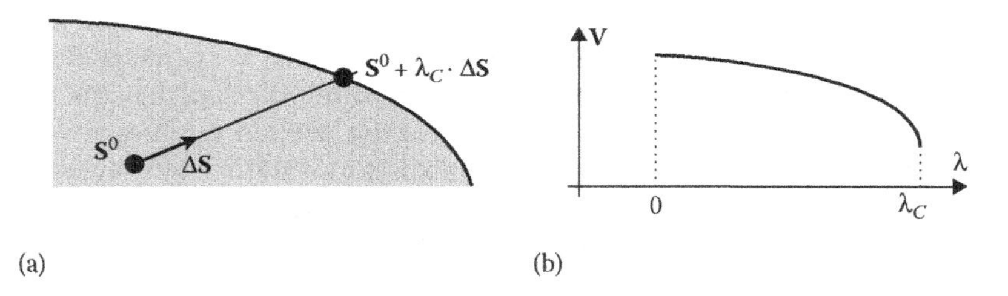

FIGURE 4.6
Bus power dispatch parameterization and concept of critical load margin. (a) Bus power space; (b) system variables space.

4.1.2 Margin to Voltage Collapse Computation Techniques

Most of the critical load factor computation techniques are based on the same idea: how high can the load factor be without crossing the feasibility boundary? Depending on the method of approaching the feasibility boundary, there are two main sets of techniques: continuation methods and direct methods. The former follows the trajectories of state variables to the voltage collapse point, whereas the latter computes the voltage collapse point by solving a set of nonlinear equations that includes the singularity condition for the Jacobian matrix. In further subsections 4.1.2.1 and 4.1.2.2, both sets will be analyzed.

From here on, the steady-state equations system is simplified to a generic notation as follows:

$$\mathbf{g}(\mathbf{x},\lambda)=\mathbf{0} \tag{4.4}$$

In Equation 4.4, **g** represents the parameterized power flow (Equation 4.3). These equations are functions of **x** and λ. Vector **x** contains the system state variables, that is, angle and module of bus voltages, and λ represents the load margin. As a consequence, for each value of λ, there will be a new solution for the state variables **x**; thus, trajectories $\mathbf{x}=\mathbf{x}(\lambda)$ can be traced. These trajectories will end as soon as λ reaches its upper limit in the voltage collapse point, that is, the critical load margin λ_C.

4.1.2.1 Continuation Methods

Continuation techniques are used to trace the trajectories of the state variables **x** while an independent parameter λ is evolving, as a succession of different roots of the state equations **g** (Seydel 2009). Concerning electric power systems, continuation techniques have been widely applied to trace voltage trajectories and detect the voltage collapse point, using the independent parameter λ to control the evolution of both the demand and the generation (Cañizares and Alvarado 1993).

The aim is to find a new root $(\mathbf{x}, \lambda)^{(K+1}$ starting from a previous equilibrium point $(\mathbf{x}, \lambda)^{(K}$. The first idea is just to increase the parameter from $\lambda^{(K}$ to $\lambda^{(K+1}$, and then to update the solution of system Equation 4.4 using a standard nonlinear equation system solver, such as Gauss–Seydel or Newton–Raphson. Nevertheless, this simple objective becomes more difficult as the system gets close to the voltage collapse point. As the system evolves toward its saddle-node bifurcation, the Jacobian matrix conditioning gets worse. This fact makes it more difficult to converge nonlinear equation system solvers.

To overcome this difficulty, continuation techniques split the new root-finding process into two steps: predictor and corrector. Figure 4.7 shows a complete continuation iteration, including predictor and corrector steps. The first approximation $(\mathbf{x}, \lambda)^{(K^*}$ of the new root is computed using the predictor step. This approximation represents a good starting point for the corrector

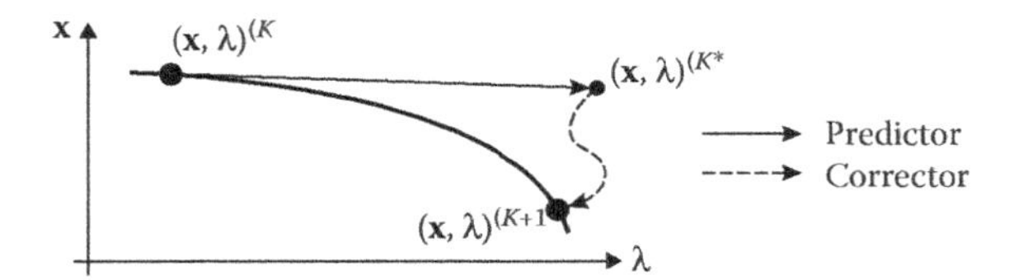

FIGURE 4.7
Complete continuation iteration.

step, where the equation system $\mathbf{g}(\mathbf{x}, \lambda) = \mathbf{0}$ is solved using a standard nonlinear equation system solver and adding additional equations that control the conditioning of the problem; thus, convergence to the new solution $(\mathbf{x}, \lambda)^{(K+1}$ is improved.

4.1.2.1.1 Predictor Step

The predictor step takes the solution $(\mathbf{x}, \lambda)^{(K}$ and uses it to provide a first approximation $(\mathbf{x}, \lambda)^{(K^*}$ of the new root $(\mathbf{x}, \lambda)^{(K+1}$ of the state equations. Most of the predictor techniques found in the literature may be classified into two families: polynomial and differential.

The polynomial predictor step is built using a polynomial approximation of the trajectories $\mathbf{x}(\lambda)$ by means of the solutions $(\mathbf{x}, \lambda)$ previously computed. The most common polynomial predictor vectors are the polynomial of degree 0 (constant predictor) and 1 (secant predictor). The constant predictor is the trivial one, where the approximation is equal to the initial solution. The secant predictor is the most used in literature, and requires two solutions, $(\mathbf{x}, \lambda)^{(K-1}$ and $(\mathbf{x}, \lambda)^{(K}$, to build it. Higher-order polynomial approximations will be more accurate, but will involve greater computational effort. In addition, the accuracy of a polynomial approximation is related to the proximity of the solutions used to build it; thus, they are inadequate to predict large perturbations on the load margin λ.

The differential predictor step is built using the Taylor approximation of the trajectories $\mathbf{x}(\lambda)$, truncated to a concrete degree. In contrast to polynomial techniques, the coefficients of the differential predictor step are the successive derivatives of the state variables $\mathbf{x}$ with respect to the load margin λ. The main handicap of these methods is the computing of derivatives of degree higher than one. However, the first-degree differential predictor, also known as the *tangent predictor*, is easy to compute, since the first derivative of the trajectories $\mathbf{x}(\lambda)$ can be obtained by directly deriving the system Equation 4.4:

$$\cancelto{0}{\mathbf{g}(\mathbf{x},\lambda)} + \mathbf{g}_\lambda \cdot \Delta\lambda_p + \mathbf{g}_\mathbf{x} \cdot \Delta\mathbf{x}_\mathbf{p} = \mathbf{0} \tag{4.5}$$

In Equation 4.5, $\mathbf{g}_\mathrm{x} = [\partial g_\mathrm{i}/\partial x_\mathrm{j}]$ represents the Jacobian matrix of state equations $\mathbf{g}$ with respect to state variables $\mathbf{x}$, while $\mathbf{g}\lambda = [\partial g_\mathrm{i}/\partial\lambda]$ is the parameterization vector of both the active and the reactive power system evolution.

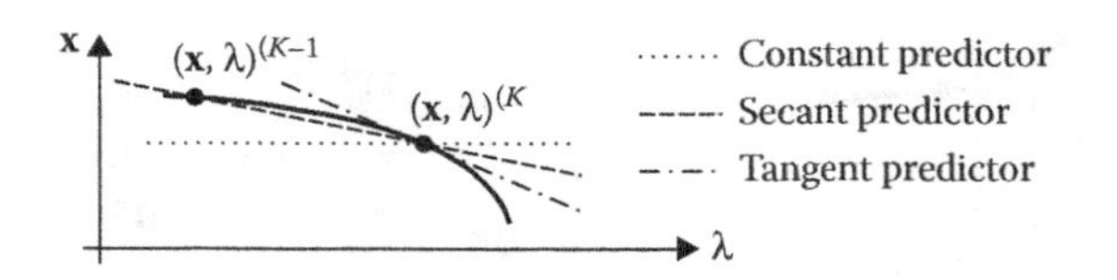

FIGURE 4.8
Predictor steps comparison.

From Equation 4.5, a direction in the state variables space which is tangent to trajectories $\mathbf{x}(\lambda)$ is computed. Figure 4.8 illustrates the difference between the constant, the secant, and the tangent predictor vectors.

Since the load margin represents an extra degree of freedom, an additional definition is necessary to complete the definition of the predictor step. This can be done by defining the module of the predictor vector or by fixing one of its components. This second option, known as *local parameterization*, is the most used. The load margin is usually the system variable controlled by the predictor step when the system is far enough from the voltage collapse point, whereas if the system is close to the feasibility boundary, it is better to use a bus voltage to control the predictor step size.

4.1.2.1.2 Corrector Step

The corrector step converges system Equation 4.4 to obtain the new root $(\mathbf{x}, \lambda)^{(K+1}$, using the approximation $(\mathbf{x}, \lambda)^{(K^*}$ provided by the predictor step as a starting point. The method employed to achieve the new root may be any of the numerical methods used to solve systems of nonlinear equations. In addition, since the load margin λ represents an extra variable, there exists a degree of freedom that has to be compensated by adding an extra equation.

In the most basic version of the corrector step, the system compensation is carried out by fixing one of the variables at its value computed by the predictor step, typically the load margin.

$$\begin{gathered}\mathbf{g}(\mathbf{x},\lambda)=\mathbf{0} \\ \lambda-\lambda^{(K^*}=0\end{gathered} \tag{4.6}$$

It can be deduced that system Equation 4.6 represents the standard power flow equations with a bus power dispatch defined by the fixed value of the load margin. This alternative is very practical, since it can be easily programmed using commercial power flow solvers. The disadvantage is the convergence problems that standard power flow algorithms present when the point of work is close to the voltage collapse boundary.

Instead of the load margin, one of the state variables x_i can be fixed at the value provided by the predictor step.

$$\mathbf{g}(\mathbf{x},\lambda)=\mathbf{0}$$
$$x_i - x_i^{(K^*} = 0 \tag{4.7}$$

The fixed state variable is usually a bus voltage V_i. This alternative provides a better convergence than fixing the load margin and thus is to be recommended when the system is close to the voltage collapse point. If the load margin fixation is out of bounds, the corrector step may lead to an unsolvable power flow, while the bus voltage fixation guarantees the existence of the solution.

From a geometrical point of view, the fixation of one of the system variables is represented by a hyperplane whose intersection with the trajectories $\mathbf{x}(\lambda)$ is the new root $(\mathbf{x}, \lambda)^{(K+1}$. Figure 4.9 represents the approximation of $(\mathbf{x}, \lambda)^{(K^*}$ provided by the predictor step, and some alternatives for the corrector step.

As is shown in Figure 4.9, neither of the alternatives for the fixation, the load margin (point A) and one of the state variables (point B), is optimal. To improve the convergence of the corrector step, the additional hyperplane may be reoriented. The aim is to minimize the distance between the approximation $(\mathbf{x}, \lambda)^{(K^*}$ and the trajectories $\mathbf{x}(\lambda)$. To achieve this, the additional hyperplane must be perpendicular to the trajectories $\mathbf{x}(\lambda)$, as Figure 4.9 depicts (point C). Unfortunately, it is impossible to compute the optimal direction, because it is the tangent vector to the trajectories $\mathbf{x}(\lambda)$ at the new root $(\mathbf{x}, \lambda)^{(K+1}$ that we're trying to obtain.

The best alternative consists of using the tangent vector corresponding with the previous solution $(\mathbf{x}, \lambda)^{(K}$, that is, the tangent predictor step $(\Delta\lambda_p, \Delta\mathbf{x}_p)$. Therefore, the additional equation will be the perpendicularity of the tangent predictor step and the corrector step.

$$\mathbf{g}(\mathbf{x},\lambda)=\mathbf{0}$$
$$\Delta\lambda_p \cdot \left(\lambda - \lambda^{(K^*}\right) + \Delta\mathbf{x}_\mathbf{p}^\mathrm{T} \cdot \left(\mathbf{x} - \mathbf{x}^{(K^*}\right) = 0 \tag{4.8}$$

The difference between the corrector step described in Equation 4.8 and the optimum represented by point C in Figure 4.9 will be proportional to the length of the predictor step, as Figure 4.10 shows.

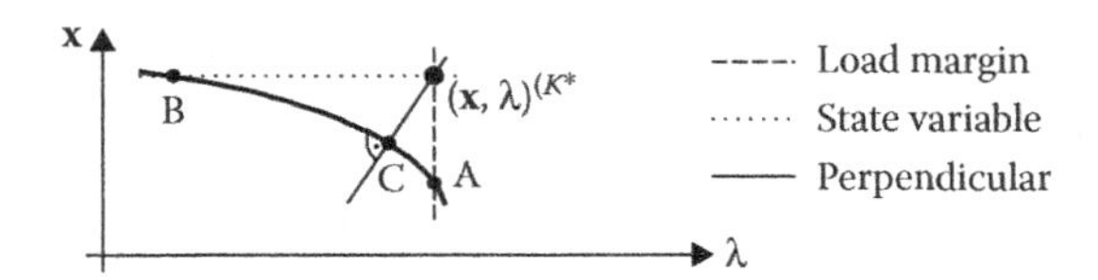

FIGURE 4.9
Corrector step alternatives.

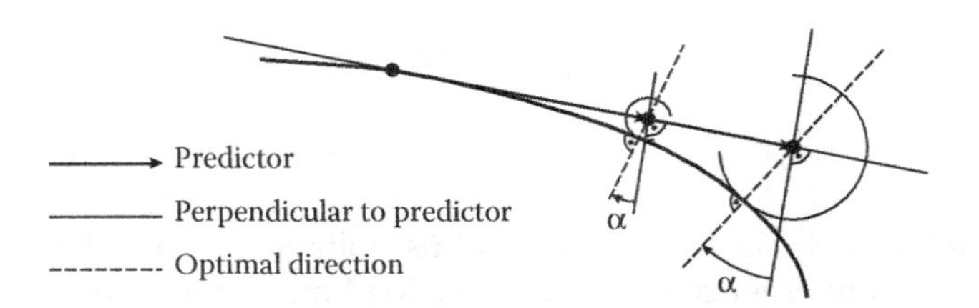

FIGURE 4.10
Difference between corrector perpendicular to predictor and corrector perpendicular to variables' trajectories.

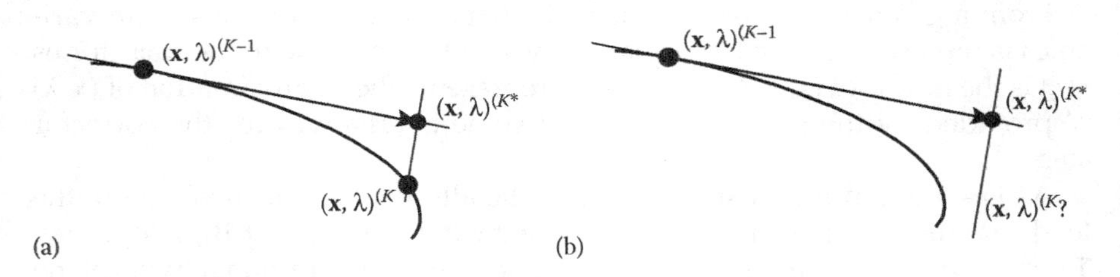

FIGURE 4.11
Corrector step feasibility as a function of predictor step size.

In Figure 4.10, the distance between the corrector steps grows with the size of the predictor step. Even when the predictor step size is inadequate, the corrector step may diverge because the resulting power flow is unfeasible.

In Figure 4.11a, the size of the predictor step is short enough to run into a feasible corrector power flow, since the corrector step and the state variables' trajectories intersect each other. However, in Figure 4.11b, for a larger predictor step, the corrector step and the state variables trajectories are separated; as a result, the corrector power flow is unfeasible. As a consequence, it is very important to include convergence controls, such as step length modulation control, to identify unsolvable power flows during the corrector step. In the case of unsolvability, the predictor step may be reduced as much as necessary to obtain a feasible power flow during the corrector step.

4.1.2.2 Direct Methods

Direct methods are closely related to the mathematical characteristics of the voltage collapse point, that is, the singularity of the Jacobian matrix. From a mathematical point of view, the voltage collapse point represents a saddle-node bifurcation in the power flow equations. In a saddle-node bifurcation, two points of work collide and annihilate each other; thus, the system jumps from being feasible with two roots to being unfeasible. At the saddle-node bifurcation, the system will have only one root, and the Jacobian matrix of the state equations system will be singular.

To compute the voltage collapse point using direct methods, a set of additional equations is required to force the Jacobian matrix to be singular. To achieve this, the state equations system is augmented by forcing the Jacobian matrix to have an eigenvector associated to a singular eigenvalue (Cañizares 1995). This could be either the left or the right eigenvector.

$$\begin{aligned} \mathbf{g}(\mathbf{x},\lambda) &= \mathbf{0} \\ \mathbf{w}^{\mathrm{T}} \cdot \mathbf{g}_{\mathrm{x}} &= 0 \\ \|\mathbf{w}\| &\neq 0 \end{aligned} \tag{4.9}$$

System of Equations 4.9 is formed by the parameterized power flow Equations 4.4, plus the singularity condition. The singularity condition is defined by making zero the product of the Jacobian matrix and a vector **w**, which represents the left eigenvector. Finally, vector **w** has to be set as a non-zero vector, controlling either its module or one or more of its components. The solution of Equation 4.9 will hence represent a saddle-node bifurcation of the system, that is, the voltage collapse point, and **w** will be the corresponding left eigenvector associated with the singular value of the Jacobian matrix. A system equivalent to Equation 4.9 may be formulated using the right eigenvector.

The direct methods can also be formulated as a set of optimality conditions (Echavarren et al. 2011a,b). The computation of the margin to the voltage collapse point can be formulated as an optimization problem, where the objective is to maximize the load margin, subject to the system of parameterized steady-state Equation 4.4:

$$\begin{aligned} &\max \quad \lambda \\ &s.to \quad \mathbf{g}(\mathbf{x},\lambda) = \mathbf{0} \ : \ \mu \end{aligned} \tag{4.10}$$

In Equation 4.10, vector μ represents the Lagrange multipliers associated with the equality constraints, that is, the steady-state equations of the system. The solution of the optimization problem Equation 4.10 is obtained by formulating the Lagrangian of the problem:

$$\mathcal{L} = \lambda - \mathbf{g}^{\mathrm{T}} \cdot \mu \tag{4.11}$$

The Karush–Kuhn–Tucker (KKT) optimality conditions are thus formulated by deriving the Lagrangian Equation 4.11 with respect to the variables of the system: the load margin λ, the state variables **x**, and the Lagrange multipliers μ.

$$\partial\mathcal{L}/\partial\lambda: \qquad 1-\mathbf{g}_\lambda^{\mathrm{T}}\cdot\mu=0 \tag{4.12}$$

$$\partial\mathcal{L}/\partial\mathbf{x}: \qquad -\mathbf{g}_\mathbf{x}^{\mathrm{T}}\cdot\mu=\mathbf{0} \tag{4.13}$$

$$\partial\mathcal{L}/\partial\mu: \qquad -\mathbf{g}=\mathbf{0} \tag{4.14}$$

Beginning with the last of the optimality conditions, Equation 4.14 represents the parameterized power flow Equation 4.4. The other two optimality conditions define the equilibrium point to be a saddle-node bifurcation of the system. On the one hand, Equation 4.12 forces the Lagrange multipliers vector μ to be a nonzero vector. On the other hand, Equation 4.13 requires the product of the Lagrange multipliers vector μ and the Jacobian matrix $\mathbf{g}_\mathbf{x}$ results to be zero; thus, μ is the left eigenvector of the Jacobian matrix $\mathbf{g}_\mathbf{x}$ corresponding to its singular eigenvalue.

Direct methods hence consist of converging a nonlinear equations system whose solution is the voltage collapse point. This convergence may be achieved using standard techniques for nonlinear equations systems, such as Gauss–Seydel or Newton–Raphson. However, convergence of KKT optimality conditions presents some difficulties. Because of the topology of the Lagrangian, if the system is initially far from the voltage collapse point or the Lagrange multipliers have an inadequate starting value, then the convergence may be difficult.

To run a complete iteration of the Newton–Raphson on KKT optimality conditions, the system presents twice the number of variables of the system, that is, the load margin, the state variables, and the Lagrange multipliers.

$$\begin{bmatrix} -\sum_k \mu_k\cdot\left(\nabla^2 g_k\right) & \begin{matrix}-\mathbf{g}_\lambda^{\mathrm{T}}\\ -\mathbf{g}_\mathbf{x}^{\mathrm{T}}\end{matrix} \\ \begin{matrix}-\mathbf{g}_\lambda & -\mathbf{g}_\mathbf{x}\end{matrix} & \cdot \end{bmatrix}\cdot\begin{bmatrix}\Delta\lambda\\ \Delta\mathbf{x}\\ \Delta\mu\end{bmatrix}=-\begin{bmatrix}1-\mathbf{g}_\lambda^{\mathrm{T}}\cdot\mu\\ -\mathbf{g}_\mathbf{x}^{\mathrm{T}}\cdot\mu\\ -\mathbf{g}\end{bmatrix} \tag{4.15}$$

A complete iteration of the Newton-Raphson on KKT optimality conditions requires the second derivatives of the power flow equations. In Equation 4.15, $\nabla^2 g_k$ represents the Hessian of the parameterized power flow Equation 4.4, with respect to the load margin λ and the state variables $\mathbf{x}$. In the System of System Equations 4.15 also reveals the importance of the Lagrange multipliers, since a bad choice of their initial values may result in numerical stability problems for the matrix in the left-hand side, or even make it singular if the starting values for the Lagrange multipliers are set to zero.

4.1.2.3 Other Voltage Stability Indices

As well as the load margin, many other voltage stability indices have been defined in the literature (Cañizares 2002; de Souza et al. 1997). Most of them are based on the fact that the voltage collapse point corresponds to a saddle-node bifurcation, and hence, the Jacobian of the state equations is singular. Those indices are usually easy and fast to compute. However, the drawback is that they don't provide any information about the variables' trajectories; thus, any perturbation in the system, such as the saturation of the reactive power production in generation units, won't be taken into account. In addition, some of these indices provide useful mathematical information about the distance to the singularity, but with no physical interpretation of the power system's operation.

One set of alternative indices is those based on the algebraic decomposition of the Jacobian matrix of the state equations, such as the Eigen decomposition (ED) or the singular value decomposition (SVD) (Lof et al. 1992; Dobson 1992). The ED consists of factorizing the Jacobian matrix of the state equations in a set of associated values called *eigenvalues*. If the matrix decomposed is singular, at least one of the eigenvalues will be zero. Therefore, if a power system is getting close to voltage collapse, the Jacobian matrix will present an eigenvalue tending to zero. In addition, eigenvectors corresponding to the singular value provide useful information about critical buses:

- The largest elements of the left eigenvector correspond to the buses where power injection will cause the greatest perturbation of state variables.
- The largest elements of the right eigenvector correspond to the buses that present the state variables with the highest sensitivity to power system injections.

Concerning the singularity of the Jacobian matrix in the voltage collapse point, some authors propose to work with nonsingular reduced versions of it (Chen et al. 2003). By selecting a suitable bus, the corresponding rows and columns in the Jacobian matrix are removed and the new Jacobian is recalculated. From that point, the determinant is computed and compared against the determinant of the original Jacobian matrix. Also, methods exist to reduce the bus impedance matrix of a network into its two-bus equivalent model at a referred bus (Lee 2016); thus, equations of the equivalent model could then be derived to facilitate an equivalent nodal analysis at the referred bus to determine its voltage stability limit. Other approaches, after convenient Jacobian matrix manipulation on the critical bus, use this reformulated Jacobian to build test functions. These test functions give an approximation of the maximum load margin but without state variables' trajectories (Chiang and Jean-Jumeau 1995).

Other methods compute the distance to voltage collapse by means of Lyapunov functions, also known as *energy functions* (Irisarri et al. 1997; Hiskens and Hill 1989). Lyapunov functions are scalar functions used to prove the stability of an equilibrium point of an ordinary differential equation system. Lyapunov functions have been widely used not only for small-signal stability analysis, but also for the voltage stability problem. The idea is to associate the power system state equations with a dissipation function that represents the gradient of the energy function; thus, the equilibrium point represents a minimum energy point for the Lyapunov function. In addition, energy functions allow state variables' trajectories to be traced, but both the computational effort and the mathematical difficulties of energy function construction make these energy function–based methods less suitable than others, such as the continuation techniques (Overbye and Klump 1998).

Finally, most recent works on the holomorphic embedding load-flow method (HELM) present it as a very useful tool to trace the trajectories of system variables (Trias and Marin 2016). HELM consists of solving the complex power flow equations by defining complex voltages as holomorphic functions of a complex parameter. This complex parameter is used to expand the voltages as polynomial functions. Therefore, the computing of each coefficient of the polynomial will be carried out by solving a linear equation system involving all the coefficients with lower degrees. To date, no work using HELM with voltage collapse point detection has been carried out, but the results on the power flow problem solution indicate that HELM has great potential.

4.2 Control Actions against Voltage Collapse

Voltage collapse situations can be forecast by system operators by means of load flow techniques and stability analysis. Indicators such as the load margin provide useful information about the future behavior of the system. If the system operator considers that the system may be dangerously close to voltage collapse, then control or preventive actions have to be taken. The most common reasons for a voltage collapse situation are large energy transfers and network bottlenecks. Generation units can be redispatched to reorganize network power flows (Capitanescu and Wehenkel 2011). In addition, voltage collapse situations are associated with insufficient reactive power margin. Generation units and other reactive power sources can provide that reactive support.

However, short-term voltage collapse situations can also appear, mostly when some outage occurs. In these situations, dynamic reactive power sources are required to recover the voltage stability. Even in the case when dynamic resources are insufficient or exhausted, an emergency load shedding has to be applied.

Voltage collapse mitigation resources not only have to supply reactive support to the system to guarantee the stability of network voltages, but also have to provide an appropriate mix of static and dynamic support.

4.2.1 Synchronous Machines

Generation units play a key role in the voltage stability control and mitigation problem, mainly due to being the principal active power sources in the network, but also due to being capable of generating or aborting reactive power to regulate bus voltages. By means of the synchronous machine excitation system, the generation unit presents a reactive power margin used to maintain the voltage of a selected bus, typically the high-voltage bus in the substation, where the plant is connected to the transmission grid (van Cutsem 2007).

The reactive power margin in synchronous machines is often considered as invariant, that is, a minimum and a maximum limit for the reactive power. However, it is important to take into account that the reactive power limits depend on the active power production of the generation unit (Nilsson and Mercurio 1994). In Figure 4.12, typical synchronous generator capability curves are depicted.

In drawing the capability curves, there exist two main operational limits: the maximum rotor current and the maximum stator current (Adibi and Milanicz 1994). Both limits are closely related to the heating of the machine. It can be shown that the reactive power margin becomes narrower as the active power production increases. The curves in Figure 4.12 have to be completed with the turbine maximum and minimum active power production. Due to this increase in reactive power capability at lower real power output, system planners and operators may choose to generate less than rated real power to have a more reactive power margin from generators.

In the case when the reactive power production of a generation unit gets saturated, the generator will not be able to sustain the scheduled bus voltage. This fact always causes the stability of the system voltages to deteriorate, and sometimes this saturation may even bring the system to an immediate voltage collapse. A deeper analysis of the effects of reactive power limits on the system voltage stability is presented in Section 4.3.

4.2.2 Renewable Energy Sources

Distributed generation can be an effective resource to prevent the system from suffering a voltage collapse, especially in distribution and islanded

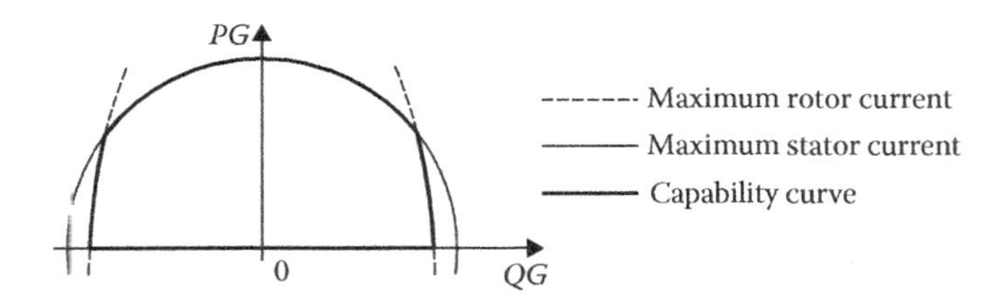

FIGURE 4.12
Synchronous generator capability curves.

networks. On the one hand, radial or weakly meshed networks are sensitive to voltage instability, especially under contingency situations, since the alternatives to reconfigure the power flows through network branches are limited. On the other hand, if energy resources are far from the consumers, the voltage drops along the network branches become larger as the demand grows (Kim et al. 2016).

The main advantage of considering renewable energy sources is that they can be installed close enough to the demand buses. Due to this fact, power flows and voltage drops in network branches get reduced along the network. In this way, the presence of renewable energy sources reduces the probability of the voltage collapse phenomenon, especially in isolated systems (Daher et al. 2008). Moreover, the presence of sufficient renewable generation units plus an efficient communication system provides several possibilities for controlling the voltage profile of the network (Močnik and Žemva 2014).

Renewable energy sources are normally interfaced to the grid through power electronics and energy storage systems (Carrasco et al. 2006). These power electronics have the ability to control the system's reactive power; thus, they can control the system voltage profile. This ability is especially useful for isolated networks with no point of common coupling where the voltage is controlled by a synchronous machine or the transmission grid (Daher et al. 2008).

4.2.3 Reactive Power Sources

Reactive power support is indispensable to maintain acceptable bus voltages and guarantee the system voltage stability (Taylor and International Conference on Large High Voltage Electric Systems 1995). Static reactive power sources, such as shunt capacitors, can provide an additional reactive power margin in zones of the system where it would be needed. Static sources are key control devices in short-term operational planning studies. Unfortunately, static reactive power sources are unable to provide reactive power support in the case of short-term voltage collapse situations. On the one hand, the switching operation takes too much time, and on the other hand, rapid depression of the bus voltages results in less reactive power being provided, since it is proportional to the square of the applied voltage.

In the short term, it's better to consider dynamic reactive power sources. These dynamic sources are flexible ac transmission systems (FACTS) devices for shunt compensation, such as static synchronous compensators (STATCOM) and static VAR compensators (SVC). Both devices are capable of controlling bus voltage within limits by modulating the reactive power they inject into or absorb from the network (Cañizares and Faur 1999; Hiskens and McLean 1992). To achieve this, static reactive power sources are connected through power electronics that allows them to be regulated continuously. To illustrate the differences between them, Figure 4.13 shows their corresponding operating characteristics.

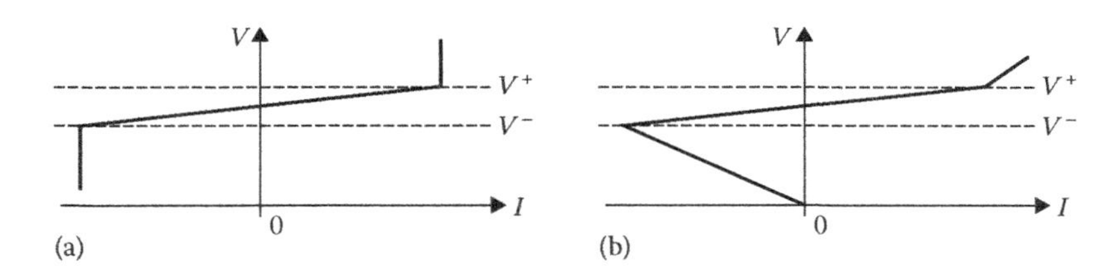

FIGURE 4.13
(a) STATCOM and (b) SVC operating characteristics.

As can be seen in Figure 4.13, the difference between them is how they work when the voltage is outside the control boundaries. Whereas the STATCOM maintains the current invariant, the SVC works as constant impedance. Due to this fact, the reactive power from a STATCOM decreases linearly with the voltage, whereas in the case of the SVC, the reactive power is quadratic with respect to bus voltage; thus, the STATCOM provides better reactive power support at low voltages. To get all the advantages from both the static and dynamic reactive power sources, they are installed and operated simultaneously. By using the static VAR sources from the operational planning, it is possible to maximize the dynamic performance of the dynamic VAR sources.

The strategy of installing many small distributed reactive power sources located close to the most critical buses results in being most efficient (Chang 2012). The performance of reactive power sources controlling bus voltage stability is very local, and it is greatly reduced when the controlled bus is distant from the reactive power source. In addition, the cost associated with the installation of distributed reactive power sources may be higher, but it is widely compensated with a better performance under *N*-1 situations that helps to meet the reliability criterion (Kincic et al. 2005). The coordination of reactive power sources can also include the distributed generation resources. This combined control is designed to cover momentary situations of low renewable energy production; hence, all the reactive power sources and distributed energy resources are coordinated. However, if some distributed energy resources are unable to support the system voltages, each reactive power source acts considering local measurements to support the voltage and increase the reactive power margin in its own area of influence (Wang et al. 2014).

4.2.4 Transformers

In the transmission grids, substation transformers can control the voltage of a selected bus by means of changing its tap. This control variable gives the system operator a useful tool to improve the stability of the voltages (Vournas and Karystianos 2004; Zhu et al. 2000). In the case of distribution grids, the automatic load tap changer (LTC) transformers help to maintain acceptable voltage levels at the loads. Typically, as load increases, the LTCs move the tap position to maintain the voltage level if it remains under its minimum limit long enough. This control protocol is designed to maintain customers' voltages, but may be dangerous for the stability of voltages.

During the voltage deterioration previous to the collapse, the LTC control detects a low voltage, and, if the process is slow enough, the control will time out and begin to raise the transformer tap position. As a consequence, the line current will rise as the tap does, increasing the voltage drop in the branches and making the voltage collapse more severe. Due to this fact, sometimes the best option against voltage collapse is to switch the automatic LTC control system off (Ohtsuki et al. 1991). Automatic blocking systems may be added to the LTC control to identify those situations when the LTC may cause the voltage stability in the system to deteriorate.

4.2.5 Load Shedding

Demand reduction is a very effective control action to improve the stability of the system. In the case of underfrequency scenarios, emergency load shedding is used to restore the synchronism (Anderson and Mirheydar 1992), especially in island power systems, which are especially sensitive to power imbalances (Sigrist et al. 2013). However, load shedding is also a useful control action to restore the stability of system voltages when the power system is getting close to the load margin (Echavarren et al. 2006). Load shedding contributes to decreasing the power flows through the network and power losses, improving not only the bus voltage profile, but also the stability of the system voltages. However, load shedding results in high costs to electricity suppliers and consumers; thus, it is considered a last resort in situations when the available control actions may have been exhausted, or they may not be fast enough to prevent the system voltage collapse. As a consequence, load-shedding schemes should be designed to be as effective as possible, optimizing both the location of the loads to be shed and the amount of demand to be reduced. In addition, load shedding should be applied in combination with an active power generation redispatch to control the generation–demand balance in the system. Finally, it is important to consider the reactive power demand, since during the load shedding, the power factor of the loads may be kept invariant.

The load-shedding schemes can be either automatic or manual (Nikolaidis and Vournas 2008). The automatic load-shedding schemes can be very useful for short-term voltage collapse situations, but the appropriate adjustment of settings for the undervoltage detectors and time delays can be a very difficult task. On the contrary, manual load-shedding schemes can be applied successfully if the system is approaching voltage collapse slowly, but may be insufficient in the case when an unexpected network situation brings the system to a sudden voltage collapse. Manual shedding schemes are adequate when short-term operational planning studies forecast possible voltage collapse situations due to expected load and actual contingencies, but a fast voltage collapse detection and mitigation system should also be considered for unexpected scenarios.

However, it is important to consider that, far from large-scale networks, an emergency load shedding could result in a voltage collapse situation (Nguyen and Turitsyn 2015). The increasing levels of penetration of distributed energy resources will cause the distribution grids to operate in unconventional conditions. The flow of active or reactive power may become reversed in certain realistic situations. In addition, active participation of future distribution-level power electronics in reactive power compensation may also lead to the local reversal of reactive power flows. As a consequence, existing emergency control actions such as load shedding not only fail to restore the system to normal operating conditions, but also aggravate the situation.

4.3 Sensitivity Analysis

For all the control devices described in the previous subsections, there exist several works on how to maximize their efficiency, considering all the reliability constraints and the operation limits of each device. Those works employ different techniques to achieve the optimal operating point, such as linear, quadratic, or even nonlinear programming, dynamic programming, neural networks, and so on. Many of them employ first-order sensitivities of the load margin with respect to the corresponding control variable. Sensitivity analysis provides first-order approximations of how much the load margin will vary when any of the control devices of the system are used (Echavarren et al. 2006; Dobson and Lu 1992a). Sensitivity analysis is thus a key tool to optimize control variables to improve the load margin of the system.

Let's start modifying system Equation 4.4 to introduce a vector $\mathbf{u}$ representing system control variables (demand, generation, transformers taps, shunts, etc.).

$$\mathbf{g}(\mathbf{x},\lambda;\mathbf{u})=\mathbf{0} \tag{4.16}$$

For a given vector $\mathbf{u}$ and the corresponding voltage collapse point $(\mathbf{x}, \lambda)$, a perturbation $\Delta\mathbf{u}$ of the control variables is considered. As a consequence, corresponding perturbations in the load margin $\Delta\lambda$ and the state variables $\Delta\mathbf{x}$ arise. Expanding system Equation 4.16 using first-order Taylor series, we obtain

$$\mathbf{0}\approx \cancel{\mathbf{g}(\mathbf{x},\lambda;\mathbf{u})}^{0}+\mathbf{g}_{\lambda}\cdot\Delta\lambda+\mathbf{g}_{\mathbf{x}}\cdot\Delta\mathbf{x}+\mathbf{g}_{\mathbf{u}}\cdot\Delta\mathbf{u} \tag{4.17}$$

In Equation 4.17, $\mathbf{g}_u = [\partial g_i / \partial u_k]$ represents the derivative of state equations $\mathbf{g}$ with respect to control variables $\mathbf{u}$. First-order sensitivity analysis usually consists of obtaining the inverse of the Jacobian matrix $\mathbf{g}_x$. Nevertheless, as long as the point $(\mathbf{x}, \lambda)$ corresponds to the voltage collapse point, which is a saddle-node bifurcation, the Jacobian matrix $\mathbf{g}_x$ is singular, and its inverse thus does not exist. To overcome this difficulty, the Lagrange multipliers vector μ will be required. Multiplying Equation 4.17 and the Lagrange multipliers vector μ, we obtain

$$\mu^T \cdot \mathbf{g}_\lambda \cdot \Delta\lambda + \mu^T \cdot \mathbf{g}_x \cdot \Delta\mathbf{x} + \mu^T \cdot \mathbf{g}_u \cdot \Delta\mathbf{u} = 0 \tag{4.18}$$

Now, if Equations 4.13 and 4.14 from the optimality conditions set are considered, then the load margin improvement $\Delta\lambda$ can be formulated as

$$\Delta\lambda + \mu^T \cdot \mathbf{g}_u \cdot \Delta\mathbf{u} = 0 \tag{4.19}$$

First-order sensitivities of load margin with respect to control actions are thus obtained as

$$\frac{d\lambda}{d\mathbf{u}} = -\mathbf{g}_u^T \cdot \mu \quad \rightleftharpoons \quad \frac{d\lambda}{du_k} = -\sum_j \mu_j \cdot \frac{\partial g_j}{\partial u_k} \tag{4.20}$$

Sensitivities (Equation 4.20) provide very useful information to deal with power system scenarios where the margin to voltage collapse is considered unsafe. Considering Equation 4.3, the different control actions can be classified according to how the corresponding state equation g_i has to be derived with respect to the control action u_k:

- On the one hand, if the control variable corresponds to a voltage control device of the network, such as transformer taps, shunt elements, or generation unit–controlled voltages, then the derivative is applied to the net power S injected into the network.

$$\frac{\partial g_i}{\partial u_k} = -\frac{\partial S_i}{\partial u_k} \tag{4.21}$$

- On the other hand, if the selected control variable is related to the bus power dispatch, then the corresponding derivative is applied to the net specified power S^0. Besides, the evolution vector ΔS is usually built up as proportional to the initial power dispatch. Therefore, the corresponding derivative has to be applied to the evolution vector net specified power ΔS.

$$\frac{\partial g_i}{\partial u_k} = \frac{\partial S_i^0}{\partial u_k} + \lambda_C \cdot \frac{\partial \Delta S_i}{\partial u_k} \tag{4.22}$$

To illustrate the performance of first-order sensitivities to identify optimal measures against voltage collapse, a corrective load shedding has been carried out in two hourly scenarios of the operation of the Lanzarote–Fuerteventura (LF) power system, a peak demand and an off-peak demand. The LF power system is divided into two areas, Lanzarote and Fuerteventura, connected through a 65 MVA submarine cable.

Tables 4.1 and 4.2 present the active power balance in the LF system for the peak and the off-peak scenarios, respectively. In both of them, Fuerteventura is the exporting area and Lanzarote the importing area. In the off-peak scenario (Table 4.2), Lanzarote imports 12 MW, almost 15% of its demand, while in the peak scenario (Table 4.1), the energy imported is 30.6 MW, 22% of its demand.

The voltage collapse margin has been computed considering a bus power dispatch evolution vector $\mathbf{\Delta S}$ proportional to the corresponding initial dispatch $\mathbf{S}^0$ for the active power ($\mathbf{\Delta P} = \mathbf{PG}^0 - \mathbf{PD}^0$) and zero for the reactive power ($\mathbf{\Delta Q} = 0$). After the voltage collapse point has been computed, first-order sensitivities with respect to the active power demand that the optimal buses to shed load or adjust the active power generation dispatch should be identified. Sensitivities with respect to the reactive power demand are also computed, since the load shedding will be applied considering the power factor of loads as invariant. To check the performance of the load shedding and the accuracy of the sensitivities, a load of 1 MW has been shed in each scenario, distributed among the three buses with the highest sensitivity with respect to demand. After the demand reduction, the voltage collapse margin is recomputed. Tables 4.3 and 4.4 show the results on the voltage collapse

TABLE 4.1

Active Power Balance in the LF Peak Scenario (MW)

Area	PG	PD	Ploss	Pexport
Fuerteventura	103.7	70.6	2.6	30.6
Lanzarote	118.3	147.6	1.3	–30.6
Total	222.0	218.2	3.9	0.0

TABLE 4.2

Active Power Balance in the LF Off-Peak Scenario (MW)

Area	PG	PD	Ploss	Pexport
Fuerteventura	53.1	40.4	0.6	12.0
Lanzarote	73.2	84.6	0.7	−12.0
Total	126.3	125.0	1.3	0.0

TABLE 4.3

Impact of Load Shedding on the Voltage Collapse Margin for the LF Peak Scenario

	Initial	Final	Diff
Base case demand (MW)	218.2	217.2	−1.0
VCM (MW)	141.9	150.8	8.9
	65.0%	69.4%	

TABLE 4.4

Impact of Load Shedding on the Voltage Collapse Margin for the LF Off-Peak Scenario

	Initial	Final	Diff
Base case demand (MW)	125.0	124.0	−1.0
VCM (MW)	213.7	218.8	5.1
	170.9%	176.4%	

margin before and after the load shedding, for the peak and the off-peak LF scenarios, respectively.

The first result to be noted in Tables 4.3 and 4.4 is the voltage collapse margin (VCM). The VCM in the off-peak scenario is 214 MW, much greater than the VCM of the peak scenario, which is equal to 142 MW. In terms of the critical load margin λ_C, it increases from 65% in the peak scenario to 171% in the off-peak scenario. These results show that the voltages are more stable in the off-peak scenario than in the peak scenario. Due to this fact, the load shedding is more effective in the peak scenario, where 1 MW of load shed increases the VCM by almost 9 MW, whereas for the off-peak scenario, the same load shed only increases the VCM by little more than 5 MW.

Tables 4.5 and 4.6 show, for each scenario, the loads where the shed has been applied. For each load, the different columns represent the initial load (PD^0/QD^0) and the load shed ($\Delta PD/\Delta QD$), in megawatts and MVAr; the sensitivity of the voltage collapse margin, in megawatts/megawatts and megawatts/MVAr; and the expected improvement in the voltage collapse margin, in megawatts, as a result of the product of sensitivity and the load shed. Finally, the last column is the total expected improvement in the voltage collapse margin, in megawatts.

Comparing the expected VCM improvement in Tables 4.5 and 4.6 with the actual improvement in Tables 4.3 and 4.4, it can be appreciated that the accuracy of the first-order sensitivities is better for the peak scenario, in which the difference between expected and actual improvements is much less than for the off-peak scenario.

Focusing on the sensitivities, it can be appreciated that the reactive power demand is as important in the load shedding as the active power demand. However, despite the fact that in both tables, the sensitivities ∂VCM with

TABLE 4.5

Loads Selected for the Load Shedding in the LF Peak Scenario

		Active Power Demand				Reactive Power Demand				Total
Name	Area	PD^0	ΔPD	∂VCM	ΔVCM	QD^0	ΔQD	∂VCM	ΔVCM	ΔVCM
MATASBL	F	4.79	−0.52	−9.21	4.83	1.66	−0.18	−9.70	1.76	6.60
GTARAJ1	F	0.21	−0.02	−5.99	0.14	0.08	−0.01	−6.46	0.06	0.19
GTARAJ2	F	4.13	−0.45	−5.99	2.71	0.00	0.00	−6.46	0.00	2.71
Total		9.13	−1.00	−7.68	7.68	1.74	−0.19	−9.56	1.82	9.50

TABLE 4.6

Loads Selected for the Load Shedding in the LF Off-Peak Scenario

		Active Power Demand				Reactive Power Demand				Total
Name	Area	PD^0	ΔPD	∂VCM	ΔVCM	QD^0	ΔQD	∂VCM	ΔVCM	ΔVCM
MATASBL	F	0.71	−0.24	−8.27	1.97	0.25	−0.08	−7.28	0.60	2.56
GTARAJ1	F	0.12	−0.04	−6.08	0.24	0.05	−0.01	−5.17	0.08	0.32
PLAYABL	L	2.17	−0.72	−5.45	3.94	0.44	−0.15	−4.67	0.69	4.63
Total		3.00	−1.00	−6.15	6.15	0.73	−0.24	−5.58	1.36	7.51

respect to the active and reactive power demand are similar, the amount of reactive power load to be shed is less than the active; thus, the effect of ΔVCM on the VCM is less important.

To show how the load shedding has influenced the voltage stability, pilot buses' trajectories for both scenarios are plotted next.

Figures 4.14 and 4.15 show a comparison between the nose curves before (dotted line) and after (continuous line) the load shedding. Figure 4.14 (peak scenario) depicts how the voltage collapse point moves from 360 to 368 MW, that is, 9 MW of improvement minus 1 MW of load shed. In the case of the off-peak scenario of Figure 4.15, the voltage collapse point moves from 339 to 343 MW, that is, 5 MW of improvement minus 1 MW of load shed.

As Table 4.5 shows, all the loads shed belong to the Fuerteventura subsystem. Due to that fact, the load-shedding effect is more remarkable in Figure 4.14a than in Figure 4.14b. For the Fuerteventura pilot buses, the trajectories after the load shedding present not only higher voltage levels but also less negative gradients with respect to active power demand. On the contrary, in Figure 4.14b, the Lanzarote pilot buses' trajectories before and after the load shedding present voltage levels and gradients very similar to each other.

In the off-peak scenario, as Table 4.6 shows, the loads shed have been distributed among both subsystems. In Fuerteventura, 0.28 MW of load has been shed, whereas in Lanzarote, the load shed has been 0.72 MW. Due to that fact, the load-shedding effect in Figure 4.15a and b is equivalent. All the trajectories after the load shedding present similar improvements in voltage levels and voltage gradients with respect to active power demand.

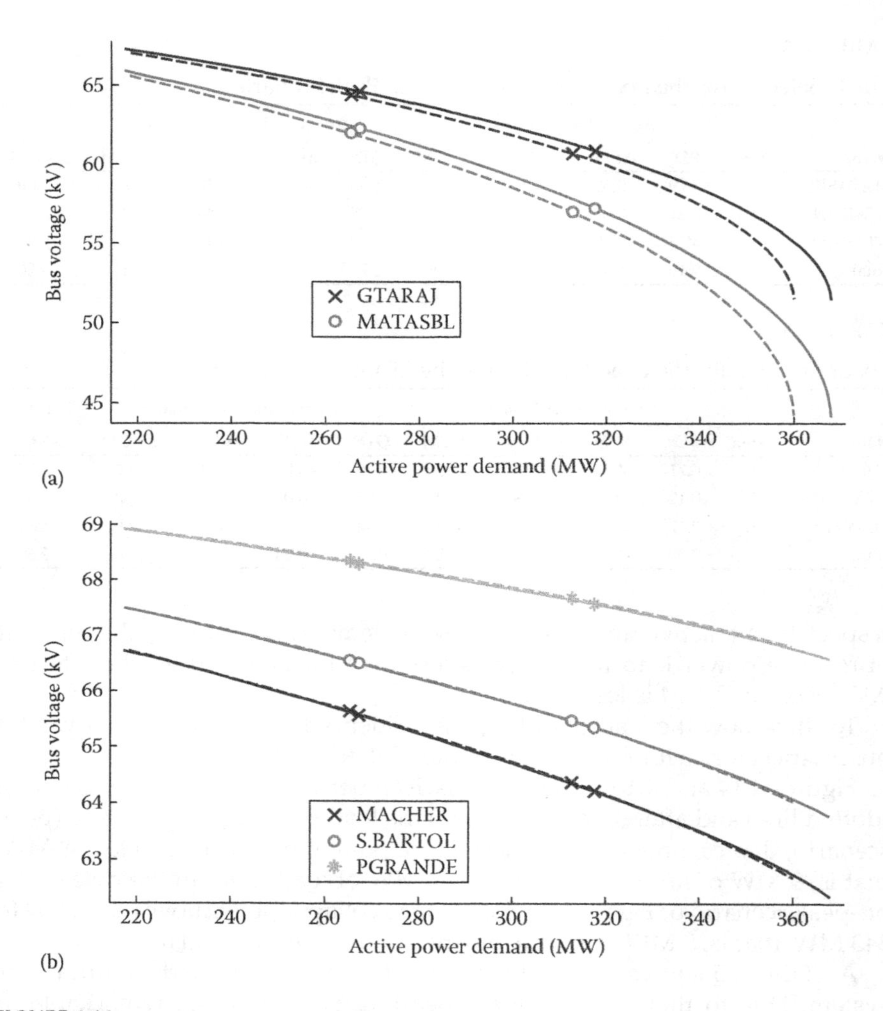

FIGURE 4.14
Pilot buses voltage evolution in the peak scenario during demand increase: (a) FUERTEVENTURA pilot buses, (b) LANZAROTE pilot buses.

These load-shedding results may be extrapolated to the distributed energy resources allocation problem. Since the sensitivities with respect to active power demand and generation are opposite in sign, these can be employed to check the performance of installing distributed energy resources to enhance the stability of system voltages. Therefore, all the previous results on load shedding can be interpreted as results on injecting active power from distributed energy resources.

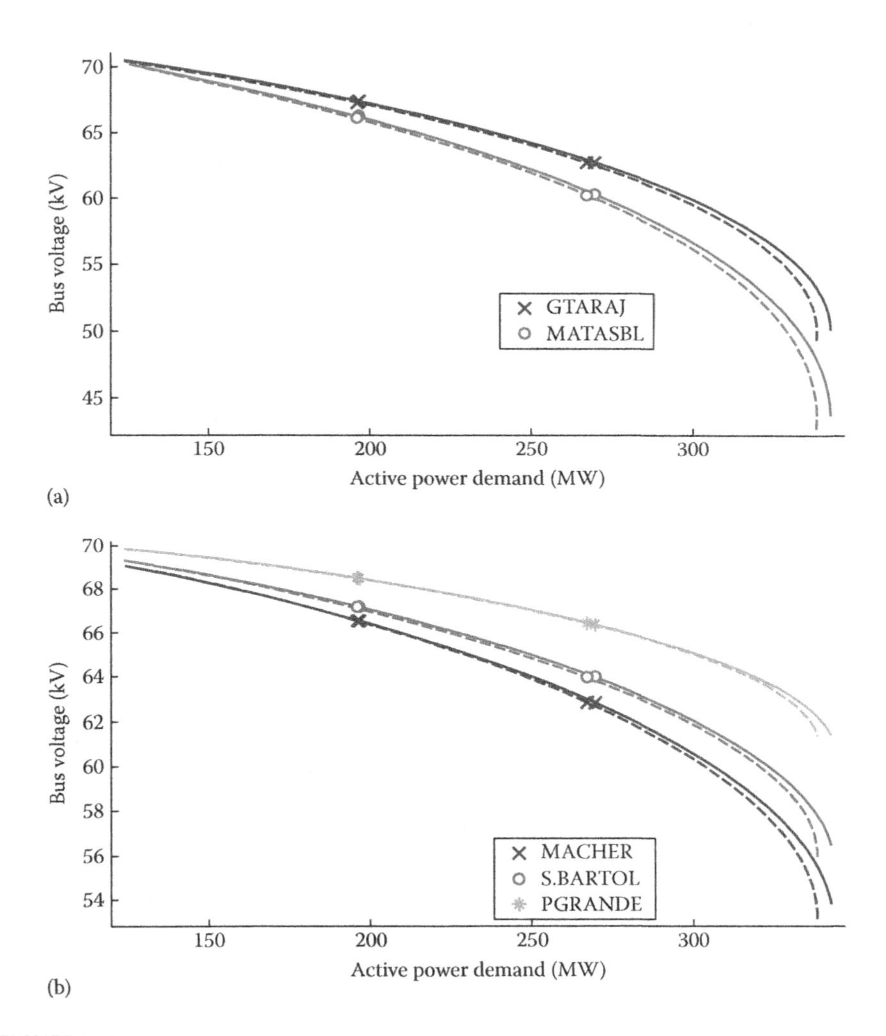

FIGURE 4.15
Pilot buses voltage evolution in the off-peak scenario during demand increase: (a) FUERTEVENTURA pilot buses, (b) LANZAROTE pilot buses.

4.4 Effect of Reactive Generation Limits on Voltage Stability

The stability margin of a power system is affected by a number of features of the power network, such as the generator voltages, the load of the system, and the generation–demand imbalance between the power system areas. However, one of the most important variables when analyzing voltage collapse phenomena is the reactive power limits of generation units.

Each time a machine saturates its reactive power limits, either minimum or maximum, the voltage stability of the system undergoes deterioration (Dobson and Lu 1992b). In most of those cases, the gradient of voltages with respect to the load margin becomes more pronounced, and the load margin to voltage collapse is reduced, but the voltage stability remains. However, in highly loaded cases, the stability margin may exhibit a discontinuous increase when a generator reaches a reactive power generation limit, but the equilibrium point is on the unstable branches of the nose curves. As a consequence, the power system becomes immediately unstable due to any inevitable small perturbation, and a dynamic voltage collapse leading to blackout may follow.

Considering the system $\mathbf{g}(\mathbf{x}, \lambda) = 0$, as the independent parameter λ increases from the starting point, the state variables $\mathbf{x}$ approach the voltage collapse point. Along that manifold, if a unit reaches its reactive power generation limit, the power system state equations must be modified to set the reactive power generation of the unit equal to its limit. The generator bus type changes from PV to PQ. This structural perturbation in the power system model results in a change in the initial unlimited reactive power generation manifold $\mathbf{x}(\lambda)$, that is, the nose curves. This change may result in one of two qualitatively different consequences: (a) a voltage stability deterioration or (b) an immediate voltage collapse, both depicted in Figure 4.16.

Considering the first case scenario, Figure 4.16a, the system variables $\mathbf{x}$ (i.e., bus voltage angle and module) follow the unlimited manifold till a reactive power limit is reached. From that point, the system variables change to the limited manifold. This limited manifold presents a more pronounced gradient with respect to the load margin λ and is closer to the voltage collapse point than the unlimited manifold. Therefore, the generation unit saturation has implied a deterioration of the system voltage stability, but the system remains stable, and the actual trajectory can reach the voltage collapse point of the limited manifold.

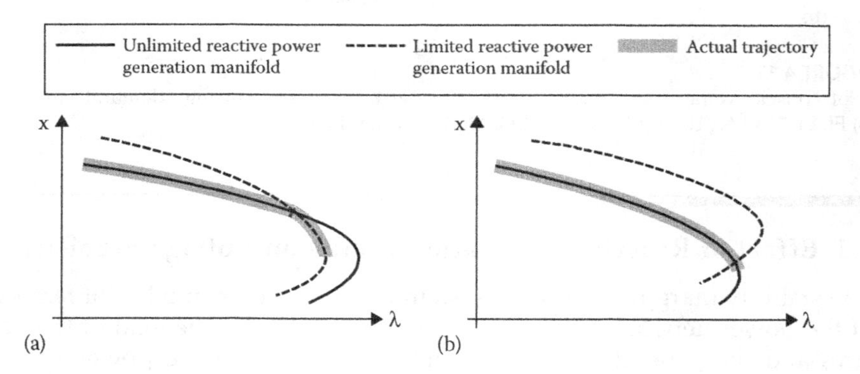

FIGURE 4.16
Unlimited and limited manifolds, and actual trajectory of system variables.

In the second case scenario, Figure 4.16b, the system variables also change from the unlimited to the limited manifold when a reactive power limit is reached. However, in this case, the limited manifold corresponds to the unstable branch of the nose curve. Now, the limited manifold is formed by a succession of unstable equilibrium points. In practice, this situation leads to an immediate voltage collapse due to the occurrence of any small perturbation in the system. That's why in Figure 4.16b, the actual trajectory ends as soon as the system changes from the unlimited to the limited manifold. This end point is known as *limit-induced collapse*.

This stability exchange phenomenon between two equilibrium points shown in Figure 4.16b indicates that the system has passed through a point called *transcritical bifurcation* (Echavarren et al. 2009). Using bifurcation analysis, it has been well established that this possible limit-induced collapse can be predicted before it happens by monitoring the sign of the sensitivity of reactive power generation QG with respect to its own bus voltage VG.

$$\frac{dQG}{dVG} = QG_{VG} - \mathbf{QG}_{\mathbf{x}}^{\mathrm{T}} \cdot \mathbf{g}_{\mathbf{x}}^{-1} \cdot \mathbf{g}_{VG} \tag{4.23}$$

where:

$\mathbf{g}_{VG} = [\partial g_i / \partial VG]$ is the derivative of state equations $\mathbf{g}$ with respect to the generator voltage VG

$\mathbf{QG}_{\mathbf{x}} = [\partial QG / \partial x_j]$ is the derivative of reactive power QG with respect to state variables $\mathbf{x}$

$QG_{VG} = [\partial QG / \partial VG]$ is the derivative of reactive power QG with respect to the generator voltage VG

When the reactive power is close to saturation, if the sensitivity dQG/dVG is positive, then the saturation will just cause the stability of the system voltages to deteriorate. However, if the sensitivity dQG/dVG is negative, then the saturation will bring the system to a limit-induced collapse and a possible blackout.

To illustrate this, an hourly scenario of the operation of the LF power system has been analyzed, focusing on the SALINAS power plant and its reactive power production. The LF power system is divided into two areas, Lanzarote and Fuerteventura, connected through a 65 MVA submarine cable. SALINAS belongs to Fuerteventura (Area 1). Two sets of 66 kV pilot buses have been defined, one for the Fuerteventura subsystem ("Gran Tarajal 66 kV" and "Matas Blancas 66 kV"), and one for the Lanzarote subsystem ("Macher 66 kV," "San Bartolomé 66 kV", and "Punta Grande 66 kV").

The VCM has been computed considering a bus power dispatch evolution vector ΔS proportional to the corresponding initial dispatch $\mathbf{S}^0$ for the active power ($\mathbf{\Delta P} = \mathbf{PG}^0 - \mathbf{PD}^0$) and reactive power ($\mathbf{\Delta Q} = -\mathbf{QD}^0$) to maintain the power factor of the loads invariant. In a first analysis, the system demand is increased from its initial value to the voltage collapse point without

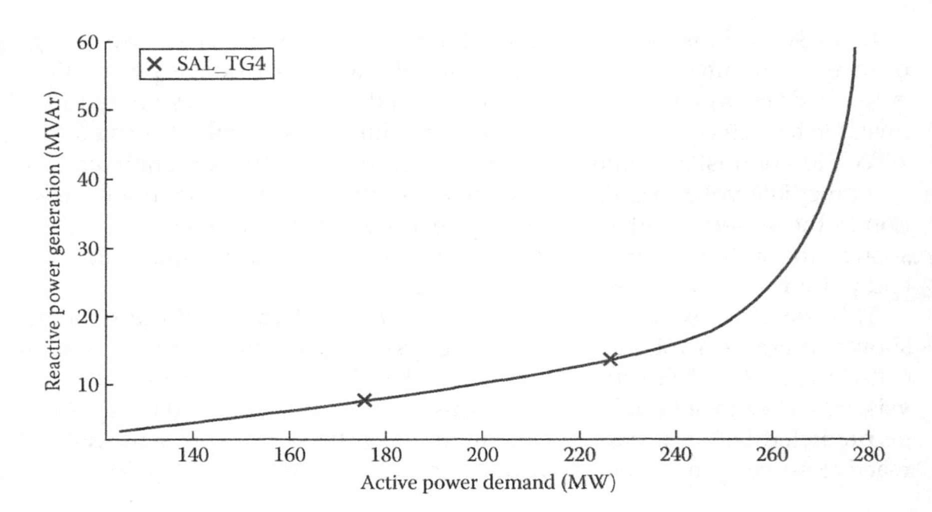

FIGURE 4.17
SALINAS reactive power generation evolution during demand increase.

considering the reactive power generation of SALINAS. Both the load and the generation evolutions are proportional to their original magnitudes. Reactive power demand has been increased considering the power factor of the loads. Figure 4.17 represents the trajectories during demand increase of the SALINAS unit reactive power generation and its sensitivity with respect to the generator bus voltage.

Figure 4.17 shows how the reactive power generation of SALINAS unit grows monotonically from 3 to 59 MVAr, as the demand evolves from an initial value of 125.5 MW to a final value at the voltage collapse point of 277.2 MW. The original reactive power generation maximum limit is 25 MVAr, which means that the unit should be saturated at 260.5 MW. To identify the critical point from where the reactive power generation saturation will result in an immediate voltage collapse, reactive power sensitivity with respect to the generator voltage has to be depicted.

Figure 4.18 details the trajectory of the reactive power sensitivity with respect to the generator voltage around the critical point. The sensitivity starts from 31.5 MVAr/kV and decreases asymptotically to minus infinite at voltage collapse (277.2 MW). The discontinuities in the trajectory, due to the reactive power generation saturation of other machines in the network, can be appreciated. Indeed, the sign of the sensitivity changes at 257.5 MW, from 9.5 to −1.1 MVAr/kV, due to the saturation of two generation units that also belong to the SALINAS plant. The corresponding value of SALINAS reactive power generation at the critical point is 22.22 MVAr.

Figure 4.19 shows the voltage trajectories during demand increase of two sets of 66 kV pilot buses, corresponding respectively to the FUERTEVENTURA (GRAN TARAJAL and MATAS BLANCAS) and LANZAROTES (MACHER,

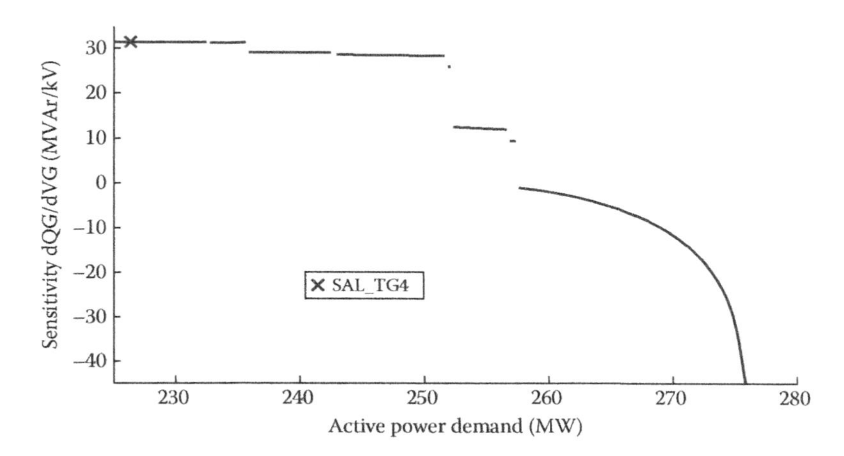

FIGURE 4.18
SALINAS unit sensitivity with respect to generator voltage evolution during demand increase.

SAN BARTOLOME, and PUNTA GRANDE) subsystems. In all of them, the sudden change in voltage gradient due to the saturation of other generation units can be appreciated.

The simulation is run again, this time considering a maximum reactive power limit for SALINAS of 20 MVAr. According to Figures 4.17 and 4.18, the generation unit will reach that limit when the demand is 253.8 MW, corresponding to a reactive power sensitivity with respect to the generator voltage equal to 12.35 MVAr/kV.

All of the plots in Figure 4.20 show a sudden perturbation in voltage gradient when the demand reaches 253.8 MW; thus, the SALINAS generation unit's reactive power generation saturates at maximum limit. Due to this new perturbation, the new point of voltage collapse is found at a demand of 263.1 MW, 14.1 MW less than in the previous simulation without considering reactive power limits for SALINAS. In Figure 4.20a, the bus voltage of the SALINAS generation unit remains constant till its reactive power saturation. Because the new system remains stable, the SALINAS bus voltage starts to fall at that point. In Figure 4.20b and c, the pilot buses' voltages present similar behavior; thus, when the saturation occurs, the voltage gradient becomes more negative. Thus, the saturated system stability is weaker than before, but remains stable till the voltage collapse point.

The third simulation corresponds to a maximum reactive power limit for SALINAS of 30 MVAr. Figures 4.17 and 4.18 show that the generation unit will reach that limit when the demand is 265.4 MW, corresponding to reactive power sensitivity with respect to the generator voltage equal to −5.6 MVAr/kV.

Now, as Figure 4.21 shows, the sudden perturbation in voltage gradient occurs at a demand of 265.4 MW when the SALINAS generation unit reactive power generation saturates at maximum limit. In this case, the new point of

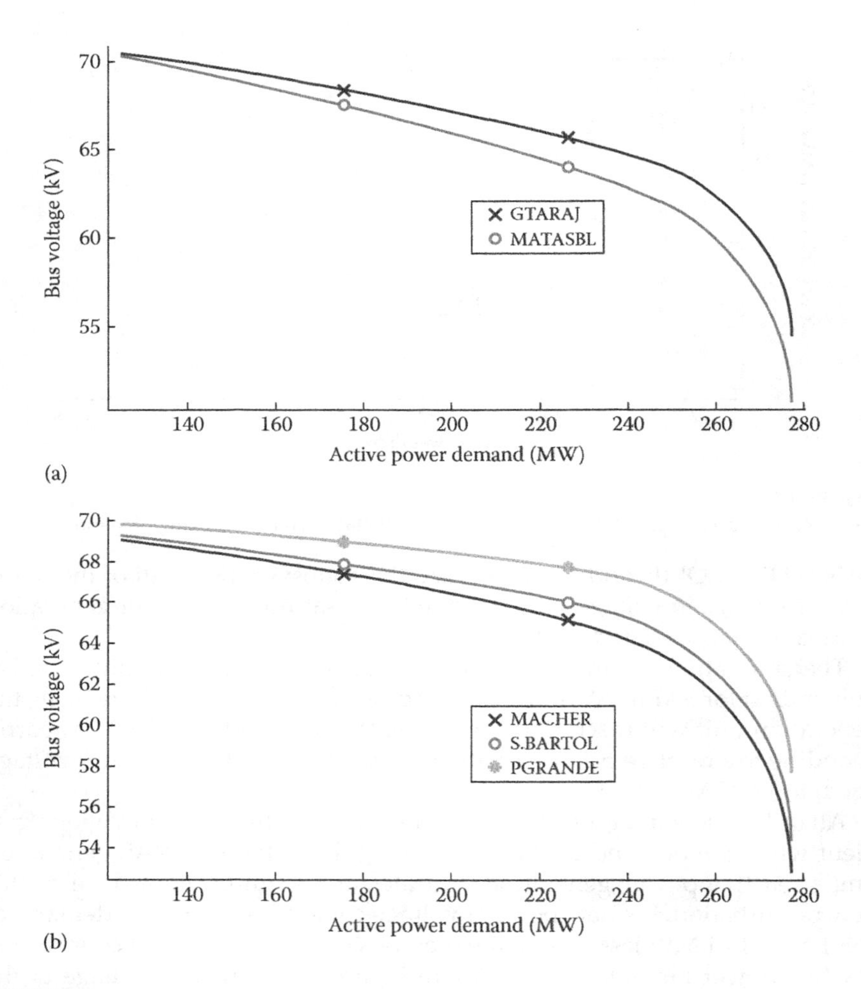

FIGURE 4.19
Pilot buses voltage evolution during demand increase: (a) FUERTEVENTURA pilot buses, (b) LANZAROTE pilot buses.

voltage collapse is found at a demand of 266.7 MW, 10.5 MW less than the initial simulation without considering reactive power limits for SALINAS. However, this new voltage collapse point can be reached, since the SALINAS saturation has undergone a change of sign in the voltage gradient; that is, the saturated system is unstable. This means that the system will suffer an immediate voltage collapse due to a limit-induced bifurcation. In Figure 4.21a, the bus voltage of the SALINAS generation unit remains constant till its reactive power saturation, when it starts to grow because the limited system is unstable. In Figure 4.21b and c, the pilot buses' voltages present similar behavior; thus, when the saturation occurs, the voltage gradient becomes positive, so the system is unable to continue its trajectory along the new manifold.

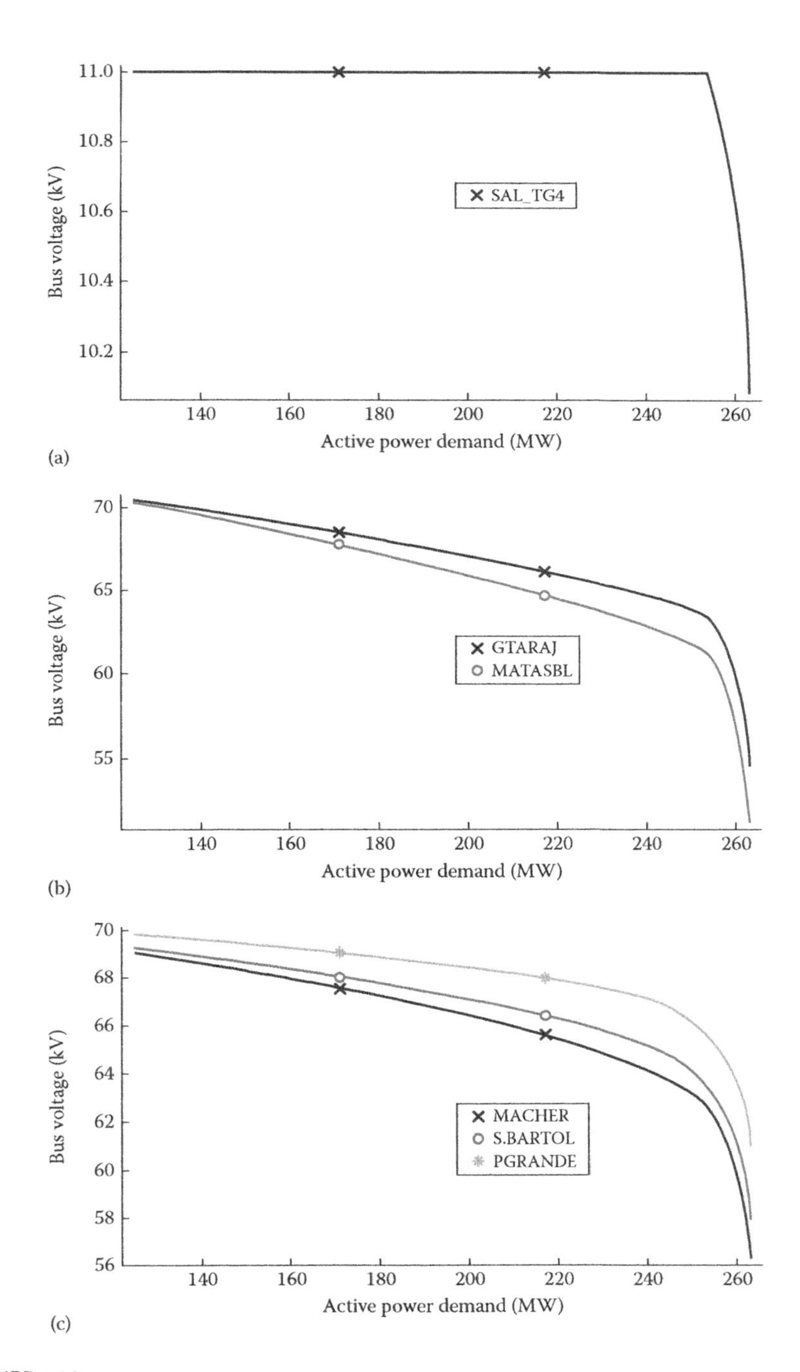

FIGURE 4.20
Pilot buses voltage evolution during demand increase: (a) SALINAS, (b) FUERTEVENTURA pilot buses, (c) LANZAROTE pilot buses.

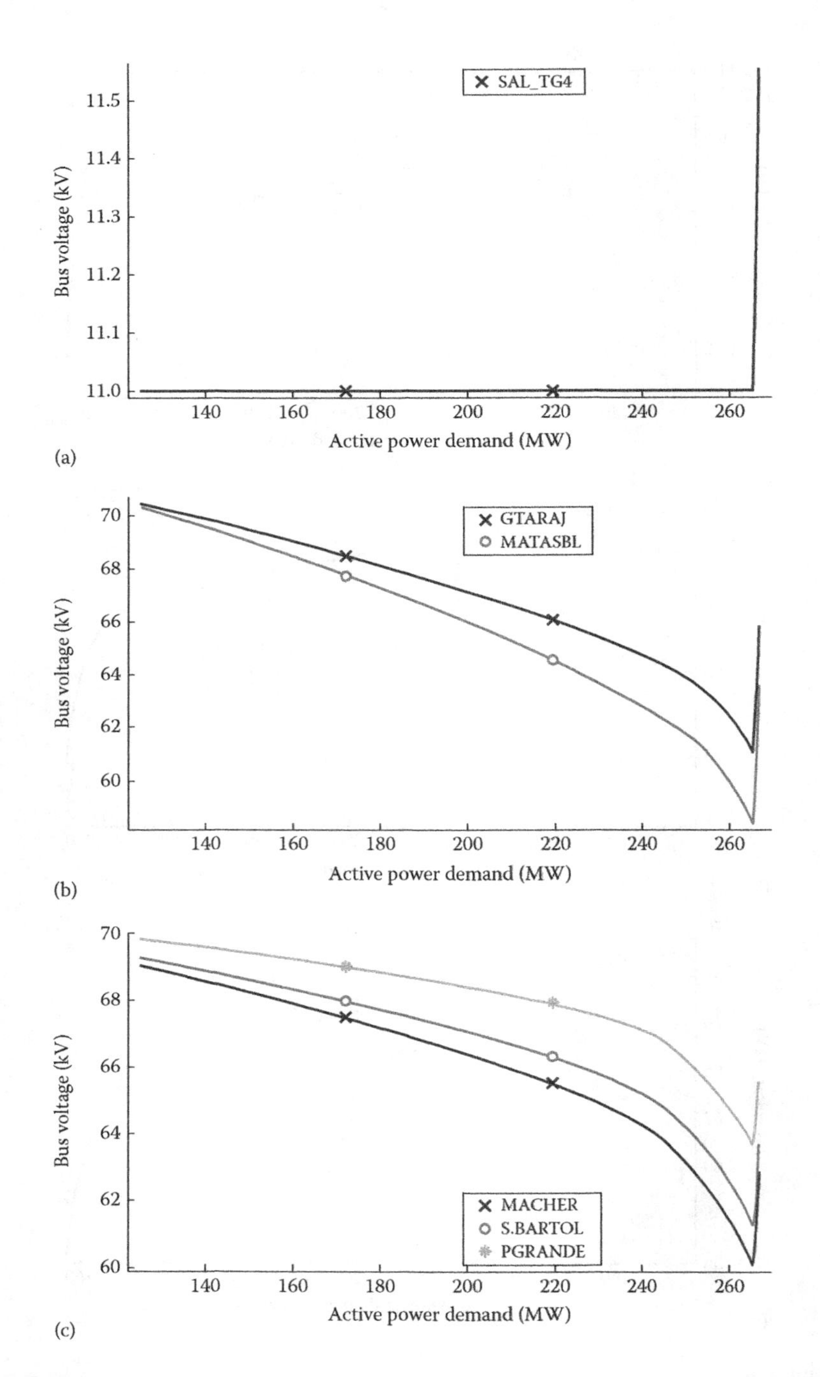

FIGURE 4.21
Pilot buses voltage evolution during demand increase: (a) SALINAS, (b) FUERTEVENTURA pilot buses, (c) LANZAROTE pilot buses.

References

Adibi, M. M., and D. P. Milanicz. 1994. Reactive capability limitation of synchronous machines. *IEEE Transactions on Power Systems* 9:29–40.

Anderson, P. M., and M. Mirheydar. 1992. An adaptive method for setting underfrequency load shedding relays. *IEEE Transactions on Power Systems* 7:647–655.

Cañizares, C. A. 1995. On bifurcations, voltage collapse and load modeling. *IEEE Transactions on Power Systems* 10:512–522.

Cañizares, C. A. 2002. Voltage stability assessment: Concepts, practices and tools. Special Publication of IEEE Power System Stability Subcommittee.

Cañizares, C. A., and F. L. Alvarado. 1993. Point of collapse and continuation methods for large AC/DC systems. *IEEE Transactions on Power Systems* 8:1–8.

Cañizares, C. A., and Z. T. Faur. 1999. Analysis of SVC and TCSC controllers in voltage collapse. *IEEE Transactions on Power Systems* 14:158–165.

Capitanescu, F., and L. Wehenkel. 2011. Redispatching active and reactive powers using a limited number of control actions. *IEEE Transactions on Power Systems* 26:1221–1230.

Carrasco, J. M., L. G. Franquelo, J. T. Bialasiewicz, E. Galvan, R. C. PortilloGuisado, M. A. M. Prats, J. I. Leon, and N. Moreno-Alfonso. 2006. Power-electronic systems for the grid integration of renewable energy sources: A survey. *IEEE Transactions on Industrial Electronics* 53:1002–1016.

Chang, Y. C. 2012. Multi-objective optimal SVC installation for power system loading margin improvement. *IEEE Transactions on Power Systems* 27:984–992.

Chen, K., A. Hussein, M. E. Bradley, and H. Wan. 2003. A performance-index guided continuation method for fast computation of saddle-node bifurcation in power systems. *IEEE Transactions on Power Systems* 18:753–760.

Chiang, H.-D., and R. Jean-Jumeau. 1995. Toward a practical performance index for predicting voltage collapse in electric power systems. *IEEE Transactions on Power Systems* 10:584–592.

Daher, S., J. Schmid, and F. L. M. Antunes. 2008. Multilevel inverter topologies for stand-alone PV systems. *IEEE Transactions on Industrial Electronics* 55:2703–2712.

de Souza, A. C. Z., C. A. Canizares, and V. H. Quintana. 1997. New techniques to speed up voltage collapse computations using tangent vectors. *IEEE Transactions on Power Systems* 12:1380–1387.

Dobson, I. 1992. Observations on the geometry of saddle node bifurcation and voltage collapse in electrical power systems. *IEEE Transactions on Circuits and Systems I: Fundamental Theory and Applications* 39:240–243.

Dobson, I., H. Glavitsch, C.-C. Liu, Y. Tamura, and K. Vu. 1992. Voltage collapse in power systems. *IEEE Circuits and Devices Magazine* 8:40–45.

Dobson, I., and L. Lu. 1992a. Computing an optimum direction in control space to avoid stable node bifurcation and voltage collapse in electric power systems. *IEEE Transactions on Automatic Control* 37:1616–1620.

Dobson, I., and L. Lu. 1992b. Voltage collapse precipitated by the immediate change in stability when generator reactive power limits are encountered. *IEEE Transactions on Circuits and Systems I: Fundamental Theory and Applications* 39:762–766.

Echavarren, F. M., E. Lobato, and L. Rouco. 2006. A corrective load shedding scheme to mitigate voltage collapse. *International Journal of Electrical Power and Energy Systems* 28:58–64.

Echavarren, F. M., E. Lobato, and L. Rouco. 2009. Steady-state analysis of the effect of reactive generation limits in voltage stability. *Electric Power Systems Research* 79:1292–1299.

Echavarren, F. M., E. Lobato, L. Rouco, and T. Gomez. 2011a. Formulation, computation and improvement of steady state security margins in power systems. Part I: Theoretical framework. *International Journal of Electrical Power and Energy Systems* 33:340–346.

Echavarren, F. M., E. Lobato, L. Rouco, and T. Gómez. 2011b. Formulation, computation and improvement of steady state security margins in power systems. Part II: Results. *International Journal of Electrical Power and Energy Systems* 33:347–358.

Hiskens, I. A., and D. J. Hill. 1989. Energy functions, transient stability and voltage behaviour in power systems with nonlinear loads. *IEEE Transactions on Power Systems* 4:1525–1533.

Hiskens, I. A., and C. B. McLean. 1992. SVC behaviour under voltage collapse conditions. *IEEE Transactions on Power Systems* 7:1078–1087.

IEEE Power System Relaying Committee. 1996. Voltage collapse mitigation.

Irisarri, G. D., X. Wang, J. Tong, and S. Mokhtari. 1997. Maximum loadability of power systems using interior point nonlinear optimization method. *IEEE Transactions on Power Systems* 12:162–172.

Kim, Y.-S., E.-S. Kim, and S.-I. Moon. 2016. Frequency and voltage control strategy of standalone microgrids with high penetration of intermittent renewable generation systems. *IEEE Transactions on Power Systems* 31:718–728.

Kincic, S., X. T. Wan, D. T. McGillis, A. Chandra, Boon-Teck Ooi, F. D. Galiana, and G. Joos. 2005. Voltage support by distributed static VAr systems (SVS). *IEEE Transactions on Power Delivery* 20:1541–1549.

Kundur, P., N. J. Balu, and M. G. Lauby. 1994. *Power System Stability and Control.* New York: McGraw-Hill.

Lee, D. H. A. 2016. Voltage stability assessment using equivalent nodal analysis. *IEEE Transactions on Power Systems* 31:454–463.

Lof, P. A., T. Smed, G. Andersson, and D. J. Hill. 1992. Fast calculation of a voltage stability index. *IEEE Transactions on Power Systems* 7:54–64.

Močnik, J., and A. Žemva. 2014. Controlling voltage profile in smart grids with remotely controlled switches. *IET Generation, Transmission and Distribution* 8:1499–1508.

Nguyen, H. D., and K. Turitsyn. 2015. Voltage multistability and pulse emergency control for distribution system with power flow reversal. *IEEE Transactions on Smart Grid* 6:2985–2996.

Nikolaidis, V. C., and C. D. Vournas. 2008. Design strategies for load-shedding schemes against voltage collapse in the Hellenic system. *IEEE Transactions on Power Systems* 23:582–591.

Nilsson, N. E., and J. Mercurio. 1994. Synchronous generator capability curve testing and evaluation. *IEEE Transactions on Power Delivery* 9:414–424.

Ohtsuki, H., A. Yokoyama, and Y. Sekine. 1991. Reverse action of on-load tap changer in association with voltage collapse. *IEEE Transactions on Power Systems* 6:300–306.

Overbye, T. J., and R. P. Klump. 1998. Determination of emergency power system voltage control actions. *IEEE Transactions on Power Systems* 13:205–210.

Seydel, R. 2009. *Practical Bifurcation and Stability Analysis*. New York: Springer.

Sigrist, L., I. Egido, and L. Rouco. 2013. Principles of a centralized UFLS scheme for small isolated power systems. *IEEE Transactions on Power Systems* 28:1779–1786.

Taylor, C. W., N. J. Balu, and D. Maratukulam. 1994. *Power System Voltage Stability*. New York: McGraw-Hill.
Taylor, C. W., and International Conference on Large High Voltage Electric Systems. Study Committee 38. Task Force 38.02.12. 1995. *Criteria and Countermeasures for Voltage Collapse*. Paris: CIGRE.
Trias, A., and J. L. Marin. 2016. The holomorphic embedding loadflow method for DC power systems and nonlinear DC circuits. *IEEE Transactions on Circuits and Systems I: Regular Papers* PP:1–12.
van Cutsem, T. 2007. *Voltage Stability of Electric Power Systems*. New York: Springer.
Vournas, C., and M. Karystianos. 2004. Load tap changers in emergency and preventive voltage stability control. *IEEE Transactions on Power Systems* 19:492–498.
Wang, Z., H. Chen, J. Wang, and M. Begovic. 2014. Inverter-less hybrid voltage/var control for distribution circuits with photovoltaic generators. *IEEE Transactions on Smart Grid* 5:2718–2728.
Zhu, T. X., S. K. Tso, and K. L. Lo. 2000. An investigation into the OLTC effects on voltage collapse. *IEEE Transactions on Power Systems* 15:515–521.

5

Advanced Control Devices

This chapter is devoted to the new control devices that have arisen in power systems and in particular, in island power systems. These new control devices are mainly related to the increased needs for energy storage due to the integration of renewable energy sources (RES) in power systems (Lee et al. 2012). Apart from well-known pumped storage units, different energy storage systems (ESS) have been developed in recent years based on compressed air (compressed air energy storage [CAES], ultracapacitors [UCs], batteries [battery energy storage systems (BESS)], and flywheels [flywheel energy storage systems (FESS)]) (Carpinelli et al. 2013). The current main use of ESS is to smooth the generation pattern from RES, but other applications related to frequency control and voltage control are being broadly proposed. Many characteristics, such as power rating, energy capacity, lifetime, or number of charge/discharge cycles, determine system costs and thus which ESS to use in a certain application (Poonpun and Jewell 2008).

Examples of the implementation of different ESS can be found as research projects and also as actually installed projects. Flywheels were already proposed in 1989 for the very small power system in Fair Isle (Scotland), to sustain the essential loads for several (up to 15) seconds' shortfalls in wind power, thus avoiding the need to start a diesel engine (Davies and Jefferson 1989). Also in 1990, another flywheel system to cover short-term fluctuations in wind speed was installed in the Greek island of Agios Efstratios as a field demonstration (Dettmer 1990).

Several Portuguese Atlantic islands have different storage systems (Vasconcelos et al. 2015). An FESS is designed to avoid frequency stability problems in Porto Santo Island. For Madeira Island, the objective relies on the exploitation of hydro-resources through the quantification of the technical benefits resulting from variable-speed hydro-pumping stations that are able to provide primary frequency regulation services in the pump operation mode. In Flores Island, the benefits of introducing electric vehicles are addressed. A wind-hydro-pump storage station has been constructed on Ikaria Island (Greece) (Papaefthymiou et al. 2010). Another power system with wind generation and a hydro-pump storage system is also in operation in El Hierro (Spain) (Hallam et al. 2012), where wind is the only source of generation except for small diesel machines that are used only in case of emergencies (Pezic and Cedrés 2013). Advanced storage systems have also been installed in other Spanish islands, such as a 4–20 MW UC bank installed in the La Palma power system, a 500 kW–18 MW flywheel installed in the

La Gomera power system, and a 1 MW–3 MWh Li-ion battery installed in the Gran Canaria power system (Egido et al. 2015b). Pilot projects for the integration of ESS in interconnected power systems can also be found, such as two 20 MW flywheel projects in the New York Independent System Operator (NYISO) and PJM power systems (Lazarewicz and Arseneaux 2006).

To implement an ESS, first, it is necessary to design its size (Bueno and Carta 2005). This sizing is related to power system size, but more significantly to the ESS intended application. Another important component in the design process is the ESS control, which is again largely dependent on its final application. To perform any design or adjustment, an adequate model of the system is crucial. All these matters related to ESS will be covered in this chapter.

The chapter is organized as follows. An overview of current developments and applications of ESS will be presented in Section 5.1, while Section 5.2 is devoted to the modeling of the energy storage devices (SDs). Finally, the sizing of ESS will be addressed in Section 5.3. The list of references can be found in Section 5.4.

5.1 Overview of Current Developments and Applications

The massive development of RES, mainly wind and solar generation, has brought many new challenges for the operation and control of power systems due to the variable nature of these sources (Bueno and Carta 2006). On the one hand, this variability continuously creates load–generation imbalances. On the other hand, they currently do not contribute to frequency control, either through their inertia, because this kind of generation is usually connected to the power system through a stage of power electronic converters, or with upward reserve, as RES are usually operated in such a way that they generate as much power as possible, and therefore, no upward reserve is available. Thus, RES replace conventional generation technologies, but they do not usually provide frequency or voltage control (as conventional technologies effectively do). This imposes more regulating effort on the remaining conventional generation, and thus, other resources are being sought to provide this service (Delille et al. 2012; Jayalakshmi and Gaonkar 2011). ESS appear to be adequate means to smooth the generation profile of RES and also to provide frequency and voltage regulation services (Serban et al. 2013).

RES and advanced control devices such as ESS have been developed in parallel with smart grids (SGs) and electric vehicles (EVs). SGs appear to be the most promising means of integrating a large number of distributed energy sources into the power grid because of their capacity to monitor and control different generation technologies and loads (Solanki et al. 2016). They are also very suited to including ESS to improve their performance.

EVs can also be used to improve RES performance, as they can be used as a distributed storage system (Lam et al. 2016). However, dedicated ESS have also been developed in large interconnected power systems, both to smooth generation and load profile, and to provide frequency and voltage control (Kurian et al. 2015). As well as EVs, examples of advanced control devices for energy storage include magnetic energy storage units (Banerjee et al. 1990), BESS (Oudalov et al. 2007; Mercier et al. 2009; Lu et al. 1995), UCs, and FESS (Cimuca et al. 2006). In general terms, batteries are more suitable for high energy, which makes them the preferred solution for grid applications. UCs are more appropriate for high power but low energy, and although their level of maturity is quite high, the technology is not fully competitive economically. Finally, FESS are very convenient for high-power applications with a high number of charge/discharge cycles. A combination of several technologies is also a good solution to achieve different goals at the same time (Navarro et al. 2015).

Small isolated power systems are especially capable of taking advantage of the implementation of ESS (Sigrist et al. 2013). The current developments and applications of these technologies are presented in the following subsections.

5.1.1 Smart Grids

Practical methods, tools, and technologies based on advances in the fields of computation, control, and communications are allowing more intelligent processes to be introduced into the power network. This is, and will continue to be, transforming electricity networks into what are known as SGs (Massoud-Amin and Wollenberg 2005). Following the definition in Farhangi (2010), the SG is the collection of all the technologies, concepts, topologies, and approaches that allow the silo hierarchies of generation, transmission, and distribution to be replaced with an end-to-end, organically intelligent, fully integrated environment in which the business processes, objectives, and needs of all stakeholders are supported by the efficient exchange of data, services, and transactions. An SG is therefore defined as a grid that accommodates a wide variety of generation options, for example, central, distributed, variable, and mobile. It empowers consumers to interact with the energy management system to adjust their energy use and reduce their energy costs. An SG is also a self-healing system. It predicts looming failures and takes corrective action to avoid or mitigate system problems. An SG uses information technology (IT) to continually optimize the use of its capital assets while minimizing operational and maintenance costs.

It is not possible or desirable to drastically change the huge current power grids into SGs. However, the future SG is expected to arise from the interconnection of smaller structures called *microgrids*. Under the definition in E2RG (2011), a microgrid is a group of interconnected loads and distributed energy resources within clearly defined electrical boundaries that acts as a single controllable entity with respect to the grid. A microgrid can connect to and disconnect from the grid to enable it to operate in both grid-connected and island mode. A microgrid provides a solution to manage local generations and loads as a single grid-level entity. It has the potential to maximize overall system efficiency, power quality, and energy surety for critical loads (Fu et al. 2013).

Thus, many island power systems are very suitable networks to be operated under the smart microgrid (disconnected from large power systems) paradigm. Newly developed microgrids are usually equipped with all the necessary technology for measurement and control of the grid (including generation and load), which facilitates the integration of new generation technologies such as RES and load control, participation in power system control, and also the integration of new control devices such as UCs, flywheels, and batteries. Small island power systems are very suitable to be turned into smart microgrids, because the cost of investments is lower, and they are good test benches to test new developments that will probably be later translated to larger power systems such as larger island power systems and interconnected power systems.

Many proposals can be found related to the integration of advanced control devices in microgrids. In these cases, the power system is designed as a whole, and the ESS is one of the components of the power system. Some of the experiences in this field are summarized here.

A generalized model of a microgrid in an island mode is proposed in Muhssin et al. (2015). The microgrid evaluated includes a diesel backup generator along with a wind turbine generator, a photovoltaic system, an FESS, and a BESS. The potential for using responsive charging EVs is also investigated with a centralized controller, showing that the complete system reduces the fluctuation effect of the wind turbine and stabilizes the system frequency.

A demonstration project for the microgrid concept has been established at the Alameda County Santa Rita Jail in California (Alegria et al. 2014). The existing system included a 1 MW fuel cell, 1.2 MW of solar photovoltaic, and two 1.2 MW diesel generators. A 2 MW–4 MWh lithium iron phosphate battery storage system was added, connected to the system through a 2 MW–2.5 MVA power converter, resulting in adequate frequency control. A 900 kvar capacitor bank was also connected to provide power factor compensation when the microgrid operates connected to the grid, and additional necessary reactive power supply when in island mode (this solution was preferred to increasing the MVA rate of the power converter). The whole system was tested, resulting in compliance with all Institute of Electrical and Electronics Engineers (IEEE) standards.

5.1.2 Battery Energy Storage Systems

There are many proposals to include BESS in power systems to perform frequency and voltage regulation. BESS are mainly proposed for use linked to photovoltaic generation (Hill et al. 2012), although their operation together with wind generation has been also suggested (Dang et al. 2014).

One crucial component of a BESS is the battery management system (BMS), which enables making the battery a safe, reliable, and cost-efficient solution for grid operation (Rahimi-Eichi et al. 2013). A smart BMS should monitor and control the battery, including an accurate and practical algorithm to estimate its state of charge (SoC) and state of health (SoH, which is usually defined as the ratio between the actual capacity and the rated capacity of the battery, and which is reduced with battery operation due to the cycling effect). This will allow proper application-oriented measures to be defined to accurately predict the remaining useful life (RUL).

Energy management with multiple ESS (e.g., multiple batteries installed in the same grid) is also a difficult task, which is addressed in Urtasun et al. (2015). In particular, an energy management strategy for a multiple battery system that makes it possible to avoid the use of communication cables is proposed. The SoC of the battery is taken into account to prevent the battery from overcharge or overdischarge (the typical operating range of a battery is between 20% and 80% of the maximum SoC) as well as high charging/discharging currents. Simulations and experiments are carried out in a power system with two batteries and two photovoltaic plants. The control of the SoC of the battery to prevent overcharge and overdischarge conditions is also a primary concern in Wu et al. (2014), where a BESS is operated with photovoltaic generation.

Several domestic and international projects promoted by the Japanese New Energy and Industrial Technology Development Organization (NEDO), related to both wind and photovoltaic generation, are introduced in Morozumi (2015). For example, a redox-flow battery system (6 MW–6 MWh) was installed at Tomamae Winvilla 30 MW Wind Park in Hokkaido, Japan, to examine the control methods and the reliability of a battery storage system to reduce short-term fluctuation in wind power, while a sodium–sulfur (NaS) battery (1.5 MW–12 MWh) was introduced at the Wakkanai City 5 MW photovoltaic plant in Hokkaido, where a planned transmission from the solar plant was achieved, reducing the negative impacts of renewable energy by balancing and stabilizing the frequency.

When ESS are integrated with wind or solar plants, the global entity is seen as a virtual power plant. The objective is that the virtual power plant (e.g., a wind farm plus a battery storage system) should be able to emulate the characteristics of a conventional power plant. The adequate selection of the ESS is crucial, because they are still expensive systems, and the final choice will depend on the intended application. A detailed comparison of different ESS (UC, flywheel, and several types of batteries) is performed in Swierczynski et al. (2014) in terms of annual accumulated cost per cycle. Li-ion batteries result in the most adequate technology to provide the primary frequency regulation service. An interesting cost analysis is performed in Alimisis and Hatziargyriou (2013), where the employment of used EV batteries to improve the performance of a wind farm is proposed. An EV battery is considered to have reached the end of its useful life for electromobility when it can no longer provide 80% of the peak power needed for car acceleration compared with a new battery. However, current markets and trends for secondhand batteries demonstrate that there is already considerable commercial exploitation of used and reconditioned batteries.

It has also been proposed to use BESS in small island autonomous systems. For example, the impact of battery characteristics on the performance of frequency and voltage regulation is investigated in Ersal et al. (2013) for a military autonomous microgrid. A new control algorithm for BESS realization of frequency and voltage control in an isolated system with wind generation is evaluated in Sharma and Singh (2011). It is also possible to take advantage of the benefits of different storage systems, as presented in Mendis et al. (2014), where the management of a hybrid storage system with a battery and a UC is analyzed.

5.1.3 Electric Vehicles

EVs have been around since the beginning of the twentieth century. However, the number of EVs was minimal, because they were slower and more expensive than their gasoline-powered counterparts. The outlook for EVs shifted dramatically in the 1990s due to the growing concern over air quality and the possible consequences of the greenhouse effect (Chan 1993). From the beginning of the twenty-first century, economic and environmental incentives, as well as advances in technology, have paid increasing attention in plug-in hybrid electric vehicles (PHEVs) and plug-in electric vehicles (PEVs) because of their low pollution emissions. Other advantages of EVs are their potential to transfer power to the grid to alleviate peak power demand and their capability to provide ancillary services to the grid (Su et al. 2012). This leads to what is called the *vehicle-to-grid* (V2G) concept, meaning that the EVs are not only consumers in a power grid, but entities that can also act as generators that provide both active and reactive power to the power grid (Yilmaz and Krein 2013). SG technology will optimize the vehicles' integration with the grid, allowing intelligent and efficient use of energy (Boulanger et al. 2011). It is true that the bidirectional use of the batteries to provide services to the grid will reduce battery life due to battery degradation. However, it may even save money for the vehicle owners (a potential net return is estimated between $90 and $4000 per year per vehicle) with adequate capacity of the electrical connections, market value, penetration number of EV, and EV battery energy capacity (Yilmaz and Krein 2013).

To implement such a system, and because each individual car could provide only 10–20 kW, much importance is placed on the development of a management system in the form of an intermediate figure called the *aggregator* (Han et al. 2010). The whole system of many EVs together with the aggregator can be seen as a huge electric battery. The main task of the aggregator will be to deal with the charging of the small-scale batteries of the vehicles while providing the regulation service for large-scale power, meeting the power requirement from the grid operator. The effect on the SoC of the batteries due to participation in frequency regulation is evaluated in Quinn et al. (2012), taking into account probabilistic vehicle travel models, frequency control signals, and pricing signals. Among other things, a high-power home charging capability and the consideration of vehicle performance degradation are found to be key aspects of the system. Maintaining the specific and guaranteed SoC demanded by each individual customer is also a crucial element in the development of the whole system (Liu et al. 2013). Decentralized controls of the aggregator–EV system are preferred, because the communications needed for the system to perform adequately are reduced (Yang et al. 2013). Many computational challenges arise when trying to operate the system, as highlighted in Galus et al. (2011) for the provision of secondary frequency control, and distributed computing is envisioned as a way to resolve such challenges.

The integration of EVs with other ESS and with RES in an isolated power system has also been investigated. For example, a centralized control scheme for islanded microgrids with a high penetration of PEVs is analyzed in Abdelaziz et al. (2014). In this case, the main objectives are the same as for interconnected power systems, that is, maintaining the customer's required SoC of PEVs and providing primary frequency regulation. Moreover, load shedding is more prone to happen in small isolated power systems due to the lack of system inertia, with a consequent reduction in the quality of service, and thus special emphasis is also put on minimizing load shedding.

EVs are also proposed in Sharma et al. (2015) to improve the system restoration service using the concept of islanding. The advantages of including the EVs to assist service restoration are highlighted in the form of an increase in priority load restored, a decrease in switching operations, and an increase in total load restored.

5.1.4 Flywheels

The use of flywheels to provide frequency and voltage regulation has been widely proposed. Many of these proposals link the flywheel operation with wind generation. For example, the technical and economic performance of a small power system with wind generation and a flywheel is presented in Thatte et al. (2011), where the control of both the wind farm and the flywheel is designed to provide frequency regulation. Direct coupling of a flywheel with a windmill double fed induction generator (DFIG) is studied in Gayathri-Nair and Senroy (2013) to provide low-voltage ride through and power smoothing. The results show that by modifying the controls of the DFIG and the flywheel, the system achieves a very smooth regulated power output and prolonged low-voltage ride through capability. The application of an FESS in an island power system with a wind farm to improve the frequency quality is also proposed in Takahashi and Tamura (2008). An FESS was also introduced in a remote island in Okinawa prefecture, Japan, in 2009 (Sakamoto et al. 2011). The island has four diesel generators with a total rated capacity of 900 kW and two 245 kW wind turbine generators. The 200 kW FESS was equipped with both frequency- and voltage-stabilizing control systems to reduce the fluctuations due to wind energy.

Flywheels can be also combined with flexible industrial loads to provide primary frequency control with economic benefits, as presented in Wandelt et al. (2015). A 1 MW battery is compared with a 1 MW flywheel. The relatively small capacity of the battery (1 MWh, as batteries usually have a power to capacity ratio of one) and the flywheel (0.25 MWh, as flywheels have typically a power to capacity ratio of four) is overcome by operating the flexible load when necessary. Thus, the storage system has to be appropriately sized, so that the industrial plant's production is not affected by the provision of primary frequency control.

Finally, flywheels are also proposed to independently participate in regulation services, especially for frequency regulation. For example, a flywheel is proposed in Zhang et al. (2014) as particularly suitable to closely follow the continuously changing Automatic Generation Control (AGC) signals because of its rapid response, fast ramping capability, high efficiency, and ability to undergo a large number of charge/discharge cycles. An independent FESS for frequency regulation, voltage regulation, and oscillation damping is proposed in Lazarewicz and Rojas (2004) as an attractive cost solution compared with the generator-based approach. Two pilot projects (1 MW, 250 kWh) developed in California and New York State are summarized in Lazarewicz and Arseneaux (2006), where the preliminary data of real operation show that the system can follow AGC signals very quickly while providing reactive power to improve the local power factor. The results for more than 1 year's operation of one pilot plant (1 MW, 250 kWh) in the New England power system are presented in Lazarewicz and Ryan (2010). One of the main conclusions is that, on a daily basis, there are between 7 and 13 roughly 80% depth-of-discharge cycles and many smaller ones. This leads to an average of 180 kWh per hour injection into the system, resulting in approximately 125,000 full charge/discharge cycles over a 20-year life of the system (much of this operation is at full power).

Due to the limited storage of FESS, it is difficult to offer a guaranteed frequency regulation service in any situation, such as is provided by conventional generators (Zhang et al. 2014). Thus, to allow FESS participation, certain independent system operators (ISOs) (such as NYISO and the Independent System Operator of New England [ISO-NE]) have introduced rules to allow energy-limited resources to offer a frequency regulation service regardless of their storage levels. A simple schedule strategy is proposed in Lazarewicz and Ryan (2010), based on the SoC of the FESS, to maximize the reserve that the FESS can provide. For example, when the system is at half charge, it is scheduled at 0 MW, because it can provide its nominal power to both up and down reserve.

Improved offer strategies to reserve markets are obtained in Zhang et al. (2014) via a robust optimization approach ensuring that the FES does not violate its storage capability limits while the frequency regulation service requests are fully satisfied. These strategies are tested for a 20 MW, 6 MWh, FESS system, which resembles the two 20 MW pilot systems that have been developed in Stephentown (New York) and Chicago (Lazarewicz and Ryan 2010).

5.1.5 Ultracapacitors

UC technology has been developed since the 1980s, representing an improved type of double-layer capacitor based on a ceramic with an extremely high specific surface area and a metallic substrate (Bullard et al. 1989). Nowadays, UCs are becoming a very interesting solution to storing electrical energy due to their high efficiency, robustness, simplicity, and increasing energy density. Compared with batteries, the main advantage of UCs is their high power density (del Toro Garcia et al. 2010). Moreover, UCs have a better cycling performance, reaching a high cycle life of up to 500,000 cycles, according to the manufacturer, because of the chemical and electrochemical inertness of the activated carbon electrodes. However, in practice, UCs exhibit performance fading when they are used for months, depending on the particular application they are involved in, which has been carefully studied based on different aging tests (El Brouji et al. 2009).

Due to their characteristics, UCs can be used in conjunction with RES. For example, a UC is used in Cho and Hong (2010) together with a solar plant to supply a stand-alone building microgrid. Appropriate power management controllers and power converters are needed to overcome the high dependency of solar energy on environmental conditions with the integration of a UC in the system. Another proposal for a solar and UC system is found in Xue et al. (2009), where the UC is preferred to more traditional proposals using batteries. The design of the conditions for charge and discharge of the UC, as well as different charging modes, are analyzed as crucial for the adequate behavior of the whole system.

However, UCs have usually been proposed to be used in conjunction with other ESS because of their fast response. The use of a hybrid system with fuel cells and a UC for a stand-alone system together with a wind generator has been proposed in Gyawali and Ohsawa (2009), where the fuel cells are intended for slower transients and the UC for faster transients such as generator tripping. The whole scheme can provide frequency and voltage regulation by using an adequate control. The system can operate connected to the grid and also in islanded mode (Gyawali et al. 2010). A fuel cell and UC system for a stand-alone residential application is simulated in Uzunoglu and Alam (2006). UCs are a good complement to fuel cells, as UCs can supply a large amount of power but cannot store a significant amount of energy. The combined use of the two technologies has the potential for better energy efficiency, reduction in the cost of fuel cell technology, and improved fuel usage.

A turbine-level storage system is also studied in Esmaili et al. (2013) with the objective of mitigating short-term wind energy variations and thus supporting wind generation. In this case, the storage system consists of a battery and a UC. The combination of UCs with batteries is also analyzed in Hredzak et al. (2014a), where special care is taken to maintain the operation of the hybrid power system within all important operational limits (battery within its SoC limits and UC voltage at a predefined value). Fast current changes are allocated to the UC, while the battery responds mainly to slow current changes, using a low-complexity control system with very satisfactory operational results. The operation of the system is improved if a more complex model predictive control system is used (Hredzak et al. 2014b).

The use of dedicated UCs for the provision of frequency or voltage control in small isolated power systems has also been studied. The impact of a UC on the frequency stability of Guadeloupe Island has been analyzed for two scenarios of different wind and solar photovoltaic penetration levels (Delille et al. 2012). In Jayalakshmi and Gaonkar (2011), a UC has been used, among other things, to provide frequency control for an isolated power system with RES by means of a proportional-plus-integral (PI) controller. The design of the adequate size of a dedicated UC to reduce load shedding in an isolated power system is detailed in Sigrist et al. (2015), and results of real operation of the system can also be found in Egido et al. (2015a).

5.2 Modeling of Advanced Control Devices

Many different models of batteries, flywheels, and UCs can be found, depending on the final application of the model. For example, a model of a battery to evaluate its aging is different from an adequate model of the same battery to develop an analysis of its performance as a frequency regulation provider. In general, models can be found in the literature for almost every application proposal, as the initial stage before any system implementation is the simulation of the system to assess its benefits and drawbacks. In that sense, the references in Section 5.1 usually present a model of the system that includes the model of the storage system.

An adequate control of the device must be always implemented in order to get the desired response while assuring that all its physical and safety limits are met. In this context, it is important to take into account that not only a model of the device itself is necessary, but an accurate model of the control is usually even more significant when evaluating ESS applications. This is because the control of the system mainly determines its final behavior, while the limiting characteristics of each device are met.

Several models for the advanced control devices under study will be highlighted here with a focus on the operation of the storage systems as frequency and voltage providers.

5.2.1 Battery Energy Storage Systems

Electrical models of the battery can be found in many references. For example, an electrical model based on Thévenin equivalent is used in Swierczynski et al. (2014), where the model parameter values are adjusted based on the results of dedicated tests. The model allows the battery response to transient load events at a determined SoC to be predicted. The Thévenin direct current (dc) voltage is assumed to be constant. This basic Thévenin-based battery model is improved by adding additional components to better capture the DC response, to predict runtime, or to consider the effect of aging and temperature. In this case, as in other recently developed models, a controlled voltage source is used to reflect the fact that the voltage–SoC relationship is a static nonlinear characteristic of the battery. In practice, experimental look-up tables are used to map open circuit voltage (OCV) to SoC with different curves for charging and discharging cycles to represent the hysteresis effect (Rahimi-Eichi et al. 2013). Performance degradation models are also built in Swierczynski et al. (2014) using two different approaches: Resistance-capacitor (RC), Parallel Networks, and ZARC and Warburg elements. Another example of a Thévenin model of the battery can be found in Mendis et al. (2014), following the general model in Mathworks (2016).

A grid-scale BESS consists of a battery bank, a control system, a power electronics (PE) interface for alternating current (ac)/dc power conversion, and a transformer to convert the BESS output to the transmission or distribution system voltage level (Hill et al. 2012). A simple diagram of the system is presented in Figure 5.1.

In power system analysis, a simpler model can be used to take into account the dynamics of the BESS between the reference issued by the control (active and reactive power reference) and the BESS output. A first-order transfer function is used in Chen et al. (2016), where the time delay in communication and control signal processing is modeled as a time constant. Moreover, the BESS should also satisfy the constraints related to the power limits and the stored energy. A similar approximation is used, for example, in Kottick et al. (1993).

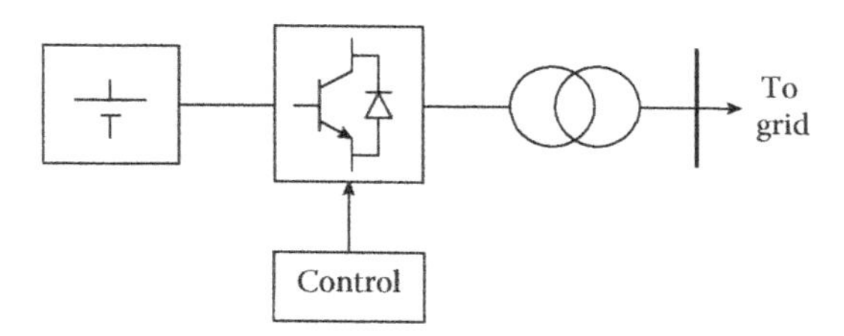

FIGURE 5.1
Simple diagram for the connection to the grid of a BESS.

5.2.2 Flywheel Energy Storage System

Flywheel energy is stored in kinetic form due to the rotation of the flywheel mass. The basic FESS consists of a flywheel driven by an electrical machine, which functions as a motor to store energy from the grid into the flywheel (charging mode), increasing its rotational speed, and as a generator to release energy from the flywheel to the grid (discharging mode), decreasing its rotational speed. The electrical machine is connected to the grid through two back-to-back power electronic converters and a DC link (Awadallah and Venkatesh 2015). A transformer is used if necessary, depending on the voltage of the network. A control system is implemented to operate the FESS complying with the active and reactive requirements from the grid. Figure 5.2 shows a simple diagram of the system.

When the energy is released from the flywheel, the speed and the available energy decrease. Moreover, depending on the type of electrical machine and control used, the maximum available power can also be reduced. For example, a switched reluctance machine with no flux weakening strategy is used in Navarro et al. (2015). In this case, the maximum torque is almost constant independently of the rotational speed, and thus, as speed decreases, maximum output power also decreases. In this case, not only the available energy but also the available maximum power must be taken into account in the FESS model.

A detailed model of the whole FESS system is presented in Suvire et al. (2012). Special emphasis is put on the control of the system to allow frequency and voltage control. The flywheel is modeled in terms of its remaining stored energy. A similar approach can also be found in Takahashi and Tamura (2008).

However, for system integration studies, a simpler model of the flywheel can be used. For example, in Xie et al. (2011), only the model of the remaining energy is used, whereas a very fast ramp rate limiter is used as a dynamic constraint in Lu et al. (2010). A first-order linear model is used in Ray et al. (2010) to represent the dynamics between the power reference to the FESS and its power output.

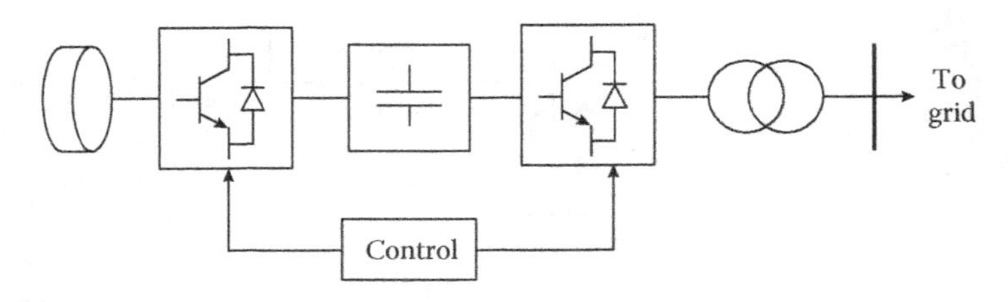

FIGURE 5.2
Simple diagram for the connection to the grid of an FESS.

5.2.3 Ultracapacitor Energy Storage System

UC storage system configuration is quite similar to that of a BESS. In this case, the charging and discharging of a capacitor bank are performed through a PE interface operated by a dedicated control system. As in BESS and FESS, a transformer is used if necessary, depending on the voltage level of the grid. A simple model of the complete system is depicted in Figure 5.3.

Several models of UCs have been presented in the literature that focus on the electrical and thermal behavior of the UC. For example, different RC network models have been compared in Shi and Crow (2008). An RC series-parallel network is also used in Chiang et al. (2013) for the electrical characteristics of the UC, while an equivalent thermal model is presented to address UC temperature. A third-order model (three RC networks) has been found to be accurate enough in Yang et al. (2008), as the rest of the time constants can be neglected. A similar objective is pursued in Eddahech et al. (2013), where a linear recursive model has been employed. Also, an electrical model is used in El Brouji et al. (2008) to assess the aging of a UC by means of evaluating the evolution of the estimated parameters of the model. A very simple model of the UC with only a series resistance added to the capacitor is used in Mendis et al. (2014). However, important limitations are included in the model, such as the safe operating voltage limits, maximum possible peak current, and maximum allowable power output.

However, as UCs are connected to power systems through a stage of power electronic converters, and because an appropriate control is applied to these converters to obtain the desired response from the UC, a simple model of the UC can be used for grid integration studies. In this case, the dynamic of the UC response to the control reference is represented by a time delay. For example, this simple model with a delay in the range of hundreds of milliseconds is used in Goikoetxea et al. (2010), where different control strategies are presented that drive the response of the UC storage system.

5.2.4 Generic Dynamic Model of Energy Storage Systems for Power System Analysis

As concluded in Sections 5.2.1 through 5.2.3 many detailed models have been used for batteries, flywheels, and UCs to take into account detailed

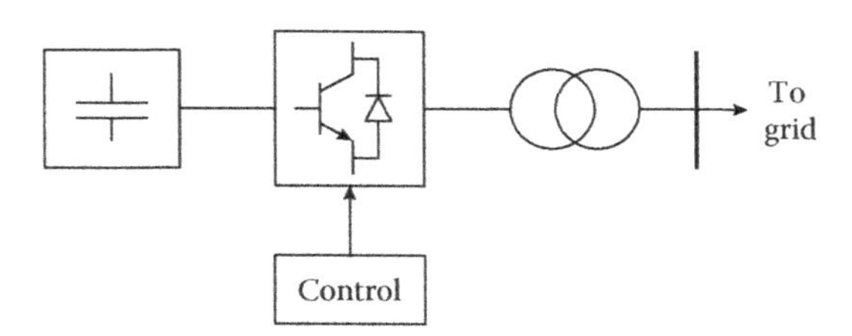

FIGURE 5.3
Simple diagram for the connection to the grid of a UC ESS.

transients, aging, and heating of the system. However, a simpler model is possible for the three devices to be used for grid integration studies. This simple model is preferred because detailed models have many parameters that are difficult to tune, and also because simple models are easier to understand and have a small number of parameters that are easier to tune, and simulation time is reduced (Rowen 1983).

A simplified model should always be focused on its application, including the interesting dynamics and the operational limits of the system regarding the application under study (Egido et al. 2004). In the case of ESS, the main dynamics and limits are the same for all of them, and thus, a general model can be established whose parameter values need to be adjusted for the specific device under study (Egido et al. 2015a).

A general model intended to reflect the response of an ESS as seen by the power system is presented in Figure 5.4. The model includes the control of the ESS (block Control), the dynamics of the PE and the SD (block PE & SD), and the SoC calculation.

The dynamic response of the PE and the SD are modeled with a first-order time constant (see block PE & SD). The inputs to this block are the setpoint reference for active (P_{ref}) and reactive (Q_{ref}) power as calculated by the control, and the outputs are the active (P_g) and reactive (Q_g) power generation. As the dynamics could be different for active and reactive power generation (due to the device characteristics or intentionally forced by the power electronics control), separate time constants (T_P and T_Q) are included in the model.

The control of the ESS includes frequency and voltage control. The details of these controls depend on the application. For example, a proportional voltage control could be used, while droop control and synthetic inertia could be established for frequency control. The inputs to the frequency control are frequency reference (f_{ref}) and measured frequency (f), while the inputs for voltage control are voltage reference (V_{ref}) and measured voltage (V). The output of each control is limited by the maximum/minimum active (P_{max}/P_{min}) and reactive (Q_{max}/Q_{min}) power available, thus obtaining the

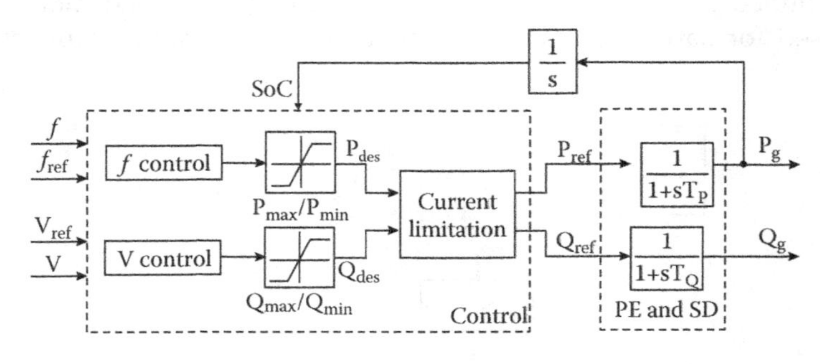

FIGURE 5.4
General model of an ESS appropriate for system analysis.

desired active (P_{des}) and reactive (Q_{des}) power. A current-limitation block is added to the control to avoid an active (P_{ref}) and reactive (Q_{ref}) power reference violating the maximum current rating of the PE. The exact implementation of the current-limiting block depends on the desired performance when the current limit is reached (e.g., the priority could be set on frequency or voltage control if the current limit is exceeded, reducing the other output as necessary, or both outputs could be reduced proportionally). The current-limiting block also needs the voltage measurement (V) to calculate the current associated with active and reactive power references for the actual voltage. This current-limiting block could also be integrated in the PE block. The outputs of the control block are the active (P_{ref}) and reactive (Q_{ref}) power reference sent to the PE.

Finally, the SoC of the SD is computed by integrating the power generation (P_g). The SoC can be used by the control in many ways. For example, the active power limits (P_{max}/P_{min}) can depend on the SoC for a certain device, or the control can take different actions if the charge of the device is reaching its limits.

The model presented in Figure 5.4 has been simulated using real data recorded from a UC (4 MW/20 MWs) in real operation with active power control. The parameters of the model (especially time constant $T_P=200$ ms) have been adjusted to represent the response of the UC under study. The frequency control block includes synthetic inertia ($H=10$ s) and droop control ($R=2\%$), the same control parameters used in the real device. As the UC is operating in a small island, frequency excursions up to 100 mHz are frequent. Thus, a 200 mHz dead band is used to avoid the continuous operation of the UC and to limit its actuation to severe events. The frequency measured during a generator disconnection contingency has been used as input to the model, and the simulated output has been compared with the real output of the UC. The results are presented in Figure 5.5, where the measured frequency is depicted in Figure 5.5a, and the simulated and real UC power output is displayed in Figure 5.5b. As can be seen in the figure, the real and simulated results coincide well. The initial delay in the UC response in both real and simulated data is due to the 200 mHz dead band.

5.3 Sizing of Advanced Control Devices

The appropriate sizing of advanced control devices is crucial to be able to economically provide regulation services in island power systems. Thus, the sizing procedure must be linked to the intended objective of the ESS in the specific power system. For that reason, many design methods have arisen depending on the particular system characteristics and intended use of the ESS.

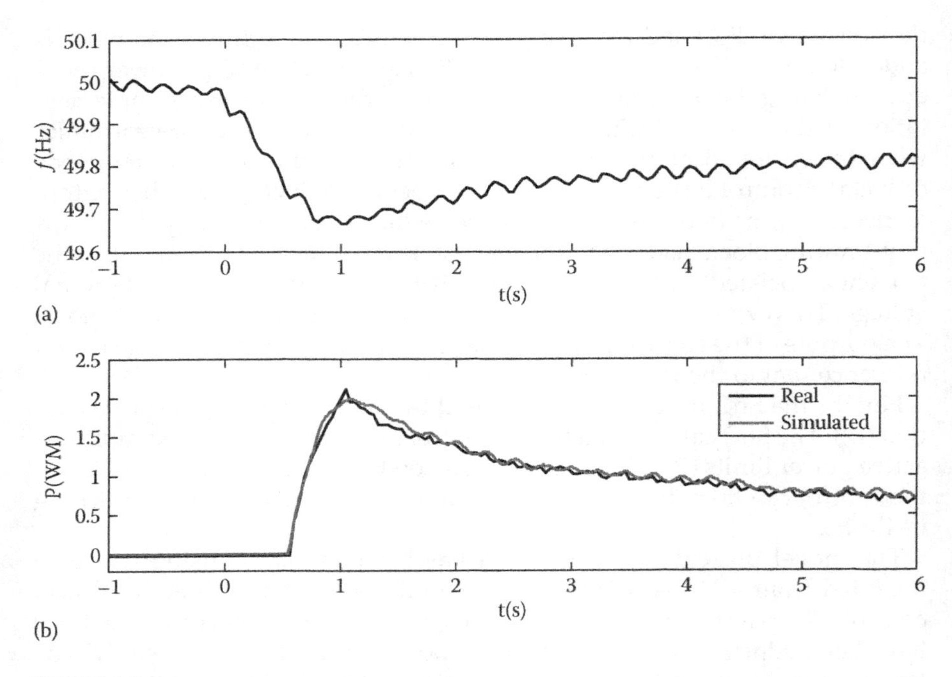

FIGURE 5.5
Real and simulated response of a UC to a generator tripping event. (a) Measured frequency; (b) simulated and real UC power output.

5.3.1 Sizing for Wind Power Farm Output Smoothing

As presented in Section 5.1, one of the applications of ESS is to smooth the output power of a wind farm. For example, the sizing of a hybrid system with a BESS and a flywheel to smooth the output of a 10 MW wind farm is presented in Lee et al. (2012). In such a case, as two different SDs are used to take advantage of their different response characteristics, an integrated sizing of both devices is performed. The design is decoupled in terms of speed of response, because flywheel actions are faster than BESS actions, and thus BESS takes care of slow transients while the flywheel tackles fast transients. Following this procedure, BESS sizing is first accomplished, and then the flywheel is sized to avoid the remaining fast transient power variations. Two different approaches are followed in terms of desired behavior. First, full compensation of wind farm output power variations is pursued, with the objective of having an output power as flat as possible. Then, partial compensation is addressed, allowing some oscillations in output power.

5.3.2 Sizing for Very Small System Operation

The sizing of a whole hybrid system for a very small island power system (1 MW) is performed in Vrettos and Papathanassiou (2011). Such a small island power system is currently usually fed by diesel generators, but if a certain penetration level of renewable generation is desired, the introduction of ESS is helpful to keep the performance of the system. In this case, wind and photovoltaic generation is introduced into the system, and a BESS is used as the ESS. Installation and operational cost of both diesel generators and BESS are taken into account in an optimization problem. CO_2 emissions cost is also included in the optimization. The entire lifetime of the project is addressed, and thus a battery life model is included. An optimization problem including all the technologies allows the complete system to be designed at one time.

5.3.3 Sizing for Continuous Frequency Control

Another usual application of ESS is frequency control. The sizing of FESS for frequency control is addressed in Li et al. (2014) with an emphasis on alleviating the regulation efforts of conventional generators. In this case, an optimization problem is stated, with the objective of minimizing not only the deviation between the regulation signal and the provided power, but also the total cost of providing the service. The total cost includes investment and operational costs of both conventional generation and FESS. Constraints are imposed to ensure that the SoC of the FESS device is kept between allowable limits and that the power generation of both conventional generators and FESS does not surpass their limits.

The sizing of a BESS to participate in AGC regulation is addressed in Chen et al. (2016). In this case, the objective is to improve the AGC performance at an affordable cost. AGC performance is evaluated by means of the North American Electric Reliability Corporation (NERC)'s CPS1 performance index, while BESS investment costs are used. BESS penetration is incremented step by step, and both performance and cost are evaluated, selecting the maximum penetration level at which the improvement in the performance does not compensate the cost increment.

5.3.4 Sizing for Reduction of Frequency Excursion after a Generator Trip

A major concern in frequency control for island power systems is the response to generator trips. Island power systems are especially sensitive to generation–load imbalances and specifically to contingencies whereby a sudden generation or load trip occurs, generator trips being more common. In this case, the frequency can experience a fast decay that reaches unallowable values (e.g., a system blackout can be expected if the frequency falls below 47.5–48 Hz). This occurs when the dynamics of the generators in the system are not fast enough to rapidly compensate the power imbalance. To avoid this situation, underfrequency load-shedding (UFLS) plans are usually implemented as a last resort tool to avoid system blackout by disconnecting a predefined amount of load. Fast-response ESS such as FESS and UCs can help the system in this situation by rapidly providing the necessary power while giving extra time for the conventional generating units to produce their response. The sizing procedure of a UC-based ESS with this objective is presented in Sigrist et al. (2015). Avoiding extreme frequency excursions can be stated as the technical objective, while minimizing the cost of the ESS is the economic objective. A more demanding technical objective would be not only avoiding extreme frequency excursions, but also reducing or even avoiding load shedding, resulting in an increase in ESS size and cost.

An example of the operation of an ESS to avoid extreme frequency excursions is presented in Figure 5.6. A small island power system (40 MW, 50 MWA) with low inertia (H=1 s) is used for the example (R=4% droop is used for the conventional generation). A 7 MW generator trips at t=1 s while a load of 30 MW is being supplied (primary reserve is 10 MW). A very simple model that comprises a first-order model (T=1.5 s) and maximum and minimum power limits is used for conventional generation. As can be seen in the figure (thin lines in every plot), if the ESS is not in use, the frequency nadir reaches almost 47 Hz (Figure 5.6a). The increase in conventional generation (PGen) is shown in Figure 5.6b. After some oscillations, a final steady state (not shown in the figure) is reached with a 7 MW increase in conventional generation and a 49.65 Hz frequency.

The size of an ESS for the example system has been designed to reduce frequency decay with the objective of preventing the frequency nadir from reaching 48 Hz. A 4 MW/2 MW ESS manages to comply with this objective for the imposed generator trip. The results of the behavior of the system with the ESS in operation are also presented in Figure 5.6. The model in Figure 5.4 is used for the ESS. Voltage control is deactivated, while a frequency control with synthetic inertia and droop control is used (H=1 s and R=4%, equal to conventional generation). A 200 mHz dead band is used to avoid ESS operation for small frequency deviations, and a 200 ms time constant is fixed to model the dynamic response of the ESS. As can be seen in Figure 5.6a, the frequency nadir stays above 48 Hz. It can be observed in Figure 5.6b that both the conventional generation (Pgen) and the ESS (PESS) increment their generation after the generator trip (t=1 s). The energy stored in the ESS is depleted at t=2.2 s (see Figure 5.6c), causing the output power of the ESS to go to zero (PESS in Figure 5.6b). After some oscillations, a final steady state (not shown in the figure) is reached with a 7 MW increase in conventional generation and a 49.65 Hz frequency, exactly the same as with the ESS not in use. Note that a 2 MW ESS would have been sufficient, as output power is less than 2 MW most of the time. However, a 4 MW device is used to better illustrate the following comments.

The control of the ESS power output while providing power to the system is also important. A synthetic energy and droop control has been used for the simulation in Figure 5.6, but different shapes have also been proposed. For example, several power injection shapes are evaluated in Goikoetxea et al. (2010). As the energy stored in the ESS becomes exhausted, this can

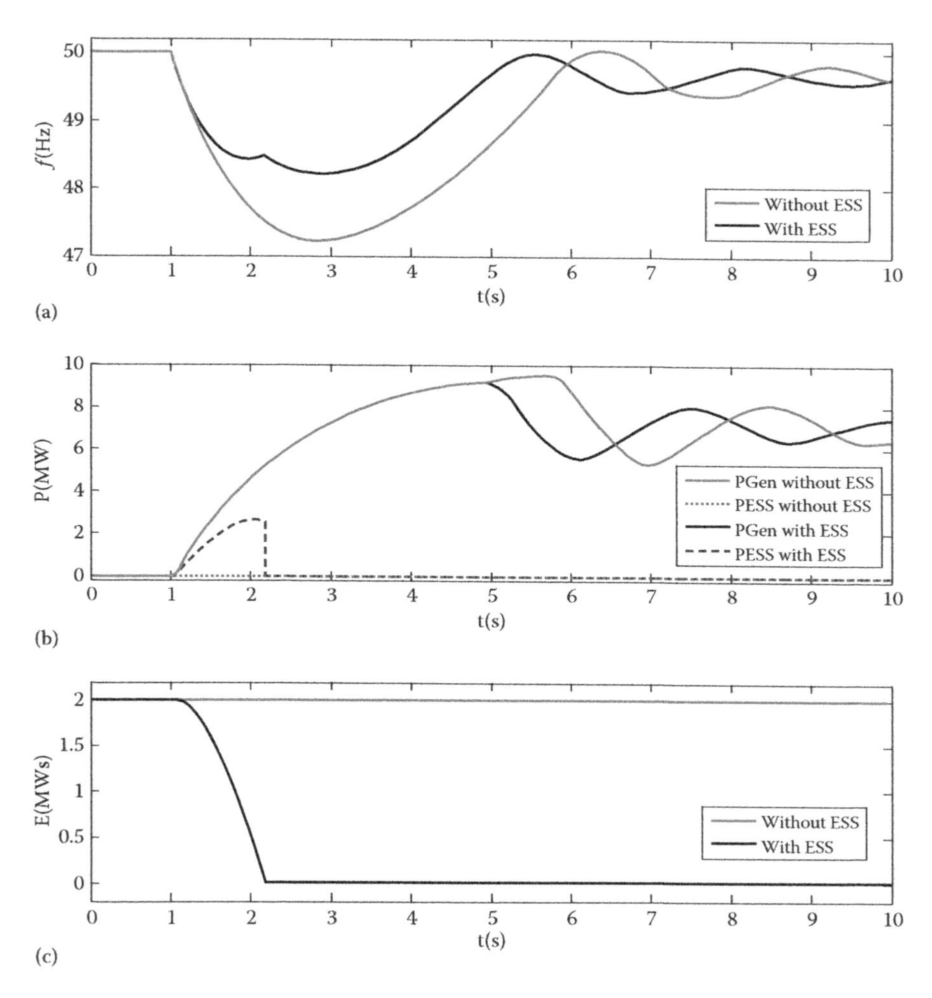

FIGURE 5.6
Example of the operation of an ESS to reduce the frequency excursion in the case of a generator trip. (a) Frequency; (b) conventional generation; (c) energy stored in the ESS.

result in the sudden loss of ESS power in the system, with the subsequent new frequency transient. In the example shown in Figure 5.6, the reduction in the ESS power output due to ESS energy being depleted causes a new frequency decay with a slight decrease up to t=3 s, when the increment in conventional generation surpasses the reduction in ESS power, and the frequency starts increasing again. The more sudden the loss, the more severe is the effect on the system frequency transient. To cope with this problem, a square-shaped power output has been evaluated in Sigrist et al. (2015), where the most appropriate moment to withdraw the ESS power is analyzed, and a tail control of the injected power is proposed.

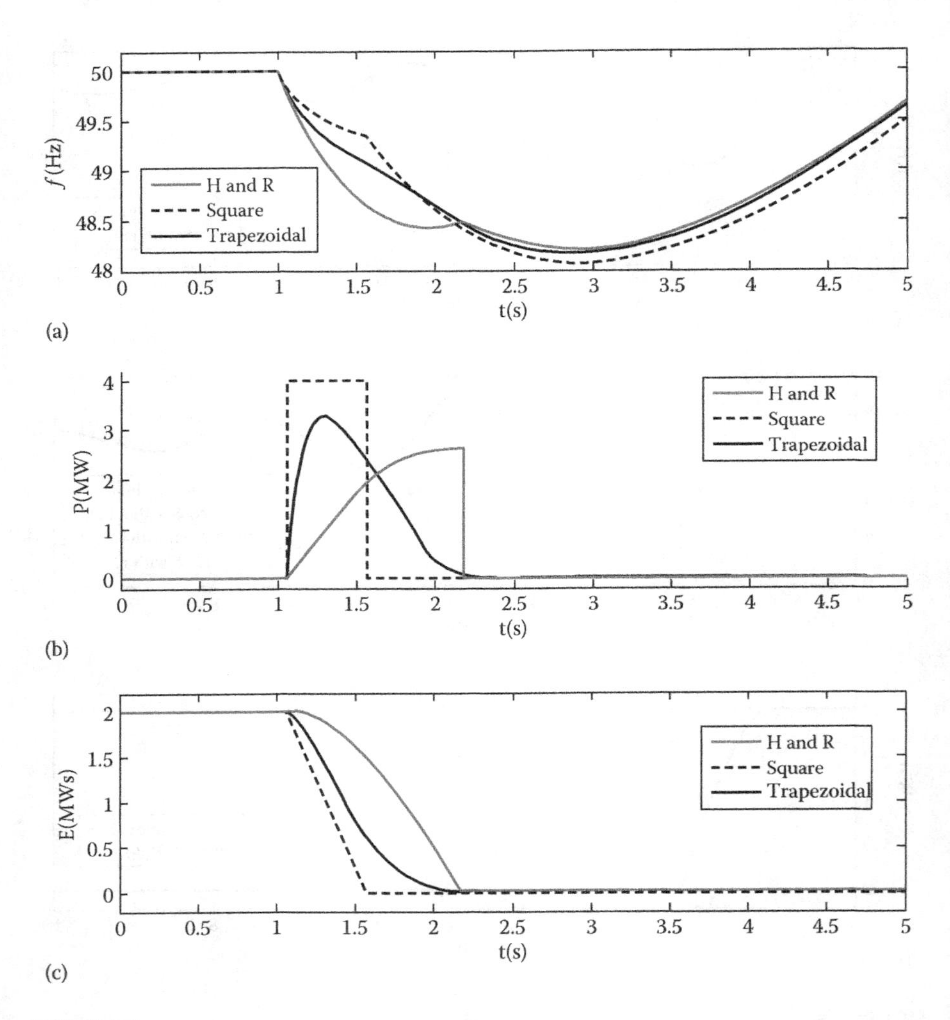

FIGURE 5.7
Examples of different control actions of an ESS after a generator trip. (a) Frequency; (b) conventional generation; (c) energy stored in the ESS.

Examples of different control actions related to ESS power injections after a generator trip are provided in Figure 5.7. Synthetic inertia and droop control (H and R) response are compared with a squared response (in this case, only PESS is presented in Figure 5.7b, and conventional generation output is not depicted). As can be seen in the figure, the square injection keeps the frequency closer to 50 Hz at the beginning of the transient (as more megawatts are generated in the system), but a great change in frequency behavior takes place when ESS energy gets depleted (t = 1.6 s). The final frequency nadir is lower than with the H and R control. An intermediate response is obtained if a nearly trapezoidal injection shape is used. ESS power is rapidly increased up to 3 MW, then briefly kept at this value, and finally slowly reduced to zero (the edges of the trapezoidal shape have been smoothed). With this trapezoidal shape, the frequency response is smoother, with no change in frequency derivative during the transient. The initial frequency decay is slightly faster than with the square injection, but the frequency nadir is similar to that with H and R control.

As previously noted, the characteristics of the control of the ESS power injection will affect the sizing of the ESS. As an example, only a 2 MW ESS is necessary in the previous example if synthetic inertia and droop control are to be used, or a 3 MW ESS would be enough if the trapezoidal shape is preferred. On the other hand, a square wave for a 2 MW ESS (keeping the 2 MW energy) would provide a different frequency, probably between those for trapezoidal and inertia and droop, as 2 MW will be provided for 1 s. Thus, an appropriate sizing procedure must take into account the kind of control used, or even must include this control as part of the design variables.

References

Abdelaziz, M. M. A., M. F. Shaaban, H. E. Farag, and E. F. El-Saadany. 2014. A multistage centralized control scheme for islanded microgrids with PEVs. *IEEE Transactions on Sustainable Energy* 5 (3):927–937.

Alegria, E., T. Brown, E. Minear, and R. H. Lasseter. 2014. CERTS microgrid demonstration with large-scale energy storage and renewable generation. *IEEE Transactions on Smart Grid* 5 (2):937–943.

Alimisis, V., and N. D. Hatziargyriou. 2013. Evaluation of a hybrid power plant comprising used EV-batteries to complement wind power. *IEEE Transactions on Sustainable Energy* 4 (2):286–293.

Awadallah, M. A., and B. Venkatesh. 2015. Energy storage in flywheels: An overview. *Canadian Journal of Electrical and Computer Engineering* 38 (2):183–193.

Banerjee, S., J. K. Chatterjee, and S. C. Tripathy. 1990. Application of magnetic energy storage unit as load-frequency stabilizer. *IEEE Transactions on Energy Conversion* 5 (1):46–51.

Boulanger, A. G., A. C. Chu, S. Maxx, and D. L. Waltz. 2011. Vehicle electrification: Status and issues. *Proceedings of the IEEE* 99 (6):1116–1138.

Bueno, C., and J. A. Carta. 2005. Technical-economic analysis of wind-powered pumped hydrostorage systems: Part II—Model application to the island of El Hierro. *Solar Energy* 78 (3):396–405.

Bueno, C., and J. A. Carta. 2006. Wind powered pumped hydro storage systems, a means of increasing the penetration of renewable energy in the Canary Islands. *Renewable and Sustainable Energy Reviews* 10 (4):312–340.

Bullard, G. L., H. B. Sierra-Alcazar, H.-L. Lee, and J. L. Morris. 1989. Operating principles of the ultracapacitor. *IEEE Transactions on Magnetics* 25 (1):102–106.

Carpinelli, G., G. Celli, S. Mocci, F. Mottola, F. Pilo, and D. Proto. 2013. Optimal integration of distributed energy storage devices in smart grids. *IEEE Transactions on Smart Grid* 4 (2):985–995.

Chan, C. C. 1993. An overview of electric vehicle technology. *Proceedings of the IEEE* 81 (9):1202–1213.

Chen, S., T. Zhang, H. B. Gooi, R. D. Masiello, and W. Katzenstein. 2016. Penetration rate and effectiveness studies of aggregated BESS for frequency regulation. *IEEE Transactions on Smart Grid* 7 (1):167–177.

Chiang, C.-J., J.-L. Yang, and W.-C. Cheng. 2013. Dynamic modeling of the electrical and thermal behavior of ultracapacitors. In *2013 10th IEEE International Conference on Control and Automation (ICCA)*. Hangzhou: IEEE.

Cho, J.-H., and W.-P. Hong. 2010. Power control and modeling of a solar-ultra capacitor hybrid energy system for stand-alone applications. In *2010 International Conference on Control Automation and Systems (ICCAS)*. Gyeonggi-do: IEEE.

Cimuca, G. O., C. Saudemont, B. Robyns, and M. M. Radulescu. 2006. Control and performance evaluation of a flywheel energy-storage system associated to a variable-speed wind generator. *IEEE Transactions on Industrial Electronics* 53 (4):1074–1085.

Dang, J., J. Seuss, L. Suneja, and R. G. Harley. 2014. SoC feedback control for wind and ESS hybrid power system frequency regulation. *IEEE Journal of Emerging and Selected Topics in Power Electronics* 2 (1):79–86.

Davies, T. S., and C. M. Jefferson. 1989. Wind-power flywheel integration. In *Energy Conversion Engineering Conference, 1989. IECEC-89, Proceedings of the 24th Intersociety*. Washington, DC: IEEE.

del Toro Garcia, X., P. Roncero-Sanchez, A. Parreno, and V. Feliu. 2010. Ultracapacitor-based storage: Modelling, power conversion and energy considerations. In 2010 *IEEE International Symposium on Industrial Electronics (ISIE)*. Bari: IEEE.

Delille, G., B. Francois, and G. Malarange. 2012. Dynamic frequency control support by energy storage to reduce the impact of wind and solar generation on isolated power system's inertia. *IEEE Transactions on Sustainable Energy* 3 (4):931–939.

Dettmer, R. 1990. Revolutionary energy: A wind/diesel generator with flywheel storage. *IEE Review* 36 (4):149–151.

E2RG, Energy-And-Environmental-Resources-Group. 2011. 2011 microgrid workshop report held by the U.S. Department of Energy (DOE) Office of Electricity Delivery and Energy Reliability (OE) 30-31 August. San Diego, CA. Report accessed at http://energy.gov/sites/prod/files/Microgrid%20Workshop%20Report%20August%202011.pdf

Eddahech, A., O. Briat, M. Ayadi, and J.-M. Vinassa. 2013. Ultracapacitor performance determination using dynamic model parameter identification. In *2013 IEEE International Symposium on Industrial Electronics (ISIE)*. Taipei, Taiwan: IEEE.

Egido, I., F. Fernandez-Bernal, L. Rouco, E. Porras, and A. Saiz-Chicharro. 2004. Modeling of thermal generating units for automatic generation control purposes. *IEEE Transactions on Control Systems Technology* 12 (1):205–210.

Egido, I., L. Sigrist, E. Lobato, L. Rouco, and A. Barrado. 2015a. An ultra-capacitor for frequency stability enhancement in small-isolated power systems: Models, simulation and field tests. *Applied Energy* 137:670–676.

Egido, I., L. Sigrist, E. Lobato, L. Rouco, A. Barrado, P. Fontela, and J. Magriñá. 2015b. Energy storage systems for frequency stability enhancement in small-isolated power systems. In *International Conference on Renewable Energies and Power Quality (ICREPQ'15)*, La Coruña, Spain, March 25–27, 2015.

El Brouji, E.-H., O. Briat, J.-M. Vinassa, N. Bertrand, and E. Woirgard. 2009. Impact of calendar life and cycling ageing on supercapacitor performance. *IEEE Transactions on Vehicular Technology* 58 (8):3917–3929.

El Brouji, H., J.-M. Vinassa, O. Briat, W. Lajnef, N. Bertrand, and E. Woirgard. 2008. Parameters evolution of an ultracapacitor impedance model with ageing during power cycling tests. In *2008 Power Electronics Specialists Conference*. PESC 2008. Rhodes: IEEE.

Ersal, T., C. Ahn, D. L. Peters, J. W. Whitefoot, A. R. Mechtenberg, I. A. Hiskens, H. Peng, A. G. Stefanopoulou, P. Y. Papalambros, and J. L. Stein. 2013. Coupling between component sizing and regulation capability in microgrids. *IEEE Transactions on Smart Grid* 4 (3):1576–1585.

Esmaili, A., B. Novakovic, A. Nasiri, and O. Abdel-Baqi. 2013. A hybrid system of Li-ion capacitors and flow battery for dynamic wind energy support. *IEEE Transactions on Industry Applications* 49 (4):1649–1657.

Farhangi, H. 2010. The path of the smart grid. *IEEE Power and Energy Magazine* 8 (1):18–28.

Fu, Q., A. Hamidi, A. Nasiri, V. Bhavaraju, S. B. Krstic, and P. Theisen. 2013. The role of energy storage in a microgrid concept: Examining the opportunities and promise of microgrids. *IEEE Electrification Magazine* 1 (2):21–29.

Galus, M. D., S. Koch, and G. Andersson. 2011. Provision of load frequency control by PHEVs, controllable loads, and a cogeneration unit. *IEEE Transactions on Industrial Electronics* 58 (10):4568–4582.

Gayathri-Nair, S., and N. Senroy. 2013. Wind turbine with flywheel for improved power smoothening and LVRT. In *2013 IEEE Power and Energy Society General Meeting (PES)*. Vancouver, Canada: IEEE

Goikoetxea, A., J. A. Barrena, M. A. Rodriguez, and F. J. Chivite. 2010. Frequency restoration in insular grids using ultracaps ESS. In *2010 International Symposium on Power Electronics Electrical Drives Automation and Motion (SPEEDAM)*. Pisa, Italy: IEEE.

Gyawali, N., and Y. Ohsawa. 2009. Effective voltage and frequency control strategy for a stand-alone system with induction generator/fuel cell/ultracapacitor. In *2009 CIGRE/IEEE PES Joint Symposium Integration of Wide-Scale Renewable Resources into the Power Delivery System*. Calgary, AB, Canada: IEEE

Gyawali, N., Y. Ohsawa, and O. Yamamoto. 2010. Dispatchable power from DFIG based wind-power system with integrated energy storage. In 2010 *IEEE Power and Energy Society General Meeting*. Minneapolis, MN: IEEE

Hallam, C. R. A., L. Alarco, G. Karau, W. Flannery, and A. Leffel. 2012. Hybrid closed-loop renewable energy systems: El Hierro as a model case for discrete power systems. In 2012 *Proceedings of PICMET '12: Technology Management for Emerging Technologies (PICMET)*. Vancouver, Canada: IEEE

Han, S., S. Han, and K. Sezaki. 2010. Development of an optimal vehicle-to-grid aggregator for frequency regulation. *IEEE Transactions on Smart Grid* 1 (1): 65–72.

Hill, C. A., M. C. Such, D. Chen, J. Gonzalez, and W. M. Grady. 2012. Battery energy storage for enabling integration of distributed solar power generation. *IEEE Transactions on Smart Grid* 3 (2):850–857.

Hredzak, B., V. G. Agelidis, and G. D. Demetriades. 2014a. A low complexity control system for a hybrid DC power source based on ultracapacitor–lead–acid battery configuration. *IEEE Transactions on Power Electronics* 29 (6):2882–2891.

Hredzak, B., V. G. Agelidis, and M. Jang. 2014b. A model predictive control system for a hybrid battery-ultracapacitor power source. *IEEE Transactions on Power Electronics* 29 (3):1469–1479.

Jayalakshmi, N. S., and D. N. Gaonkar. 2011. Performance study of isolated hybrid power system with multiple generation and energy storage units. In *2011 International Conference on Power and Energy Systems (ICPS)*. Chennai, India: IEEE.

Kottick, D., M. Blau, and D. Edelstein. 1993. Battery energy storage for frequency regulation in an island power system. *IEEE Transactions on Energy Conversion* 8 (3):455–459.

Kurian, S., S. T. Krishnan, and E. P. Cheriyan. 2015. Real time implementation of artificial neural networks-based controller for battery storage supported wind electric generation. *IET Generation, Transmission and Distribution* 9 (10): 937–946.

Lam, A. Y. S., K.-C. Leung, and V. O. K. Li. 2016. Capacity estimation for vehicle-to-grid frequency regulation services with smart charging mechanism. *IEEE Transactions on Smart Grid* 7 (1):156–166.

Lazarewicz, M. L., and J. A. Arseneaux. 2006. Status of pilot projects using flywheels for frequency regulation. *Power Engineering Society General Meeting, 2006*. Montreal, Canada: IEEE.

Lazarewicz, M. L., and A. Rojas. 2004. Grid frequency regulation by recycling electrical energy in flywheels. *Power Engineering Society General Meeting, 2004*. Denver, CO: IEEE.

Lazarewicz, M. L., and T. M. Ryan. 2010. Integration of flywheel-based energy storage for frequency regulation in deregulated markets. In *2010 IEEE Power and Energy Society General Meeting*. Minneapolis, MN: IEEE.

Lee, H., B. Y. Shin, S. Han, S. Jung, B. Park, and G. Jang. 2012. Compensation for the power fluctuation of the large scale wind farm using hybrid energy storage applications. *IEEE Transactions on Applied Superconductivity* 22 (3):5701904–5701904.

Li, X., S. Tan, J. Huang, Y. Huang, M. Wang, T. Xu, and X. Cheng. 2014. Optimal sizing for flywheel energy storage system-conventional generator coordination in load frequency control. In *2014 International Conference on Power System Technology (POWERCON)*. Chengdu, China: IEEE.

Liu, H., Z. Hu, Y. Song, and J. Lin. 2013. Decentralized vehicle-to-grid control for primary frequency regulation considering charging demands. *IEEE Transactions on Power Systems* 28 (3):3480–3489.

Lu, C.-F., C.-C. Liu, and C. J. Wu. 1995. Dynamic modelling of battery energy storage system and application to power system stability. *IEE Proceedings – Generation, Transmission and Distribution* 142 (4):429–435.

Lu, N., M. R. Weimar, Y. V. Makarov, F. J. Rudolph, S. N. Murthy, J. Arseneaux, and C. Loutan. 2010. Evaluation of the flywheel potential for providing regulation service in California. In *2010 IEEE Power and Energy Society General Meeting*. Minneapolis, MN: IEEE.

Massoud-Amin, S., and B. F. Wollenberg. 2005. Toward a smart grid: Power delivery for the 21st century. *IEEE Power and Energy Magazine* 3 (5):34–41.

Mathworks [Online]. Implement Generic Battery Model. 2016, Available: http://www.mathworks.com.au/help/physmod/powersys/ref/battery.html.

Mendis, N., K. M. Muttaqi, and S. Perera. 2014. Management of battery-supercapacitor hybrid energy storage and synchronous condenser for isolated operation of PMSG based variable-speed wind turbine generating systems. *IEEE Transactions on Smart Grid* 5 (2):944–953.

Mercier, P., R. Cherkaoui, and A. Oudalov. 2009. Optimizing a battery energy storage system for frequency control application in an isolated power system. *IEEE Transactions on Power Systems* 24 (3):1469–1477.

Morozumi, S. 2015. Japanese experience in energy storage for a distribution network with high-penetration renewable energy [Technology Leaders]. *IEEE Electrification Magazine* 3 (3):4–12.

Muhssin, M. T., L. M. Cipcigan, and Z. A. Obaid. 2015. Small microgrid stability and performance analysis in isolated island. In *2015 50th International Universities Power Engineering Conference (UPEC)*. United Kingdom: IEEE.

Navarro, G., J. Torres, P. Moreno-Torres, M. Blanco, M. Lafoz, D. Fooladivanda, C. Rosenberg, and S. Garg. 2015. Technology description and characterization of a low-cost flywheel for energy management in microgrids; energy storage and regulation: An analysis. In *2015 17th European Conference on Power Electronics and Applications* (EPE'15 ECCE-Europe); *IEEE Transactions on Smart Grid*. Geneva, Switzerland: IEEE.

Oudalov, A., D. Chartouni, and C. Ohler. 2007. Optimizing a battery energy storage system for primary frequency control. *IEEE Transactions on Power Systems* 22 (3):1259–1266.

Papaefthymiou, S. V., E. G. Karamanou, S. A. Papathanassiou, and M. P. Papadopoulos. 2010. A wind-hydro-pumped storage station leading to high RES penetration in the autonomous island system of Ikaria. *IEEE Transactions on Sustainable Energy* 1 (3):163–172.

Pezic, M., and V. M. Cedrés. 2013. Unit commitment in fully renewable, hydro-wind energy systems. In *2013 10th International Conference on the European Energy Market (EEM)*. Stockholm, Sweden: IEEE.

Poonpun, P., and W. T. Jewell. 2008. Analysis of the cost per kilowatt hour to store electricity. *IEEE Transactions on Energy Conversion* 23 (2):529–534.

Quinn, C., D. Zimmerle, and T. H. Bradley. 2012. An evaluation of state-of-charge limitations and actuation signal energy content on plug-in hybrid electric vehicle, vehicle-to-grid reliability, and economics. *IEEE Transactions on Smart Grid* 3 (1):483–491.

Rahimi-Eichi, H., U. Ojha, F. Baronti, and M. Y. Chow. 2013. Battery management system: An overview of its application in the smart grid and electric vehicles. *IEEE Industrial Electronics Magazine* 7 (2):4–16.

Ray, P., S. Mohanty, and N. Kishor. 2010. Dynamic modeling and control of renewable energy based hybrid system for large band wind speed variation. In *2010 IEEE PES Innovative Smart Grid Technologies Conference Europe (ISGT Europe)*. Gothenburg, Sweden: IEEE.

Rowen, W. I. 1983. Simplified mathematical representations of heavy duty gas turbines. *Transactions of the ASME. Journal of Engineering for Power* 105 (4):865–70.

Sakamoto, O., K. Yamashita, Y. Kitauchi, T. Nanahara, T. Inoue, T. Arakaki, and H. Fukuda. 2011. Improvement of a voltage-stabilizing control system for integration of wind power generation into a small island power system. In *2011 2nd IEEE PES International Conference and Exhibition on Innovative Smart Grid Technologies (ISGT Europe)*. Manchester: IEEE.

Serban, I., R. Teodorescu, and C. Marinescu. 2013. Energy storage systems impact on the short-term frequency stability of distributed autonomous microgrids, an analysis using aggregate models. *IET Renewable Power Generation* 7 (5):531–539.

Sharma, S., and B. Singh. 2011. Performance of voltage and frequency controller in isolated wind power generation for a three-phase four-wire system. *IEEE Transactions on Power Electronics* 26 (12):3443–3452.

Sharma, A., D. Srinivasan, and A. Trivedi. 2015. A decentralized multiagent system approach for service restoration using DG islanding. *IEEE Transactions on Smart Grid* 6 (6):2784–2793.

Shi, L., and M. L. Crow. 2008. Comparison of ultracapacitor electric circuit models. In *2008 IEEE Power and Energy Society General Meeting: Conversion and Delivery of Electrical Energy in the 21st Century*. Pittsburgh, PA: IEEE.

Sigrist, L., I. Egido, E. Lobato, and L. Rouco. 2015. Sizing and controller setting of ultracapacitors for frequency stability enhancement of small isolated power systems. *IEEE Transactions on Power Systems* 30 (4):2130–2138.

Sigrist, L., E. Lobato, and L. Rouco. 2013. Energy storage systems providing primary reserve and peak shaving in small isolated power systems: An economic assessment. *International Journal of Electrical Power and Energy Systems* 53:675–683.

Solanki, A., A. Nasiri, V. Bhavaraju, Y. L. Familiant, and Q. Fu. 2016. A new framework for microgrid management: Virtual droop control. *IEEE Transactions on Smart Grid* 7 (2):554–566.

Su, W., H. Eichi, W. Zeng, and M. Y. Chow. 2012. A survey on the electrification of transportation in a smart grid environment. *IEEE Transactions on Industrial Informatics* 8 (1):1–10.

Suvire, G. O., M. G. Molina, and P. E. Mercado. 2012. Improving the integration of wind power generation into AC microgrids using flywheel energy storage. *IEEE Transactions on Smart Grid* 3 (4):1945–1954.

Swierczynski, M., D. I. Stroe, A. I. Stan, R. Teodorescu, and D. U. Sauer. 2014. Selection and performance-degradation modeling of LiMO/Li Ti O and LiFePO/C battery cells as suitable energy storage systems for grid integration with wind power plants: An example for the primary frequency regulation service. *IEEE Transactions on Sustainable Energy* 5 (1):90–101.

Takahashi, R., and J. Tamura. 2008. Frequency control of isolated power system with wind farm by using flywheel energy storage system. In *18th International Conference on Electrical Machines*, ICEM 2008. Pattaya, Thailand: IEEE.

Thatte, A. A., F. Zhang, and L. Xie. 2011. Coordination of wind farms and flywheels for energy balancing and frequency regulation. In *2011 IEEE Power and Energy Society General Meeting*. Detroit, MI: IEEE.

Urtasun, A., E. L. Barrios, P. Sanchis, and L. Marroyo. 2015. Frequency-based energy-management strategy for stand-alone systems with distributed battery storage. *IEEE Transactions on Power Electronics* 30 (9):4794–4808.

Uzunoglu, M., and M. S. Alam. 2006. Dynamic modeling, design, and simulation of a combined PEM fuel cell and ultracapacitor system for stand-alone residential applications. *IEEE Transactions on Energy Conversion* 21 (3):767–775.

Vasconcelos, H., C. Moreira, A. Madureira, J. P. Lopes, and V. Miranda. 2015. Advanced control solutions for operating isolated power systems: Examining the Portuguese islands. *IEEE Electrification Magazine* 3 (1):25–35.

Vrettos, E. I., and S. A. Papathanassiou. 2011. Operating policy and optimal sizing of a high penetration RES-BESS system for small isolated grids. *IEEE Transactions on Energy Conversion* 26 (3):744–756.

Wandelt, F., D. Gamrad, W. Deis, and J. Myrzik. 2015. Comparison of flywheels and batteries in combination with industrial plants for the provision of primary control reserve. In *2015 IEEE Eindhoven PowerTech*. Eindhoven, Netherlands: IEEE.

Wu, D., F. Tang, T. Dragicevic, J. C. Vasquez, and J. M. Guerrero. 2014. Autonomous active power control for islanded AC microgrids with photovoltaic generation and energy storage system. *IEEE Transactions on Energy Conversion* 29 (4):882–892.

Xie, L., A. A. Thatte, and Y. Gu. 2011. Multi-time-scale modeling and analysis of energy storage in power system operations. In *2011 IEEE Energytech*. Cleveland, OH: IEEE.

Xue, J., Z. Yin, B. Wu, Z. Wu, and J. Li. 2009. Technology research of novel energy storage control for the PV generation system. *Power and Energy Engineering Conference*. APPEEC 2009. Asia-Pacific.

Yang, H., C. Y. Chung, and J. Zhao. 2013. Application of plug-in electric vehicles to frequency regulation based on distributed signal acquisition via limited communication. *IEEE Transactions on Power Systems* 28 (2):1017–1026.

Yang, W., J. E. Carletta, T. T. Hartley, and R. J. Veillette. 2008. An ultracapacitor model derived using time-dependent current profiles. In *51st Midwest Symposium on Circuits and Systems*. MWSCAS 2008. Knoxville, TN: IEEE.

Yilmaz, M., and P. T. Krein. 2013. Review of the impact of vehicle-to-grid technologies on distribution systems and utility interfaces. *IEEE Transactions on Power Electronics* 28 (12):5673–5689.

Zhang, F., M. Tokombayev, Y. Song, and G. Gross. 2014. Effective flywheel energy storage (FES) offer strategies for frequency regulation service provision. *Power Systems Computation Conference (PSCC)*. Detroit, MI: IEEE.

6

Weekly Unit Commitment Models of Island Power Systems

Due to the isolated nature and the small size of island power systems, most of them are operated under a centralized scheme. In a classical centralized scheme, generating units are programmed according to economic dispatch rules that take into account the security of supply issues. Generation is remunerated according to the standard costs of the different generation technologies. Unit commitment models over different time frames serve as the main tool for planning and operating purposes to analyze the economic impact of the different options that can be accomplished within an island power system.

The operation costs of island power systems are higher, not only because of expensive fuel transportation and lower efficiencies of power generation technologies (e.g., diesel), but also because of technical requirements on spinning reserves guaranteeing frequency stability. Power system operators of island grids keep a certain amount of generation capacity as spinning reserve to ensure that the island is able to withstand the sudden outage of any generating unit and also address unforeseen load variations.

In Section 6.1, spinning reserve constraints are discussed as the key factor driving the economic dispatch of an island power system. The traditional criterion for setting the minimum amount of spinning reserve requirements (that it should be greater than or equal to the capacity of the largest online generator) is also discussed. Section 6.1 also points out the importance of the deployment of interconnection links between two islands, which yields economic benefits for the operation of the system and also contributes to providing reserve to both connected islands, which increases the reliability of the overall multi-island power system.

In Section 6.2, a brief overview of unit commitment solution methods, unit commitment models targeted at increasing renewable penetration, and the latest research on unit commitment models, incorporating frequency dynamics criteria as a theoretical research line, will be presented. Section 6.2 justifies the proposal in this chapter to formulate in Section 6.3 a deterministic weekly unit commitment divided into hourly units to obtain the optimal operation of an island power system that satisfies the reserve requirements. An explicit formulation of the interconnection links between islands is included to take into account the reserve that the interconnectors provide to each island, and additional reserve constraints are formulated to ensure that each of the islands is able to withstand the loss of each connecting line. The unit commitment formulated in Section 6.3 reflects the common practice and security criteria of the majority of real island system operators when operating their systems. Even though the model is weekly (with a time horizon of 1 week) and deterministic (using demand and renewable as known fixed data), it can be used for a variety of mid-term and long-term island power system studies, as will be shown in Chapter 7.

To close the chapter, in Section 6.4, illustrative case studies of real Spanish island power systems of different size and complexity are proposed to show how the dispatch of generation of island power systems is implemented using the weekly unit commitment model. The small La Palma island power system, the medium-sized Lanzarote–Fuerteventura islands power system, and the large-sized Mallorca–Menorca islands power system are used to outline the importance of the spinning reserve requirements and island interconnections in island power systems.

6.1 Spinning Reserve Requirements

6.1.1 Overview of Spinning Reserve Requirements

The key technical issue that drives the economic operation of island power systems is spinning reserve requirements. Spinning reserve requirements displace cheaper units in favor of more expensive units and increase the start-up costs of generators.

To see how these spinning reserve requirements affect the economic operation, let us consider a fictitious island system comprising three units and just 1 h of operation. Table 6.1 shows the technical features of the units (maximum and minimum operating output) as well as the economic features (cost). Assuming a demand level of 120 MWh, if no reserve requirements are taken into account, the cheapest unit (UNIT1) will be programmed at its maximum power of 100 MW, and the second cheapest (UNIT2) will

TABLE 6.1

Economic Dispatch with No Reserve Requirements under Consideration

	UNIT1	UNIT2	UNIT3
*P*max (MW)	100	80	60
*P*min (MW)	20	15	10
Cost (€/MWh)	10	20	30
Demand (MWh)		120	
Production (MWh)	100	20	0
Up reserve (MW)	0	60	0
Down reserve (MW)	80	5	0
Cost (€/h)		1400	

TABLE 6.2

Economic Dispatch with Reserve Requirements Taken into Account

	UNIT1	UNIT2	UNIT3
*P*max (MW)	100	80	60
*P*min (MW)	20	15	10
Cost (€/MWh)	10	20	30
Demand (MWh)	120		
Production (MWh)	95	15	10
Up reserve (MW)	5	65	50
Down reserve (MW)	80	50	0
Cost (€/h)		1550	

produce the other 20 MW, yielding an overall hourly cost of 1400 €/h. The most expensive UNIT3 will remain unconnected. The system is not operating in a secure mode, since it is not able to withstand the outage of UNIT1 and UNIT2. If UNIT1 is lost, UNIT2 is not able to generate the lost power of 100 MW, since it has only 60 MW of reserve, and in the same way, if UNIT2 is lost, the power system will end in blackout.

Table 6.2 shows how the economic dispatch has to be modified to include spinning reserve requirements. UNIT3 has to be on, producing its minimum output and displacing cheaper units, since the up reserve that this provides to the system is required in the case of a contingency of UNIT1 or UNIT2. The overall operating cost increases to 1550 €/h, which represents an increase of 10.7% with respect to the unsecure unit commitment. The secure power system has not only a higher operating cost, but also a larger number of start-ups, with the associated start-up cost increase.

It is common practice among island system operators to establish a value of minimum spinning reserve requirements so as to be able to cover the loss of the largest generating unit (REE 2006; EDF 2007; TRANSPOWER 2009; Matsuura 2009). This restriction should be formulated within unit commitment models, not as a static minimum reserve amount equal for each hour, but as a dynamic value varying for each hour of the week, since the biggest connected unit varies with the evolution of the demand curve.

It should be noted that the covering of the outage of the largest connected unit does not guarantee in all cases that the outage of a smaller one will also be endured. This is because when a unit is lost, not only the produced power is lost, but also the reserve it provides. As an illustrative example of why reserve requirements must apply to all connected units, consider the fictitious system of Table 6.3. The system is able to withstand the outage of the biggest connected unit (UNIT1) with the reserve provided by UNIT2 and UNIT3. However, there is not enough system reserve to cover the loss of UNIT2, since it is generating 60 MW, and the reserve provided by UNIT1 and UNIT3 only sums up to 55 MW. Thus, when imposing within unit commitment models

TABLE 6.3

Illustrative Example of Why Reserve Requirements Must Apply to All Connected Units

	UNIT1	UNIT2	UNIT3
Pmax (MW)	100	105	60
Pmin (MW)	20	15	10
Cost (€/MWh)	10	20	30
Demand (MWh)		165	
Production (MWh)	95	60	10
Up reserve (MW)	5	45	50
Down reserve (MW)	80	53	0
Cost (€/h)		2450	

the condition that the system must be able to withstand the loss of every unit, it is important to exclude the reserve provided by the lost unit.

It should be noted that within unit commitment formulations found in the literature, either the reserve power of the lost unit is not excluded or the minimum spinning reserve is considered as a static fixed value for each hour (Ortega-Vazquez et al. 2007).

6.1.2 Island Interconnections

Electric interconnection by means of submarine cables (between islands or between the mainland and an isolated island) provides manifold benefits to isolated power systems: interconnections increase security of supply, reduce the island power system cost, and introduce flexibility for increasing the penetration of variable renewable sources (EUROELECTRIC 2012). Interconnection links provide an alternative to missing or costly storage, and islands can take advantage from power flow traveling in both directions.

The cost of cable interconnections can also be cheaper than that of conventional diesel generation. Typically, alternating current (ac) submarine transmission becomes more difficult for distances over 50 km, while beyond 100 km, direct current (dc) submarine interconnection becomes the preferred technology. Islands close to the mainland are often already interconnected, while remoter islands cannot afford it. An alternative for remote islands is to search for potential connections between islands. Electric island interconnections, both to the mainland and between isolated remote islands, are widely deployed all over the world.

Concerning spinning reserve requirements, submarine interconnection links between islands (both high-voltage direct current [HVDC] and ac) provide up reserve into the connected islands, reducing the need for online thermal units' reserve. However, the economic dispatch of interconnected islands must be formulated including reserve constraints guaranteeing that the islands are able to withstand the loss of each connecting line.

6.2 Island Power Systems Unit Commitment Models

Different solution methods have been proposed for unit commitment models within the literature (Hobbs et al. 2002; Padhy 2004): exhaustive enumeration, priority listing, dynamic programming, mixed-linear integer programming (MILP), Lagrangian relaxation, interior point methods, tabu search, expert systems, neural networks, simulated annealing, particle swarm optimization, fuzzy systems, and genetic algorithms. Of this vast list of solution methods, MILP is generally preferred by researchers. The existence of robust and efficient solvers of MILP problems lies behind this decision. The widespread General Algebraic Modeling Systems (GAMS) modeling language (GAMS 2015) or new options such as Julia for mathematical programming (JuMP) modeling language (JuMP 2015) are used to write in an efficient way and interact with the available commercial solvers (Bonmin, Cbc, Clp, Couenne, CPLEX, ECOS, GLPK, Gurobi, Ipopt, KNITRO, MOSEK, NLopt, Conopt).

Renewable energy sources (RES) offer an interesting solution to decrease the dependency on fossil fuels and increase island sustainability. For these reasons, extensive research has been done on methods and studies aimed at increasing RES penetration of electric power systems, for which both stochastic and deterministic approaches on mid-term and long-term time frames have been proposed in the literature (Ortega-Vazquez et al. 2009). The impact of wind on reserve requirements is of current interest to researchers and system operators. State-of-the-art reviews can be found in Holttinen et al. (2012) and Kiviluoma et al. (2012). Studies on specific countries, such as the Netherlands (Parsons et al. 2004), Quebec (Robitaille et al. 2012), the United States (Parsons et al. 2004), and Spain (Fernandez-Bernal et al. 2015), are available in the literature. The effect of wind on island reserves has also been studied in specific island power systems such as the United Kingdom (Strbac et al. 2007), Ireland (Meibom et al. 2011), and Crete (Hansen and Papalexopoulos 2012). Concerning short-term weekly operation, the current practice of island system operators uses the best available demand and wind forecasts as fixed data; a deterministic weekly unit commitment serves as the basic tool to obtain the optimal thermal dispatch.

A new line of research considers frequency constraints within the unit commitment of island power systems, arguing that due to the small size and small inertia of island power systems, dynamic evolution after a generator disturbance can cause inadmissible primary steady-state frequencies and/or dangerous minimum frequencies during the transient part of the system dynamics after a contingency. For instance, constraints to limit primary steady-state frequency are formulated (Restrepo and Galiana 2005), a mixed unit commitment and grid simulation is found (Lei et al. 2001), reserve requirements and system dynamics are coupled using load–frequency sensitivity indexes (Chang et al. 2013), and approaches to control the minimum frequency of the island power system are formulated (Sokoler et al. 2015).

This is an interesting research frontier, but can be regarded as a theoretical line of work that has so far not been applied in practice by island system operators. As already mentioned, common practice establishes a minimum amount of spinning reserve, guaranteeing that an island power system is able to withstand the loss of the largest generating unit. The fulfillment of this criterion guarantees that after the manual or automatic activation of secondary reserve, no deviation of frequency will occur, and thus, the steady-state error of the frequency after the primary reserve regulation is not considered as relevant. Concerning the minimum dynamic value of frequency, island system operators design load-shedding plans to maintain frequency well above admissible limits (Sigrist 2015). Further research should evaluate whether the increase in system cost of the unit commitment constrained by frequency restrictions compensates for the potential avoidance of load shedding that protects island power systems, and thus, whether island system operators will be willing to update their actual unit commitment models with frequency constraints.

6.3 Mathematical Formulation of the Unit Commitment

This section details the mathematical MILP of an hourly unit commitment on a weekly time frame. The model includes the possible interconnection links between two islands and the key reserve constraints that each island power system must fulfill for security reasons. The start-up cost is a key factor, which may in fact determine the unit commitment results, and this takes different values depending of the type of start-up (hot, mild, or cold). Thus, the model formulation includes a detailed representation of the start-up and shut-down processes of thermal units, taken from the tight and compact formulation of the self-thermal unit commitment problem defined (Morales-Espana et al. 2013), showing better computational performance than other possible formulations of ramping processes, as proved by the authors.

Combined-cycle plants (CCGTs) are used in some islanded power systems (usually the larger ones). The inclusion of CCGTs within unit commitment is complex, since they can be operated in different operating configuration modes based on the number of gas and steam turbines. For instance, the islanded power system of the Spanish Mallorca–Menorca islands is equipped with several 2gas + 1steam CCGTs and one 3gas + 1steam CCGT. Depending on the desired use (operation/planning/investment studies or, similarly, short-term/mid-term/long-term studies) and the accuracy of the unit commitment, CCGTs may be modeled with different ways of increasing complexity and computational burden. The simplest is *aggregated modeling*, in which a CCGT corresponds to a pseudo-unit that is treated as a regular thermal unit ignoring the different operating configurations used, for instance, by several non-islanded power systems. The second is *component or physical unit modeling*, in which each physical component of the CCGT, that is, each gas and steam turbine, is modeled with its technical characteristics. Finally, the most complex formulation is *configuration-based or mode modeling*, which models the CCGT using multiple and mutually exclusive configurations (or modes of operation). Each of the modes has its own technical characteristics and follows a different transition path. Taking into account that the focus of the formulation is the key reserve constraints that drive the operation of island power systems, and also that many islands do not have CCGTs, the latest research (Morales-España et al. 2016) (which is recommended for the accurate configuration-based modeling of CCGTs in preference to other options [Liu et al. 2009; Troy et al. 2012]) is not included in this chapter. Instead, a simplified formulation of configuration-based CCGTs that ignores the transition paths between modes is included to model islands with CCGT units.

This section starts with the definition of the nomenclature of the model, continues with the definition of the objective function, and ends with the definition of the constraints: demand balance constraint, thermal units' technical constraints, thermal units' commitment and start-up/shut-down constraints, interconnection links constraints, system reserve constraints, and combined-cycle unit constraints.

6.3.1 Nomenclature

This subsection details the main nomenclature used in the mathematical formulation of the optimization problem. For clarification purposes, symbols are classified in sets, parameters (represented by uppercase letters), and variables—binary and continuous (represented in lowercase letters).

6.3.1.1 Sets

- g, gg generator thermal unit (1 to N_g)
- h, hh hour (1–168)
- st start-up type (hot, mild, cold)
- i, ii, iii island power system (1 to N_i)
- cc combined cycle unit (1 to N_{cc})

6.3.1.2 Parameters

- C_g^{fix} fixed cost of unit g (€)
- C_g^{lin} linear component of the variable cost of unit g (€/MWh)
- C_g^{qua} quadratic component of the variable cost of unit g (€/MWh^2)

- $C_{g,st}^{\text{start-up}}$ start-up cost of type *st* of generator *g* (€)
- $C_{g}^{\text{shut-down}}$ shut-down cost of unit *g* (€)
- $D_{i,h}$ demand of island *i* in hour *h* (MW)
- $Res_{i,h}$ renewable energy production of island *i* in hour *h* (MW)
- $P_{g}^{\min}$ minimum power generation of unit *g* (MW)
- $P_{g}^{\max}$ maximum power generation of unit *g* (MW)
- R_{g}^{up} ramp-up of unit *g* (MW/h)
- R_{g}^{down} ramp-down of unit *g* (MW/h)
- $Minhdown_{g,st}$ number of hours that a unit must be off to consider a start-up of -type *st* (No.)
- $Minhup_{g}$ minimum number of hours that a unit must be over $P_{g}^{\min}$ if a unit is started up (No.)
- $Exp_{i,ii}^{\max}$ maximum transfer capacity between islands *i* and *ii* (MW)
- $Resloss_{i}$ percentage of renewable energy of island *i* that must be covered with up reserve (%)
- $Resdown_{i}^{\min}$ minimum amount of down reserve established for island *i* (MW)

6.3.1.3 Continuous Variables

- $c_{g,h}$ operation cost of unit *g* in hour *h* (€/MWh)
- $c_{g,h}^{\text{start-up}}$ start-up cost of unit *g* in hour *h* (€)
- $c_{g,h}^{\text{shut-down}}$ shut-down cost of unit *g* in hour *h* (€)
- $p_{g,h}$ thermal power generation of unit *g* in hour *h* (MW)
- $p_{g,h}^{\text{overpmin}}$ thermal power generation over $P_{g}^{\min}$ of unit *g* in hour *h* (MW)
- $exp_{i,ii,h}$ power export leaving island *i* into island *ii* in hour *h* (MW)
- $resgen_{g,h}^{up}$ up reserve provided by thermal generating unit *g* in hour *h* (MW)
- $resgen_{g,h}^{\text{down}}$ down reserve provided by thermal generating unit *g* in hour *h* (MW)
- $reslink_{i,ii,h}^{up}$ up reserve provided by island *i* into island *ii* through interconnection links in hour *h* (MW)
- $reslink_{i,ii,h}^{\text{down}}$ down reserve provided by island *i* into island *ii* through interconnection links in hour *h* (MW)

6.3.1.4 Binary Variables

- $cx_{g,h}$ start-up decision of unit *g* in hour *h* (0/1)
- $dx_{g,h}$ shut-down decision of unit *g* in hour *h* (0/1)

- $\delta_{g,h}$ state of unit g in hour h (0/1)
- $cx\_st_{g,h,st}$ start-up decision of type st of unit g in hour h (0/1)

6.3.2 Objective Function

The objective function is formulated as the minimization of the total thermal generation cost, including the start-up, shut-down, and operation costs of thermal units:

$$\min \sum_{g,h} \left(c_{g,h} + c_{g,h}^{\text{start-up}} + c_{g,h}^{\text{shut-down}} \right) \tag{6.1}$$

The thermal operation cost is a quadratic function of the thermal generated power:

$$c_{g,h} = C_g^{\text{fix}} \cdot \delta_{g,h} + C_g^{\text{lin}} \cdot p_{g,h} + C_g^{\text{qua}} \cdot p_{g,h}^2, \qquad \forall g,h \tag{6.2}$$

where the total power of the thermal unit is defined as a function of the power generation over $P_g^{\min}$:

$$p_{g,h} = P_g^{\min} \cdot \delta_{g,h} + p_{g,h}^{\text{over } p\min} \qquad \forall g,h \tag{6.3}$$

It should be noted that the convex quadratic function of the operating cost defined in Equation 6.2 must be translated to a set of linear constraints formed as tangent cuts in a predefined number of generation points of the curve, as detailed in Lobato Miguelez et al. (2007).

The start-up cost is selected according to the start-up type using the start-up decision of the variable:

$$c_{g,h}^{\text{start-up}} = \sum_{st} \left(C_{g,st}^{\text{start-up}} \cdot cx\_st_{g,h,st} \right) \qquad \forall g,h \tag{6.4}$$

The shut-down cost is computed using the shut-down decision binary variable:

$$c_{g,h}^{\text{shut-down}} = C_g^{\text{shut-down}} \cdot dx_{g,h} \qquad \forall g,h \tag{6.5}$$

6.3.3 Constraints

6.3.3.1 Demand Balance Constraint

Concerning the demand balance of an island i, the total power generation (thermal units and renewable energy production) must be equal to total load plus the exports to interconnected islands:

$$\sum_{g \in i} (p_{g,h}) + Res_{i,h} = D_{i,h} + \sum_{ii} (exp_{i,ii,h}) \quad \forall h,i \tag{6.6}$$

6.3.3.2 Thermal Unit Technical Constraints

The maximum thermal power generation over $P_g^{\min}$ of unit g is limited using

$$p_{g,h}^{\text{over}\,p\min} \leq \left(P_g^{\max} - P_g^{\min}\right) \cdot \left(\delta_{g,h} - dx_{g,h+1} - cx_{g,h}\right) \quad \forall g,h \tag{6.7}$$

It should be noted that Equation 6.3 imposes that the unit g produces exactly $P_g^{\min}$ when the unit ends the start-up process or begins the shut-down process (Morales-Espana et al. 2013). Thermal generation increase or decrease between two consecutive hours must fulfill ramping constraints:

$$-R_g^{\text{down}} \leq p_{g,h}^{\text{over}\,p\min} - p_{g,h-1}^{\text{over}\,p\min} \leq R_g^{up} \quad \forall g,h \tag{6.8}$$

6.3.3.3 Thermal Unit Commitment, Start-Up, and Shut-Down Constraints

The condition that only one type of start-up (hot, mild, or cold) can be activated in each hour is formulated as

$$cx_{g,h} = \sum_{st} (cx\_st_{g,h,st}) \quad \forall g,h \tag{6.9}$$

Next, for each thermal unit, the type of start-up is selected according to the number of hours that the unit must be down to consider the start-up hot ($st=1$), mild ($st=2$), or cold ($st=3$):

$$cx\_st_{g,h,st} \leq \sum_{hh=Minhdown_{g,st}}^{Minhdown_{g,st+1}} (dx_{g,h-hh}) \quad \forall g,h,st \tag{6.10}$$

It should be noted in Equation 6.10 that $Minhdown_{g,hot} < Minhdown_{g,mild} < Minhdown_{g,cold}$.

The minimum number of hours that the unit must be over $P_g^{\min}$ is imposed by

$$\sum_{hh=h-Minhup_g+1}^{h} (cx_{g,hh}) \leq \delta_{g,h} \quad \forall g,h \tag{6.11}$$

The relationship between start-up, shut-down, and state binary variables is formulated as

$$\delta_{g,h} - \delta_{g,h-1} = cx_{g,h} - dx_{g,h} \quad \forall g,h \tag{6.12}$$

The minimum number of hours that the unit must be down before being started up again is formulated as

$$\sum_{hh=h-Minhdown_{g,\text{hot}}+1}^{h} \left(dx_{g,hh}\right) \leq 1-\delta_{g,h} \quad \forall g,h \tag{6.13}$$

6.3.3.4 Interconnection Links Constraints

Power exports between islands are limited by the capacity limits of the associated interconnection lines:

$$-Exp_{i,ii}^{\text{max}} \leq exp_{i,ii,h} \leq Exp_{i,ii}^{\text{max}} \quad \forall i,ii,h \tag{6.14}$$

The export of island i into island ii must be equal to the import of island ii from island i:

$$exp_{i,ii,h} = -exp_{ii,i,h} \quad \forall i,ii,h \tag{6.15}$$

6.3.3.5 System Reserve Constraints

The up and down reserves provided by thermal generating units are computed as

$$resgen_{g,h}^{\text{up}} = \left(P_g^{\text{max}} - P_g^{\text{min}}\right)\cdot \delta_{g,h} - p_{g,h}^{\text{over}\,p\text{min}} \quad \forall g,h \tag{6.16}$$

$$resgen_{g,h}^{\text{down}} = p_{g,h}^{\text{over}\,p\text{min}} \quad \forall g,h \tag{6.17}$$

Concerning the island interconnections, it should be noted that, the export from island ii into island i being $exp_{ii,i,h}$, island ii is able to increase the export, providing an up reserve to island i of value $(Exp_{ii,i}^{\text{max}} - exp_{ii,i,h})$, provided that the generators of island ii together with the other island ii interconnection links have enough up reserve. Thus, interconnection reserve is defined by the following inequality constraints:

$$reslink_{ii,i,h}^{\text{up}} \leq Exp_{ii,i}^{\text{max}} - exp_{ii,i,h} \quad \forall ii,i,h \tag{6.18}$$

$$reslink_{ii,i,h}^{\text{up}} \leq \sum_{g\in ii}\left(resgen_{g,h}^{\text{up}}\right) + \sum_{\substack{iii\neq i\\ iii\neq ii}}\left(reslink_{iii,ii,h}^{up}\right) \; \forall ii,i,h \tag{6.19}$$

Similarly, if island ii changes from export to import, the down reserve provided to island i is limited by $(Exp_{i,ii}^{\text{max}} + \text{exp}_{ii,i,h}) = (\text{Exp}_{i,ii}^{\text{max}} - \text{exp}_{i,ii,h})$, meaning that island ii stops exporting and imports the maximum amount. In addition, it is limited by the down reserve of generators belonging to island *ii* plus the down reserve of the other interconnection links of island *ii*:

$$reslink_{ii,i,h}^{\text{down}} \leq Exp_{i,ii}^{\max} - exp_{i,ii,h} \quad \forall ii,i,h \tag{6.20}$$

$$reslink_{ii,i,h}^{\text{down}} \leq \sum_{g \in ii} \left(resgen_{g,h}^{\text{down}}\right) + \sum_{\substack{iii \neq i \\ iii \neq ii}} \left(reslink_{iii,ii,h}^{\text{down}}\right) \quad \forall ii,i,h \tag{6.21}$$

Each island must have enough up reserve to withstand the loss of every generating unit *g*. When unit *g* belonging to an island *i* is lost, the total system up reserve is computed as the reserve provided by the rest of the generating units belonging to the same island plus the up reserve provided by each interconnection link with other islands. The reserve of the lost unit is excluded when formulating this constraint, as stated in Section 6.1.1. Thus, for every unit *g* belonging to every island *i*, the system up reserve constraint is expressed as follows:

$$\sum_{\substack{gg \in i \\ gg \neq g}} \left(resgen_{gg,h}^{up}\right) + \sum_{ii \neq i} \left(reslink_{ii,i,h}^{up}\right) \geq p_{g,h} \quad \forall i, g \in i, h \tag{6.22}$$

Similarly, the island must be able to withstand the potential loss of renewable energy. In this case, all generators belonging to the island and the island interconnections contribute to the up reserve. $Resloss_i$ being the fraction of renewable energy of island *i* that must be covered by up reserve, this constraint is formulated as

$$\sum_{g \in i} \left(resgen_{g,h}^{up}\right) + \sum_{ii \neq i} \left(reslink_{ii,i,h}^{up}\right) \geq Resloss_i \cdot \text{Res}_{i,h} \quad \forall i,h \tag{6.23}$$

Finally, every island *i* must be able to withstand the loss of the interconnection with a neighboring island *ii*. In this case, all generators belonging to the island and the rest of the connected islands contribute to the reserve. This constraint is formulated as

$$\sum_{g \in i} \left(resgen_{g,h}^{up}\right) + \sum_{\substack{iii \neq i \\ iii \neq ii}} \left(reslink_{iii,i,h}^{up}\right) \geq exp_{ii,i,h} \quad \forall i,ii,h \tag{6.24}$$

Concerning down reserve, the island must guarantee a minimum amount defined by the island power system operator:

$$\sum_{g \in i} \left(resgen_{g,h}^{\text{down}}\right) + \sum_{ii \neq i} \left(reslink_{ii,i,h}^{\text{down}}\right) \geq Resdown_i^{\min} \quad \forall i,h \tag{6.25}$$

6.3.3.6 Combined-Cycle Thermal Unit Constraints

CCGTs are modeled in a simplified version of the configuration-based formulation that ignores the transition paths between modes. For instance, a 2+1 CCGT is modeled by six different generators, each one corresponding to a mode: (1) gas turbine GT1, (2) gas turbine GT2, (3) gas turbines GT1 and GT2, (4) gas turbine GT1 and steam turbine ST1, (5) gas turbine GT2 and steam turbine ST1, (6) full gas/steam turbines GT1, GT2 and ST1.

Exclusion constraints are formulated for the binary decisions of the commitment, start-up, and shut-down decisions of the different modes that belong to the CCGT:

$$\sum_{g \in cc} \left(cx_{g,h} \right) \leq 1 \quad \forall cc, h \tag{6.26}$$

$$\sum_{g \in cc} \left(dx_{g,h} \right) \leq 1 \quad \forall cc, h \tag{6.27}$$

$$\sum_{g \in cc} \left(\delta_{g,h} \right) \leq 1 \quad \forall cc, h \tag{6.28}$$

As explained in Section 6.3, the formulation (Morales-España et al. 2016) is recommended for the accurate configuration-based modeling of CCGTs.

6.4 Case Studies

The performance of the unit commitment will be illustrated with three real Spanish island power systems of different size and complexity (small, medium, and large): (1) the La Palma island power system, (2) the Lanzarote–Fuerteventura interconnected islands, and (3) the Mallorca–Menorca interconnected islands*.

The first two island power systems belong to the Spanish Canary archipelago, while the last belong to the Spanish Balearic archipelago. The three case studies will emphasize the impact of reserve constraints on cost and operation of an island power system. In addition, the benefits of interconnection links between islands will be clearly demonstrated.

* It should be noted that a new link between the Mallorca–Menorca and Ibiza–Formentera islands has recently been deployed and is being tested from the beginning of 2016, forming a group of four interconnected island power systems. In addition, there are plans to upgrade the existing link between Mallorca and Menorca, and also between the Lanzarote and Fuerteventura islands, and additionally to interconnect the Lanzarote–Fuerteventura system with the Gran Canaria island power system.

6.4.1 La Palma Case Study

6.4.1.1 Description of Case Study

The La Palma power system is a small system supplied by 11 diesel thermal groups. La Palma island has 96 MW of thermal installed capacity, a peak demand of 58 MW, and an amount of renewable installed capacity of 12 MW. Figure 6.1 displays the maximum and minimum limits of the thermal units of La Palma island. It should be noted that even though all are diesel fueled, the groups are very different, not only in size and costs, but also in the amount of up reserve that they are able to provide if they are connected into the system.

The unit commitment is illustrated with the operation for 1 week in January 2015. Figure 6.2 shows the evolution of the demand and renewable production throughout the week, with a peak demand of 38.8 MW and a maximum penetration of renewable sources of 22.7% (in hour *h037*). It should be noted that the first day of the week corresponds to Saturday, so that the week ends on Friday.

6.4.1.2 Unit Commitment Results

The unit commitment model imposing the system reserve constraints (base case) was run, yielding 30 thermal start-up decisions within the week. Figure 6.3 shows the distribution of the start-up orders throughout the week, varying the number of connected units in each hour.

To show the influence of reserve constraints, the unit commitment has been rerun with no reserve constraints (no RC case), that is, removing

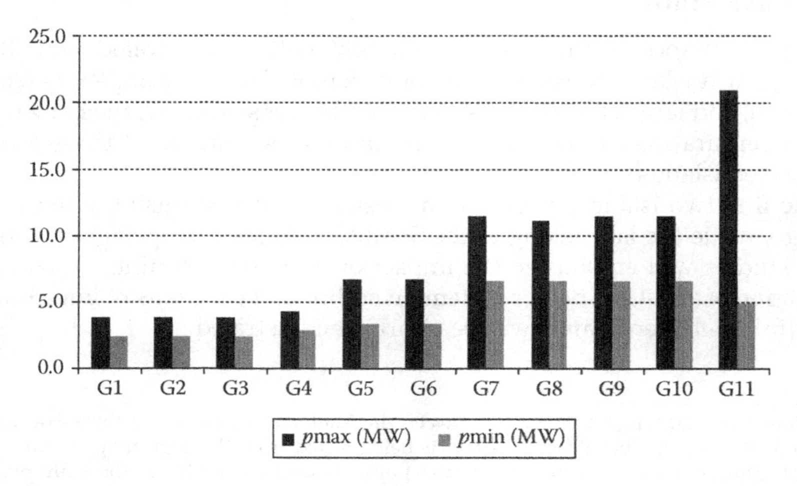

FIGURE 6.1
Maximum and minimum power of the thermal groups belonging to La Palma island power system.

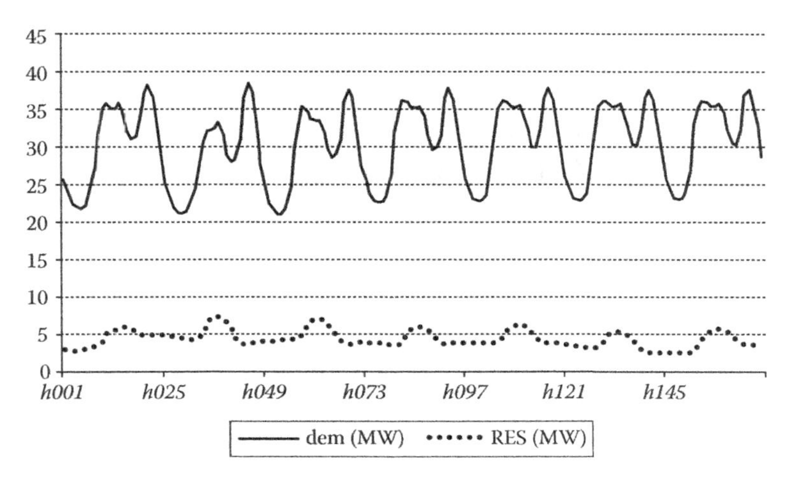

FIGURE 6.2
Demand and renewable energy in La Palma power system.

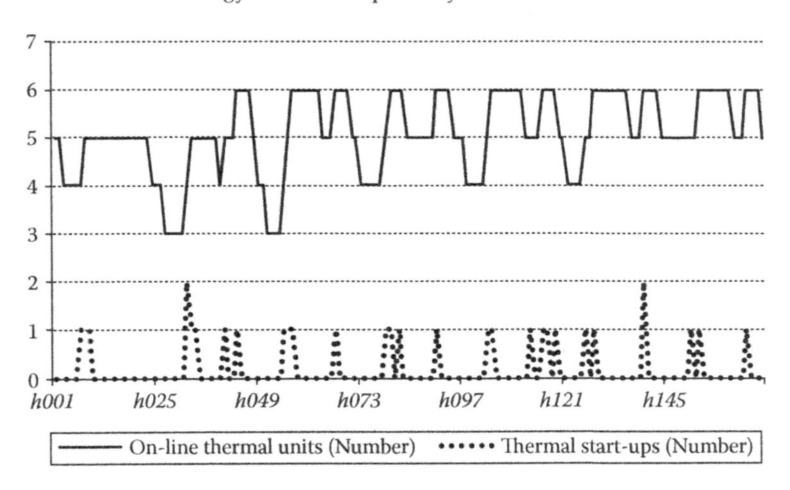

FIGURE 6.3
Number of La Palma online thermal units and thermal start-up decisions in base case (reserve constraints imposed).

Equations 6.22 through 6.25. Figure 6.4 clearly shows how the island power system requires more thermal units connected for a significant number of hours when reserve constraints have to be fulfilled. In Figure 6.5, the same conclusion is illustrated, showing the increase of start-up decisions that the imposition of reserve constraints created within the island. Since La Palma island has no interconnection links to other islands, the system up reserve is entirely provided by thermal generating units. Figure 6.6 compares the amount of the island up reserve in both cases, and in Figure 6.7, the biggest connected unit is displayed in each hour in both optimizations. The no RC

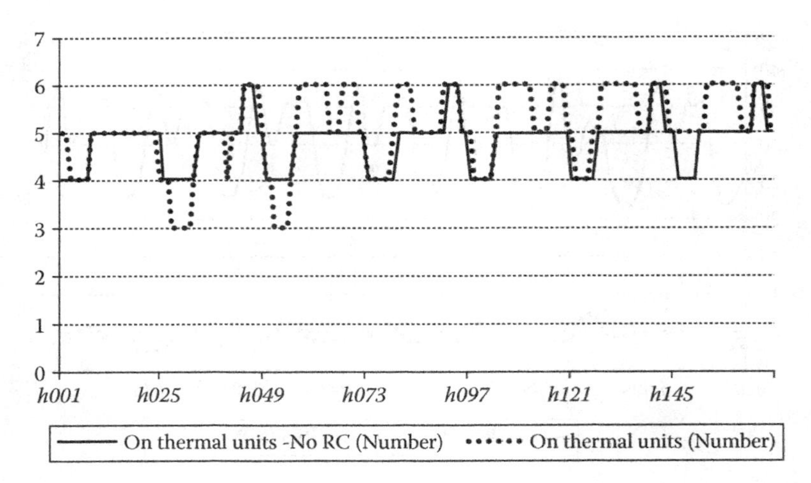

FIGURE 6.4
Comparison of La Palma thermal units in base case and no RC case.

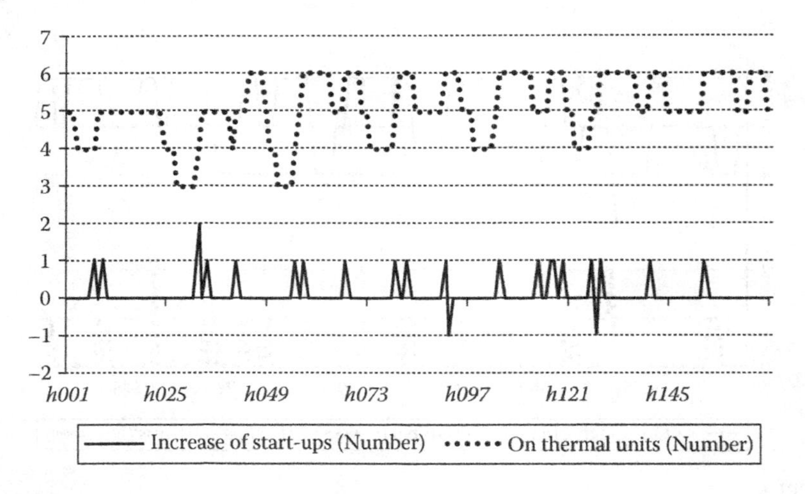

FIGURE 6.5
Increase of start-ups in La Palma in base case with respect to no RC case.

case uses G7, G8, and G9 units at full capacity (since the cost efficiency is higher when the unit is closer to Pmax), undermining the amount of system reserve. In fact, when Figures 6.6 and 6.7 are jointly analyzed, it can be seen that in the no RC case, the system is not able to withstand the loss of the biggest connected units for the majority of the week, and thus, the system is not operated in a reliable way.

Table 6.4 summarizes the impact of reserve constraints on the operation of the island power system. Reserve constraints increase the island power system cost by 5.4%, increasing both start-up costs and operation costs. The number of start-up cost decisions increases from 11 to 30, causing a 54.5%

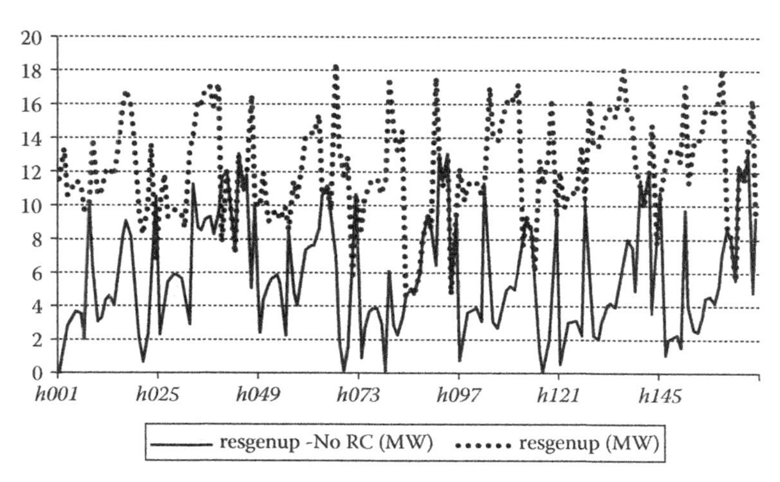

FIGURE 6.6
Comparison of La Palma up reserve in base case and no RC case.

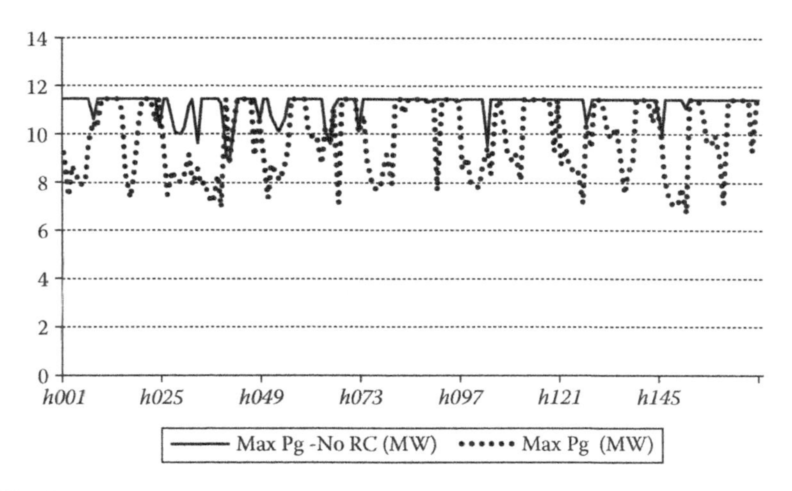

FIGURE 6.7
Greatest connected unit in base case and no RC case.

TABLE 6.4
Comparison of Base Case and No RC Case in La Palma Case Study

	Base Case	No RC Case	Difference	Difference (%)
Total cost (k€)	461.1	436.1	25.0	5.4
Operation cost (k€)	446.8	429.6	17.2	3.9
Startup cost (k€)	14.3	6.5	7.8	54.5
Number of start-ups	30	11	19	63.3
Resgenup (MW)	2031.8	966.1	1065.7	52.5

increase in the start-up costs. However, the system up reserve is increased by 52.5%.

6.4.2 Lanzarote–Fuerteventura Case Study

6.4.2.1 Description of Case Study

Lanzarote island (LZ) is interconnected with Fuerteventura (FV) island by an AC submarine link of 14.5 km and 30 MW of interconnection capacity, forming the island power system of Lanzarote–Fuerteventura (LZFV). LZ is supplied by 13 diesel groups and FV by 12 diesel units. LZ has 205 MW of thermal installed capacity, a peak demand of 166 MW, and an amount of renewable installed capacity of 58 MW. FV has 160 MW of thermal installed capacity, a peak demand of 136 MW, and an amount of renewable installed capacity of 51 MW. Figure 6.8 displays the maximum and minimum limits of the thermal units of the interconnected system, showing the variety of sizes among the diesel turbines.

The unit commitment is illustrated with the operation for 1 week in January 2015.

Figure 6.9 shows the evolution of the demand and renewable production throughout the week in the LZFV power system, with a peak demand of 219.7 MW. Compared with the La Palma power system, the renewable energy penetration is less significant, presenting a maximum of 9.6% in hour *h158*.

6.4.2.2 Unit Commitment Base Case Results

First, the results of the unit commitment model imposing the system reserve constraints (base case) are presented. Figure 6.10 shows the distribution of

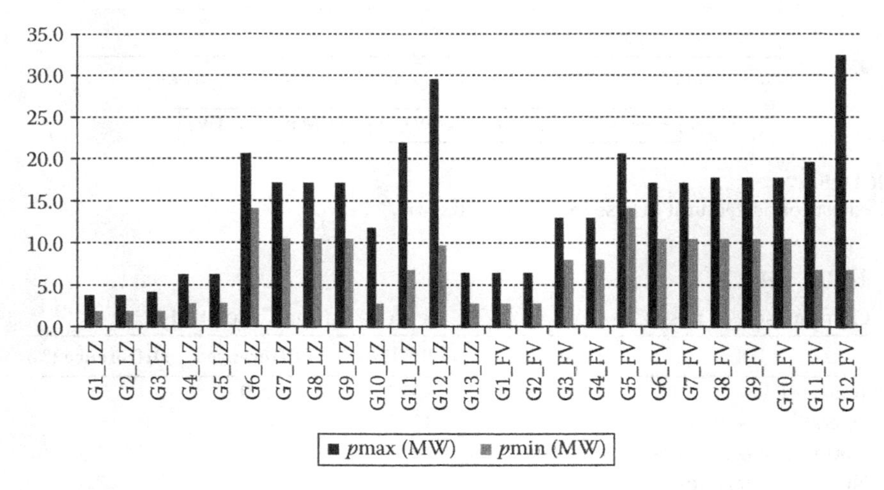

FIGURE 6.8
Maximum and minimum power of thermal groups belonging to LZFV power system.

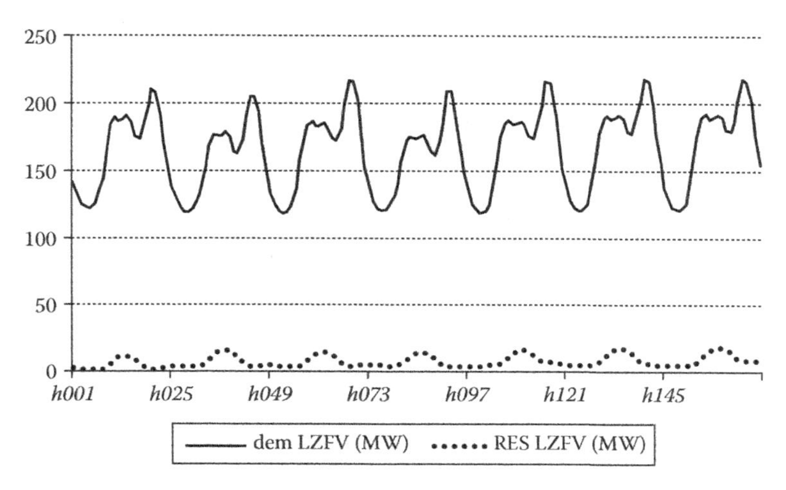

FIGURE 6.9
Demand and renewable energy in LZFV system.

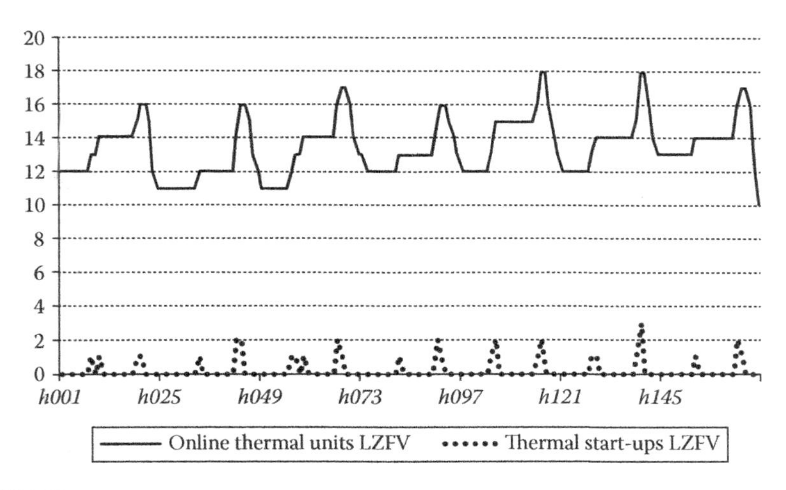

FIGURE 6.10
Number of LZFV online thermal units and thermal start-up decisions in base case.

the start-up orders and the number of thermal connected units throughout the week. The number of online thermal units oscillates between 11 and 18.

Considering LZ, the total island up reserve is formed by the reserve provided by the thermal units of LZ and the reserve provided by the interconnection link with FV. Figure 6.11 shows how both terms play a significant role in the overall island up reserve. It should be noted that the reserve provided by the AC link in LZ (reslinkup LZ term of Figure 6.11) is limited by the maximum export from FV to LZ and the amount of thermal up reserve in FV (Equations 6.18 and 6.19); Figure 6.12 shows that in the case of LZ,

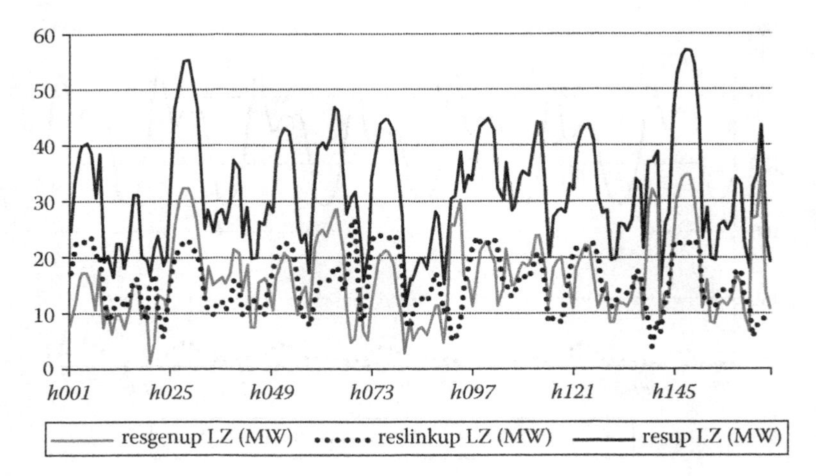

FIGURE 6.11
Total up reserve of LZ: sum of reserve of LZ generating units and interconnection link with FV.

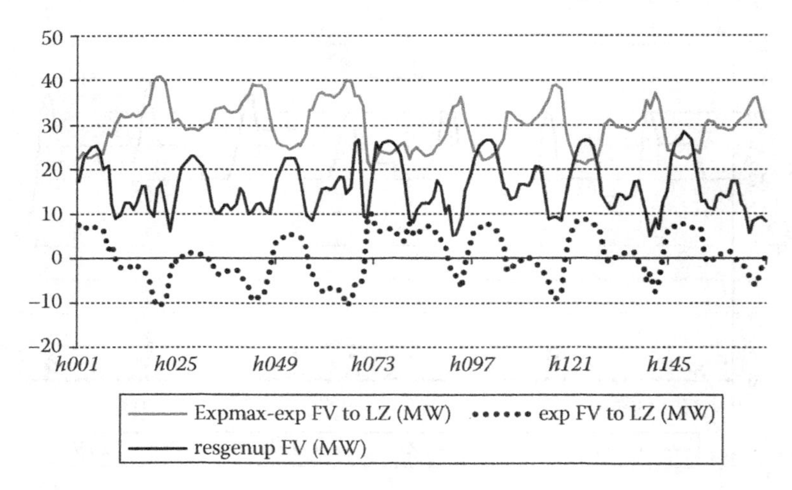

FIGURE 6.12
Up reserve link limitation in LZ.

the limiting term in the majority of the hourly periods corresponds to the amount of thermal up reserve in FV.

In Figure 6.13, it can be seen that LZ is able to withstand the loss of the interconnection with FV (Equation 6.24) in every hour; that is, the thermal units of LZ units provide an up reserve greater that the actual value of the export from FV to LZ. It should be noted that the link is used in both ways (LZ to FV and FV to LZ) during the week. Similarly, Figure 6.14 shows how the island is able to withstand the loss of the greatest connected unit (Equation 6.22); the reserve provided by the rest of the generating units

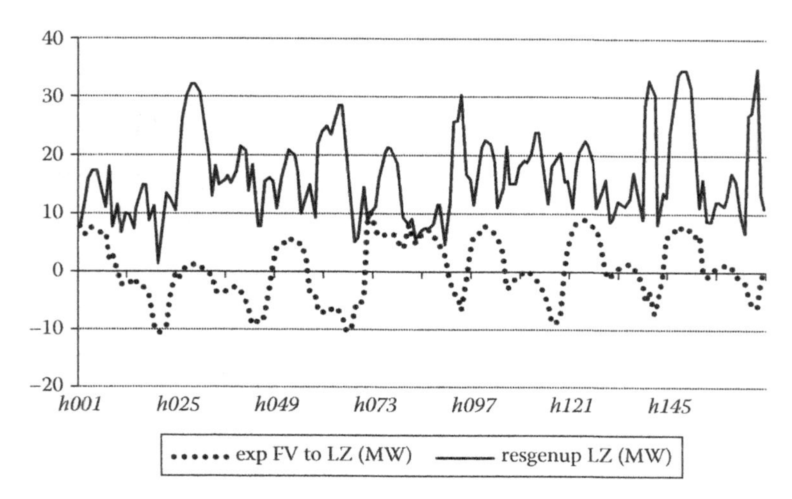

FIGURE 6.13
Export from FV to LZ and up reserve of LZ thermal units.

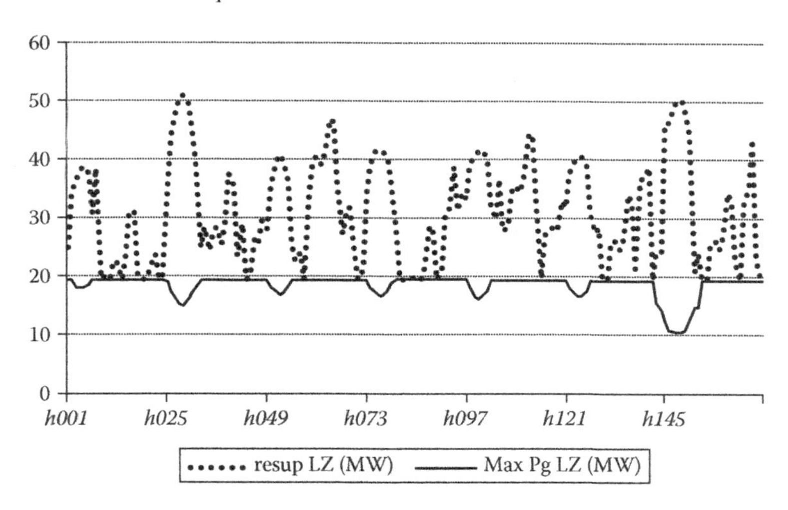

FIGURE 6.14
Total up reserve (excluding reserve of biggest unit) of LZ and maximum thermal connected unit in LZ.

and the interconnection is greater than the power produced by the biggest connected unit. Although this is not shown, reliable operation of FV is also achieved by the unit commitment.

6.4.2.3 Impact of Reserve Constraints and Interconnection Link

The impact of reserve constraints is assessed by running the unit commitment after removing the reserve constraints (no RC case). In addition, the

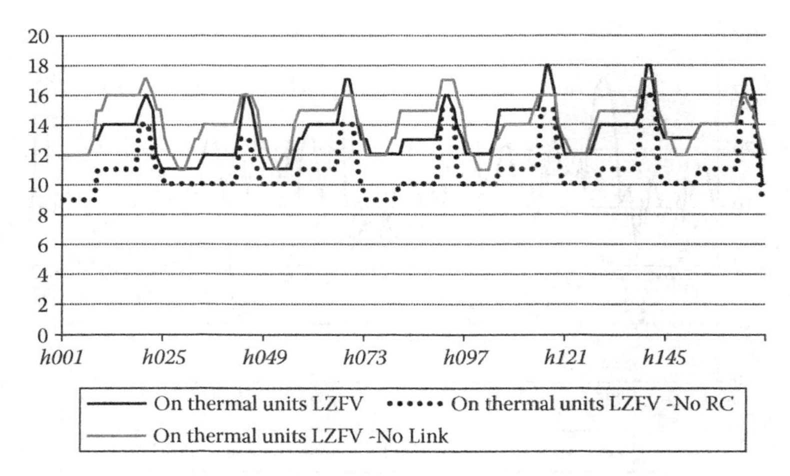

FIGURE 6.15
Thermal units in LZFV in base case, no RC case, and no link case.

technical and economic impact of the interconnection link is verified by running the unit commitment including reserve constraints but removing the interconnection capacity between the islands (no link case). Figure 6.15 illustrates the number of connected units in the three simulation conditions. If reserve constraints are incorporated, the number of connected units is greater, as has also been verified in the La Palma case study. As expected, when the interconnection link is removed, the number of connected units increases in the overall system, suggesting the important value of the link in collaborating to provide reserve in both islands.

It should be noted that reserve constraints limit the use of the interconnection at full capacity (30 MW). Figure 6.16 illustrates this fact, showing how the export from LZ to FV increases when reserve constraints are removed.

The value of the biggest connected unit in every island increases when interconnection links are used due to the reserve contribution of the interconnection link: generators can be employed closer to the maximum limit, increasing the efficiency and decreasing the cost. Figure 6.17 shows this effect in LZ.

In Table 6.5, the comparison of the three simulations is summarized for the overall LZFV system. In the LZFV system, reserve constraints increase the total cost by 3.5%; if the interconnection link is removed, the cost increases by 12.8%. It should be noted that even though the number of start-ups adds to the same figure in the three cases, the start-up cost differs significantly: reserve constraints increase base case start-up costs by 44.2%, while the removal of the interconnection link almost doubles the start-up cost. It should be noted that increasing the number of start-ups does not necessarily increase the start-up costs, since units vary significantly in size and cost.

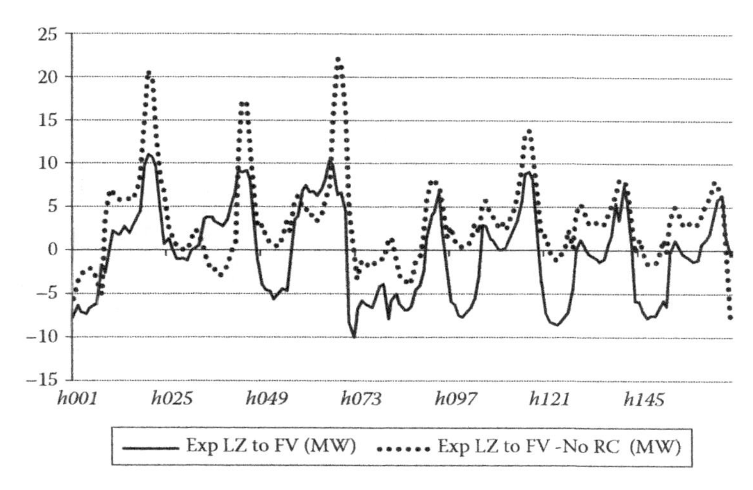

FIGURE 6.16
Export from LZ to FV in base case and no RC case.

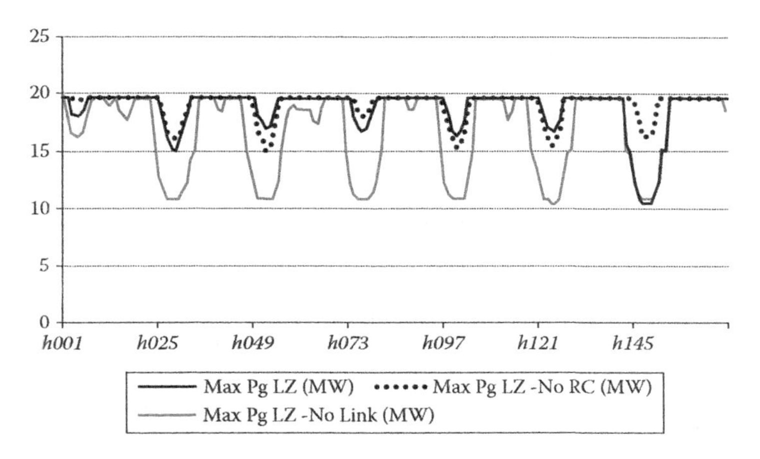

FIGURE 6.17
Biggest connected unit in base case, no RC case, and no link case.

TABLE 6.5
Comparison of Base Case, No RC Case, and No Link Case in LZFV Case Study

	LZFV				
	Base Case	No RC	Difference No RC (%)	No Link	Difference No Link (%)
Total cost (k€)	2413.1	2329.0	3.5	2721.9	−12.8
Operation cost (k€)	2362.1	2300.6	2.6	2629.2	−11.3
Startup cost (k€)	51.0	28.5	44.2	92.7	−81.7
Number of start-ups	35	35	0.0	35	0.0

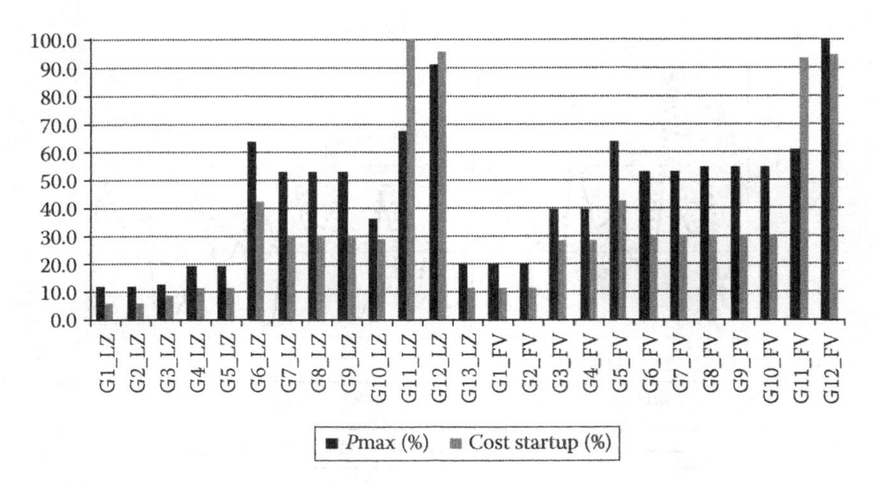

FIGURE 6.18
Relative comparison of LZFV thermal unit maximum power and start-up cost.

Figure 6.18 illustrates this fact by showing the maximum power and start-up costs of the units normalized with respect to the biggest one: the smallest start-up cost (units G1_LZ and G2_LZ) is only 6% of the biggest start-up cost (unit G11_LZ). Similarly, the smallest units (G1_LZ, G2_LZ and G3_LZ) have only 11% of the generation capacity of the biggest unit (unit G12_FV).

6.4.3 Mallorca–Menorca Case Study

6.4.3.1 Description of Case Study

Mallorca island (MALL) is interconnected with Menorca (MEN) island by an AC submarine link of around 40 km and 35 MW of interconnection capacity (a second submarine link between the islands, of capacity 45 MW, is projected before 2020), forming the islanded power system of Mallorca–Menorca (MALLMEN). MALL is also connected to the Spanish mainland by two submarine HVDC links of 237 km, 120 MW of transmission capacity each, adding to a total of 240 MW. It should be noted that each HVDC link is modeled within the unit commitment as a fictitious generator of $P_g^{max} = 120$ MW and $P_g^{min} = 0$ MW, and with an hourly variable price equal to the daily spot price of the Spanish electricity market operating in the Iberian peninsula. This modeling as fictitious generators guarantees that the MALLMEN system is able to withstand the loss of each of the HVDC links, and in addition, considers the provision of reserve of the HVDC links into the islanded system.

MEN is supplied by 11 diesel groups. The generation technologies of MALL units are much more complex and diverse: four coal units, six diesel generators, and four CCGT units. Three of the CCGT units are 2+1 units (2 gas turbines+1steam turbine) and the other corresponds to a 3+1 unit (3 gas turbines+1steam turbine). The MALLMEN system has 1794 MW of thermal installed capacity. Using the configuration-based modeling of MALL CCGT units, in which each CCGT mode of operation is represented as fictitious independent units, a total of 48 generators are found in the unit commitment model. Figure 6.19 displays the maximum and minimum limits of these real and fictitious generating units. It should be noted that the units belonging to MEN are much smaller than the MALL generators.

The unit commitment is illustrated with the operation for 1 week in October 2015. Figure 6.20 shows the evolution of the demand and renewable

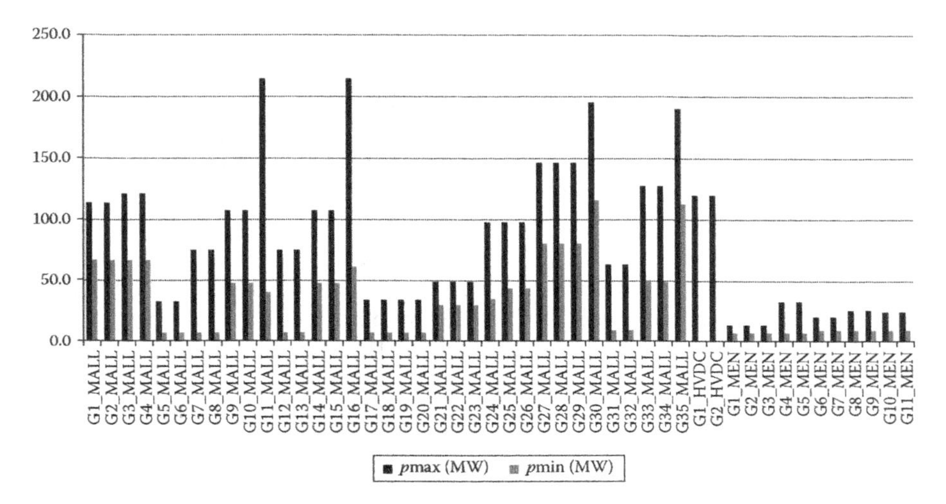

FIGURE 6.19
Maximum and minimum power of real and fictitious units belonging to MALLMEN power system.

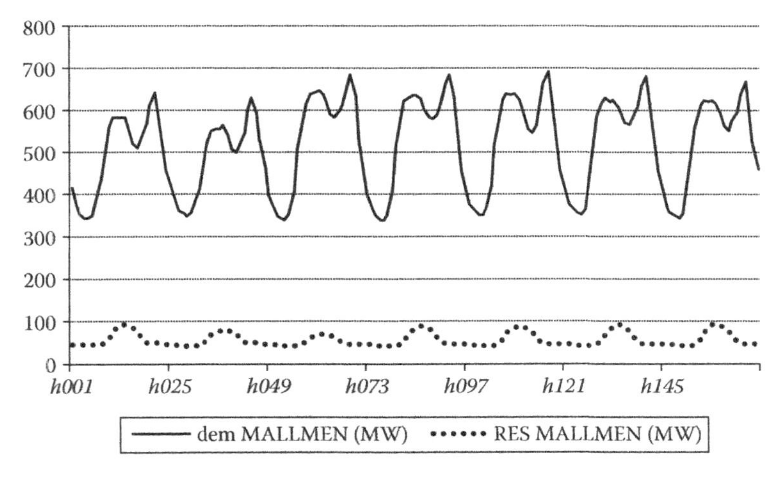

FIGURE 6.20
Demand and renewable energy in MALLMEN system.

production throughout the week in the MALLMEN power system, with a peak demand of 693 MW. Hourly renewable energy penetration has a mean value of 11.2%, presenting a maximum of 16.5% in hour *h015*. In MALLMEN, the transition from weekends (day 1 [Saturday] and day 2 [Sunday]) to working days is accompanied by an increase in afternoon and evening peak demand; in the LZFV system, this effect was not very significant (see Figure 6.9), and in the La Palma system, it was actually reversed, with weekend peaks greater than working day peaks (see Figure 6.2).

6.4.3.2 Unit Commitment Base Case Results

First, results of the unit commitment model imposing the system reserve constraints (base case) are presented. Figure 6.21 shows the distribution of the start-up orders and the number of thermal connected units during the week, varying between 7 and 10.

Figure 6.22 shows the hourly evolution of the export from MALL to MEN together with the curve of MEN demand. In contrast to the LZFV system, where the interconnection line is used in both ways, in MALLMEN, the link between the islands is always used in the same direction, exporting power from MALL to MEN. This fact is explained, on the one hand, by the small size of MEN thermal units compared with MALL thermal units (see Figure 6.19) and, on the other hand, by the cost difference of the plants. Figure 6.22 clearly states that the shape of the export follows the evolution of the MEN demand curve. In the same way, Figure 6.23 shows how the export from mainland Spain into MALL through the two HVDC links follows the MALL demand shape. It can be concluded that the HVDC links play an important role in the

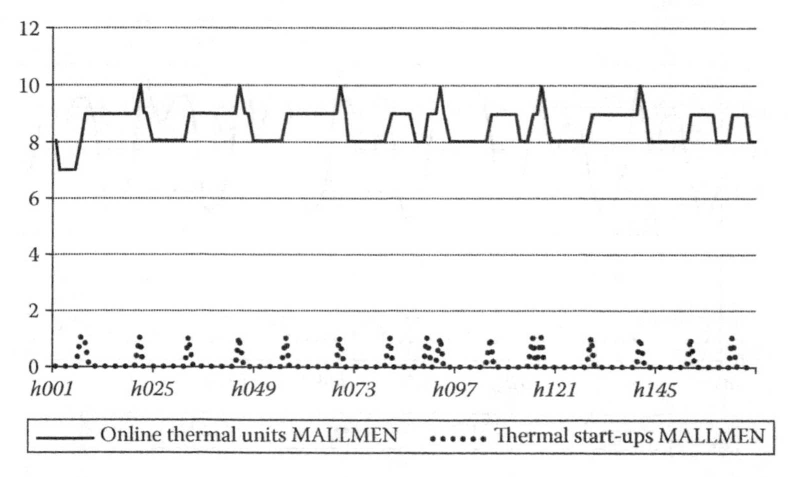

FIGURE 6.21
Number of MALLMEN online thermal units and thermal start-up decisions in base case.

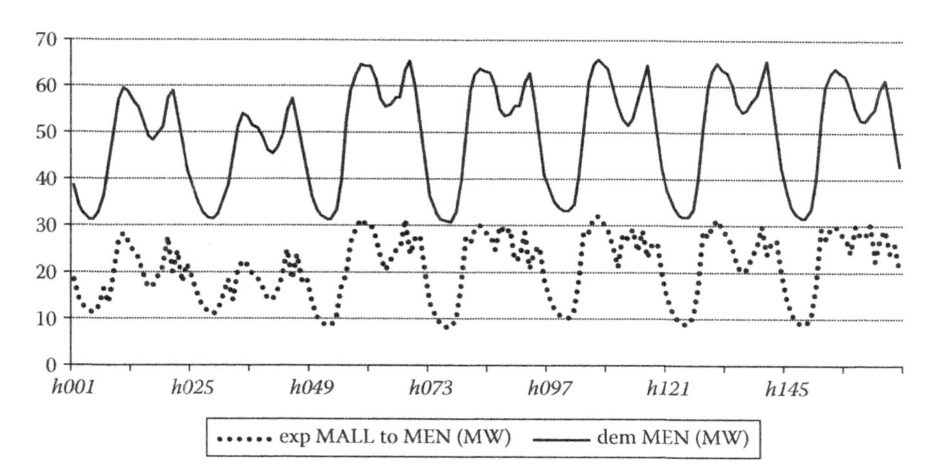

FIGURE 6.22
Hourly evolution of export from MALL to MEN and MEN demand.

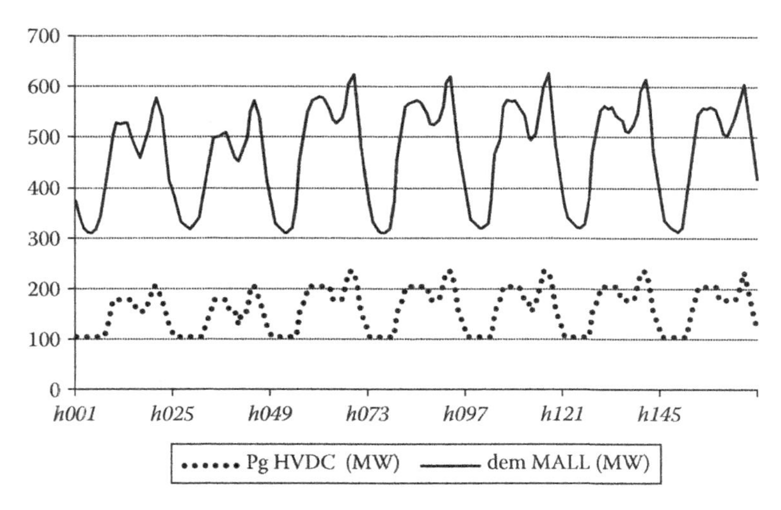

FIGURE 6.23
Hourly evolution of export from mainland Spain to MALL and MALL demand.

coverage of MALL, and similarly, the link from MALL to MEN has a crucial function in satisfying MEN demand.

Taking into account the size of the submarine AC link between the islands and the size of the generation units of the islands, the reserve provided by the AC link is small in MALL and very significant in MEN. Figure 6.24 depicts the important contribution of the up reserve provided by the link in MEN. The reserve provided by the AC link in MEN (the reslinkup MEN term of Figure 6.24) is limited by the maximum export from MALL to MEN

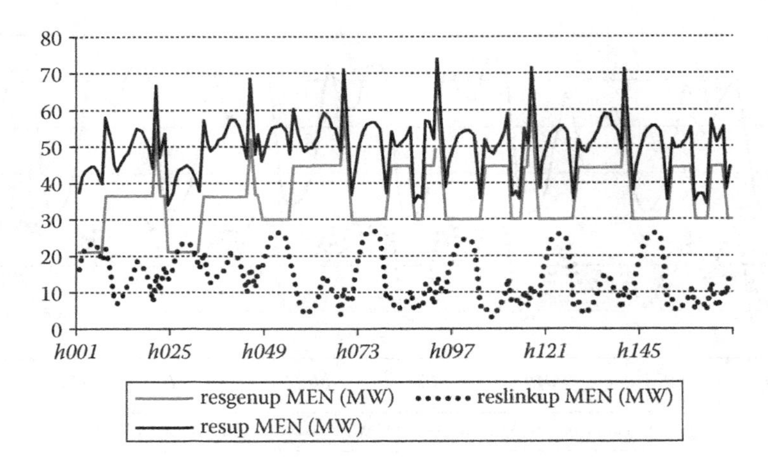

FIGURE 6.24
Total up reserve of MEN island: sum of reserve of MEN generating units and interconnection link with MALL.

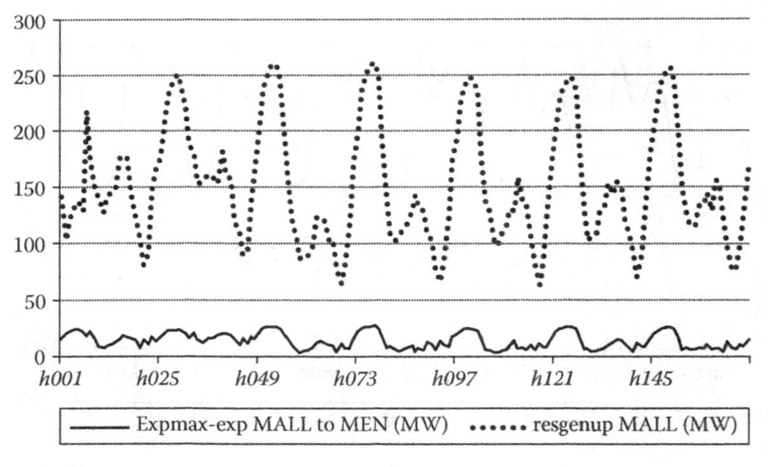

FIGURE 6.25
Up reserve link limitation in MEN.

and the amount of thermal up reserve in MALL: in contrast to LZFV, where the limiting term was thermal up reserve, in MEN island, the limiting term corresponds to the maximum export from MALL to MEN: Figure 6.25 illustrates this statement.

The unit commitment obtains a secure and reliable operation in both islands, which is illustrated for MEN in Figures 6.26 and 6.27: in every hour, the island is able to withstand both the loss of the interconnection with MALL and the loss of each thermal connected unit.

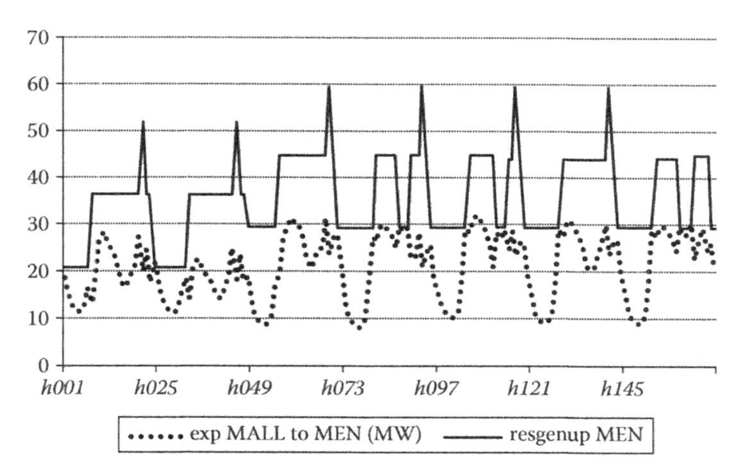

FIGURE 6.26
Export from MALL to MEN and up reserve of MEN thermal units.

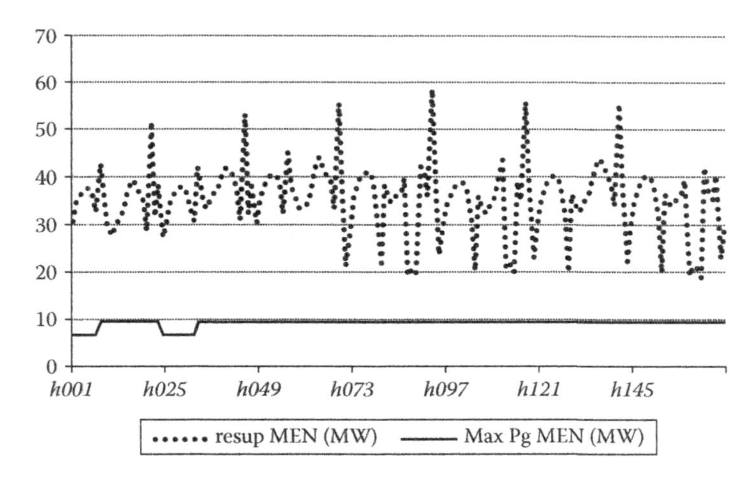

FIGURE 6.27
Total up reserve (excluding reserve of biggest unit) of MEN and maximum thermal connected unit in MEN.

6.4.3.3 Impact of Reserve Constraints and Interconnection Link

Similarly to the LZFV case study, two additional simulations (no RC case and no link case) are run to assess the impact of reserve constraints and the influence of the interconnection link. Figure 6.28 illustrates the number of connected units in the three simulation conditions. As in the case of the LZFV system, if reserve constraints are incorporated, the number of connected units is greater, and when the interconnection link is removed, the number of connected units increases. Again, reserve constraints in MEN limit the

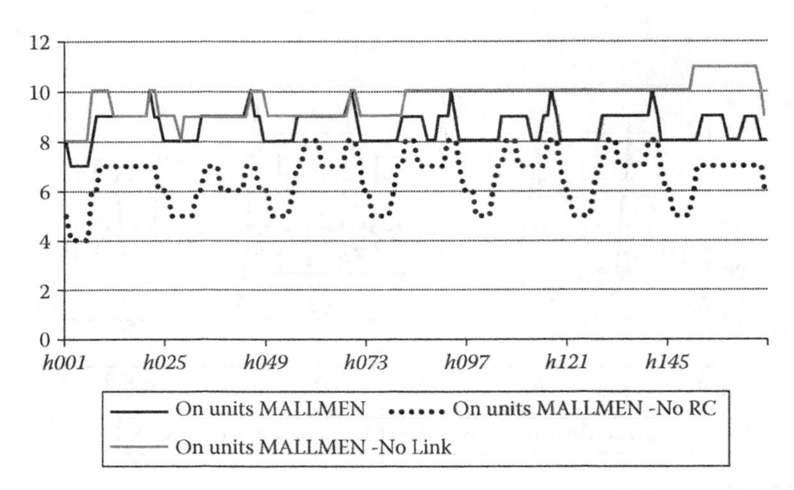

FIGURE 6.28
Thermal units in MALLMEN in base case, no RC case, and no link case.

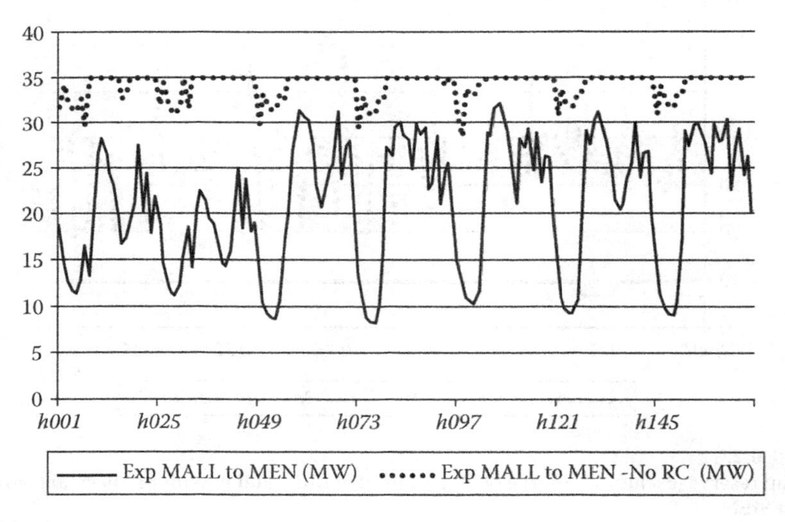

FIGURE 6.29
Export from MALL to MEN in base case and no RC case.

use of the interconnection at full capacity. Figure 6.29 indicates how the link is used at full capacity (35 MW) when reserve constraints are removed.

In Table 6.6, a comparison of the three cases is presented. In the MALLMEN system, reserve constraints increase the total cost by 9.5%, and if the interconnection link is removed, the cost increases by 22.7%. Even though the no link case presents fewer start-up decisions than the base case, the start-up cost is much greater: larger units with bigger start-up costs are connected.

TABLE 6.6

Comparison of Base Case, No RC Case, and No Link Case in MALLMEN Case Study

	MALLMEN				
	Base Case	**No RC**	**Difference No RC (%)**	**No Link**	**Difference No Link (%)**
Total cost (k€)	3492.9	3169.3	9.3	4284.4	–22.7
Operation cost (k€)	3404.6	3072.4	9.8	4124.8	–21.2
Startup cost (k€)	88.3	96.9	–9.7	159.6	–80.6
Number of start-ups	17	24	–41.2	10	41.2

6.5 Conclusions

This chapter has formulated a weekly unit commitment model that can be used to obtain the optimal secure and reliable dispatch of an island power system. Spinning reserve constraints are carefully formulated to ensure that the island power system is able to withstand the loss of every single generating unit, and also the loss of every interconnection link between islands. Three case studies of different size and complexity have been presented: the small power system of La Palma island, the medium-sized power system of the Lanzarote–Fuerteventura islands, and finally the big, complex power system of the Mallorca–Menorca islands. Although the actual economic impact of reserve constraints greatly depends on the features of the island power system, it has been shown how in all cases reserve constraints drive the economical operation of an island power system. The use of interconnection links not only enables the flow of cheaper generation power between islands, but also significantly contributes to the fulfillment of reserve constraints, which translates into a cheaper and more sustainable island operation.

References

Chang, G. W., C.-S. Chuang, T.-K. Lu, and C.-C. Wu. 2013. Frequency-regulating reserve constrained unit commitment for an isolated power system. *IEEE Transactions on Power Systems* 28 (2):578–586.

EDF. 2007. EDF SEI, Référentiel Technique (in French) (EDF Island energy systems grid Code), REF 01-07, 2005–2011, available from http://sei.edf.fr.

EUROELECTRIC. 2012. EU Islands: Towards a sustainable energy future, D/2012/12.105/24, June 2012, available from www.eurelectric.org/.

Fernandez-Bernal, F., I. Egido, and E. Lobato. 2015. Maximum wind power generation in a power system imposed by system inertia and primary reserve requirements. *Wind Energy* 18:1501–1514.

GAMS. 2015. General Algebraic Modeling System (GAMS). www.gams.com/.

Hansen, C. W., and A. D. Papalexopoulos. 2012. Operational impact and cost analysis of increasing wind generation in the Island of Crete. *IEEE Systems Journal* 6 (2):287–295.

Hobbs, B. F., M. H. Rothkopf, R. P. O'Neill, and Hung-po Chao. 2002. *The Next Generation of Unit Commitment Models*. Boston, MA: Kluwer Academic.

Holttinen, H., M. Milligan, E. Ela, N. Menemenlis, J. Dobschinski, B. Rawn, R. J. Bessa, D. Flynn, E. Gomez-Lazaro, and N. K. Detlefsen. 2012. Methodologies to determine operating reserves due to increased wind power. *IEEE Transactions on Sustainable Energy* 3 (4):713–723.

JuMP. 2015. Julia for Mathematical Programming (JuMP). https://jump.readthedocs.org/en/latest/.

Kiviluoma, J., P. Meibom, A. Tuohy, N. Troy, M. Milligan, B. Lange, M. Gibescu, and M. O'Malley. 2012. Short-term energy balancing with increasing levels of wind energy. *IEEE Transactions on Sustainable Energy* 3 (4):769–776.

Lei, X., E. Lerch, and D. Povh. 2001. Unit commitment at frequency security condition. *European Transactions on Electrical Power* 11 (2):89–96.

Liu, C., M. Shahidehpour, Z. Li, and M. Fotuhi-Firuzabad. 2009. Component and mode models for the short-term scheduling of combined-cycle units. *IEEE Transactions on Power Systems* 24 (2):976–990.

Lobato Miguelez, E., F. M. Echavarren Cerezo, and L. Rouco Rodriguez. 2007. On the assignment of voltage control ancillary service of generators in Spain. *IEEE Transactions on Power Systems* 22 (1):367–375.

Matsuura, M. 2009. Island breezes. *IEEE Power and Energy Magazine* 7 (6):59–64.

Meibom, P., R. Barth, B. Hasche, H. Brand, C. Weber, and M. O'Malley. 2011. Stochastic optimization model to study the operational impacts of high wind penetrations in Ireland. *IEEE Transactions on Power Systems* 26 (3):1367–1379.

Morales-España, G., C. M. Correa-Posada, and A. Ramos. 2016. Tight and compact MIP formulation of configuration-based combined-cycle units. *IEEE Transactions on Power Systems* 31 (2):1350–1359.

Morales-Espana, G., J. M. Latorre, and A. Ramos. 2013. Tight and compact MILP formulation of start-up and shut-down ramping in unit commitment. *IEEE Transactions on Power Systems* 28 (2):1288–1296.

Ortega-Vazquez, M. A., and D. S. Kirschen. 2007. Optimizing the spinning reserve requirements using a cost/benefit analysis. *IEEE Transactions on Power Systems* 22 (1):24–33.

Ortega-Vazquez, M. A., and D. S. Kirschen. 2009. Should the spinning reserve procurement in systems with wind power generation be deterministic or probabilistic? *In Proceedings of the International Conference on Sustainable Power Generation and Supply, 2009*. SUPERGEN '09.

Padhy, N. P. 2004. Unit commitment: A bibliographical survey. *IEEE Transactions on Power Systems* 19 (2):1196–1205.

Parsons, B., M. Milligan, B. Zavadil, D. Brooks, B. Kirby, K. Dragoon, and J. Caldwell. 2004. Grid impacts of wind power: A summary of recent studies in the United States. *Wind Energy* 7: 87–108.

REE. 2006. Ministry of industry, tourism and commerce of Spain, resolution 9613 of 26 April 2006 establishing the operation procedures for insular and extra peninsular power systems (in Spanish), official bulletin of the state nº 219 of 31 May 2006, available from www.ree.es.

Restrepo, J. F., and F. D. Galiana. 2005. Unit commitment with primary frequency regulation constraints. *IEEE Transactions on Power Systems* 20 (4):1836–1842.

Robitaille, A., I. Kamwa, A. H. Oussedik, M. de Montigny, N. Menemenlis, M. Huneault, A. Forcione, R. Mailhot, J. Bourret, and L. Bernier. 2012. Preliminary impacts of wind power integration in the Hydro-Quebec system. *Wind Engineering* 36 (1):35–52.

Sigrist, L. 2015. A UFLS scheme for small isolated power systems using rate-of-change of frequency. *IEEE Transactions on Power Systems* 30 (4):2192–2193.

Sokoler, L. E. Vinter, P., Bærentsen, R., Edlung, K., Jørgensen, J.B. 2015. Contingency-constrained unit commitment in meshed isolated power systems. *IEEE Transactions on Power Systems*. 31(5):3516–3526

Strbac, G., A. Shakoor, M. Black, D. Pudjianto, and T. Bopp. 2007. Impact of wind generation on the operation and development of the UK electricity systems. *Electric Power Systems Research* 77 (9):1214–1227.

Troy, N., D. Flynn, and M. O'Malley. 2012. Multi-mode operation of combined-cycle gas turbines with increasing wind penetration. *IEEE Transactions on Power Systems* 27 (1):484–492.

TRANSPOWER. 2009. TRANSPOWER, automatic under-frequency load shedding technical report: Appendix A: A collation of international policies for under-frequency load shedding, available from www.systemoperator.co.nz.

7

Economic Assessment of Advanced Control Devices and Renewable Energy Sources in Island Power Systems Using a Weekly Unit Commitment Model

Island power systems are facing considerable challenges to meeting their energy needs in a sustainable, affordable, and reliable way. As pointed out in Chapter 6, operation costs are higher not only because of expensive fuel transportation and lower efficiencies of the power generation technologies, but also because of the technical requirements of spinning reserves for guaranteeing frequency stability. The spinning reserve of island power systems must cover the loss of every connected unit. Since each generating unit represents a significant fraction of the total generation infeed, spinning reserve requirements displace cheaper units in favor of more expensive ones and increase the number of connected units, increasing the overall start-up cost of generators.

In this chapter, different alternative approaches that can be taken to diminish operation costs while maintaining spinning reserve security requirements are investigated to increase island sustainability: renewable energy sources (RES), energy storage systems (ESS), electric vehicles (EVs), and demand-side management measures (DSM). According to local resource availability, RES offer an interesting solution to decrease dependency on fossil fuels and increase island sustainability (Hansen and Papalexopoulos 2012). However, their intermittent behavior can affect the stability of island power systems if suitable constraints are not imposed on system operation (Rouco et al. 2008; Margaris et al. 2012). ESS and EVs offering a vehicle-to-grid operation can mitigate the impact of the intermittent behavior of RES while providing system services such as reserve provision or peak shaving (Hu et al. 2013; Ghofrani et al. 2014; Delille et al. 2012). Another way to reduce fossil fuel consumption can be provided by appropriate DSM measures (Dietrich et al. 2012; Palensky and Dietrich 2011).

In this chapter, all the mentioned options to increase island sustainability are modeled within the weekly unit commitment model presented in Chapter 6. Section 7.1 covers the modeling of ESS within the unit commitment providing island system services such as reserve and peak shaving. Section 7.3 includes the modeling of RES curtailments within the unit commitment to assess the influence of increasing RES penetration on island power systems' reliability and cost. Section 7.5 outlines the modeling of DSM and the integration of EVs in the unit commitment model.

In between the modeling sections, three case studies are presented to illustrate how the updated unit commitment can be used to make economic assessments of real island power systems for the purpose of making operating and investment decisions. In this chapter, it will be demonstrated that even though the unit commitment is a deterministic one, applied to a weekly time frame, in which demand and renewable penetration are considered as fixed input data, its use with different scenarios can serve the purpose of obtaining short-term, mid-term, and long-term conclusions. Case Study 1 in Section 7.2 will tackle the economic benefits of ESS providing primary reserve and peak shaving performed on two Spanish power system islands of different size. Case Study 2 in Section 7.4 will explore the combined effect of wind reliability and ESS availability on reserve constraints and island power system costs. Finally, using 60 real island power systems worldwide, the ambitious Case Study 3 in Section 7.6 aims at designing the most appropriate set of single and multiaction initiatives to be applied to a generic island. Initiatives fostering the deployment of RES, ESS, DSM, and EV are investigated in Case Study 3.

7.1 Energy Storage Services Modeling within Unit Commitment

7.1.1 Review

Since the beginning of the development of electric power systems, storage of electric energy has been sought to overcome technical problems and overcosts resulting from the time variation of load and generation. Current technology developments in size and costs enable ESS to provide a wide range of system security–related applications such as spinning reserve, frequency control, voltage control, power quality, and peak demand reduction (vander-Linden 2006).

Accordingly, ESS applications can be basically grouped into power and energy applications. Power applications cover those that provide relatively short power injections without the need for large energy storage, whereas energy applications are those that provide long-term power injections and hence require large energy storage. The provision of uninterruptible power supply, power quality, and primary reserve belongs to the former, whereas load shifting, peak shaving, and so on belong to the latter. For instance, NaS batteries have been used for peak shaving and backup power supply, Li-ion batteries for frequency regulation, and Ni-Cad batteries for spinning reserve and wind power firming. Always with respect to the state of the art and the technologies' modularity, pumped hydro-storage, flow, and NaS batteries are suitable for energy applications, whereas, for example, Li-ion batteries, supercapacitors, and flywheels are appropriate for power applications (Roberts 2009; Roberts and McDowall 2005; Leadbetter and Swan 2012; Chen et al. 2009; Hittinger et al. 2012). Some battery technologies, such as NaS and flow batteries, also show the ability to participate in both power and energy applications. Sensitivity analysis of storage technologies and their characteristics with respect to power and energy applications has been presented (Leadbetter and Swan 2012), where it has been found that capital cost has significant impact on all applications and for all storage technologies analyzed, whereas the improvement in efficiency seems to be a relatively insensitive parameter.

The technical impacts and benefits of ESS have been widely analyzed. In Kottick et al. (1993), the positive impact of a 30 MW/25 MWh battery ESS facility on frequency regulation in the Israeli isolated power system has been studied through a series of computer simulations of sudden demand variations. A similar analysis is carried out in Lu et al. (1995), where the emphasis, however, was on the elaboration of an incremental model of the battery ESS and the impact of governor dead bands. A method for optimal sizing and operation of a battery ESS used for primary frequency regulation in an isolated test power system is presented in Mercier et al. (2009). The state of charge limits and the nominal power are determined by using a 1 month dynamic simulation of the test system with variable wind generation and load. Note that the determination of the state of charge itself depends on the battery ESS cycling history. A fuzzy-logic-controlled supercapacitor bank for improved load frequency control of an interconnected power system has been proposed (Mufti et al. 2009). Anagnostopoulos and Papantonis (2012) studied the use of pumped storage units to increase the penetration of RES such as wind in Greece. It seems that pumped storage units become attractive for this purpose when the wind penetration level is well above the mean load-demand level. Similarly, the impact of a battery ESS on the dynamic behavior in terms of frequency and voltage of a hybrid wind–diesel system has been studied in Sebastian (2011). Applications of ESS to microgrids and control strategies and optimization of hybrid ESS have been reviewed and analyzed in Tan et al. (2013), pointing out that ESS is a promising technology for microgrids. The optimal sizing and controller settings of an ultracapacitor can be found in Sigrist et al. (2015).

Economic analysis has also been carried out. In Ross et al. (2011), a knowledge-based expert system is proposed for the ESS scheduling to optimize operation cost by diesel generation and charging/discharging cycles from wind and load profiles. Daneshi and Srivastava (2012) analyzed the impact of compressed air energy storage on operation costs and wind curtailments of a test system with respect to its location. Optimal sizing of the rated power and the energy capacity of compressed air energy storage is provided in Wang and Yu (2012), where the rated power depends on the system's security constraints and the energy capacity on the recovery of wind power curtailment. A methodology for optimal allocation and economic analysis of ESS in a hypothetical microgrid is presented in Chen et al. (2011). ESS power and energy capacities are tuned to minimize system operation costs.

In this chapter, the use of ESS in primary frequency regulation reserve and peak shaving in small isolated power systems is modeled within the weekly unit commitment model of Chapter 6. The economic operation of isolated power systems is affected by both the amount of reserve allocated to ensure frequency stability and the amount of peak-shaving generation, since their provision may require not only generators operating below their optimal operating points but also the connection of units that otherwise might not be needed. For instance, in several isolated Spanish power systems, several applications are being researched or used, ranging from a 500 kW (2.8 MWh) ZnBr battery, through a 1 MW (7.2 MWh) NaS battery, to a 162 MW pump storage.

7.1.2 Mathematical Formulation of ESS within Unit Commitment

The modeling of ESS in unit commitment includes the definition of a complete set of new equations to compute the charging and discharging levels of ESS and the reserve provision of ESS, and also the modification of the demand balance constraint and the total system reserve constraints. Only the new nomenclature, the new equations, and the equations of Chapter 7 that must be updated are presented in this chapter. As in Chapter 7, parameters start with uppercase letters, while variables start with lowercase letters.

7.1.2.1 Nomenclature

7.1.2.1.1 Sets

ess energy storage system unit

7.1.2.1.2 Parameters

$Pchar_{ess}^{max}$ maximum charging power of unit ess (MW)

$Pdisch_{ess}^{max}$ maximum discharging power of unit ess (MW)

E_{ess}^{min} minimum desired level of energy storage in unit ess (MWh)

E_{ess}^{max} maximum energy storage capacity of unit ess (MWh)

η_{ess} cycling efficiency of unit ess

7.1.2.1.3 Continuous Variables

$resess_{ess,h}^{up}$ effective ramp-up reserve provided by unit ess in hour h (MW)

$resess_{ess,h}^{down}$ effective ramp-down reserve provided by unit ess in hour h (MW)

$pchar_{ess,h}$ charging power of unit ess in hour h (MW)

$pdisch_{ess,h}$ discharging power of unit ess in hour h (MW)

$e_{ess,h}$ actual energy storage level of unit ess in hour h (MWh)

7.1.2.1.4 Binary Variables

$\delta^{char}_{ess,h}$ binary state of charging of unit ess in hour h (0/1)

7.1.2.2 Constraints

7.1.2.2.1 Demand Balance Constraint

The demand balance equation of Chapter 6 must be updated to include ess devices, forcing the total power generation (thermal units, renewable and energy storage discharge) to be equal to total load (demand and energy storage units' charge) and export to interconnected islands.

$$\sum_{g \in i}(p_{g,h}) + \sum_{ess \in i}(pdisch_{ess,h} - pchar_{ess,h}) + Res_{i,h} = D_{i,h} + \sum_{ii}(exp_{i,ii,h}) \quad \forall h, i \quad (7.1)$$

7.1.2.2.2 Reserve Constraints

The up reserve that can be used from the energy storage units in the island power system is computed by

$$resess^{up}_{ess,h} = Pdisch^{max}_{ess} - pdisch_{ess,h} + pchar_{ess,h} \quad \forall h, ess \quad (7.2)$$

Equation 7.2 assumes that no ess unit is completely discharged, but each can provide as up reserve the rated power of $Pdisch^{max}_{ess}$ (MW) independently of its actual energy level $e_{ess,h}$ (MWh). To clarify the computation of up reserve of ESS units, and how the model can be used for providing both primary reserves (a power application) and peak shaving (an energy application), let us consider, for instance, a 1 MW (7.2 MWh) NaS battery loaded in a specific hour to a certain charge (for instance, middle charge $e_{ess,h} = 3.4$ MWh). If the battery is neither charging nor discharging (which means it is not providing peak-shaving services), the maximum upward primary reserve that this battery contributes to the system is 1 MW (the battery is prepared in case of a generator trip to start discharging 1 MW to help in restoring system frequency). If the battery is charging at a power rate of 0.4 MW, the maximum contribution to upward primary reserve is 1.4 MW (stop charging 0.4 MW and start discharging 1 MW). If the battery is discharging at a power rate of 0.4 MW, the contribution is reduced to 0.6 MW.

Analogously, the down reserve that can be used from the energy storage units in the island power system is computed by

$$resess_{\text{ess},h}^{\text{down}} = pdisch_{\text{ess},h} + Pchar_{\text{ess}}^{\max} - pchar_{\text{ess},h} \quad \forall h, ess \tag{7.3}$$

Equation 7.3 assumes that the battery can contribute to down reserve when it is discharging and charging. For example, if the same battery is discharging at a power rate of 0.4 MW, the maximum contribution to downward primary reserve is 1.4 MW (stop discharging 0.4 MW and start charging 1 MW). If the battery is charging at a power rate of 0.4 MW, the contribution is reduced to 0.6 MW.

All the total system reserve constraints written in Chapter 6 (Equations 6.22 through 6.25) must be updated to take into account the reserve provided by ESS in the following way:

$$\sum_{\substack{gg \in i \\ gg \neq g}} \left(resgen_{\text{gg,h}}^{up} \right) + \sum_{ii \neq i} \left(reslink_{ii,i,h}^{up} \right) + \sum_{\text{ess} \in i} \left(resess_{\text{ess},h}^{up} \right) \geq p_{g,h} \quad \forall i, g \in i, h \tag{7.4}$$

$$\sum_{g \in i} \left(resgen_{g,h}^{up} \right) + \sum_{ii \neq i} \left(reslink_{ii,i,h}^{up} \right) + \sum_{\text{ess} \in i} \left(resess_{\text{ess},h}^{up} \right) \geq Resloss_i \cdot Res_{i,h} \quad \forall i, h \tag{7.5}$$

$$\sum_{g \in i} \left(resgen_{g,h}^{up} \right) + \sum_{\substack{iii \neq i \\ iii \neq ii}} \left(reslink_{iii,i,h}^{up} \right) + \sum_{\text{ess} \in i} \left(resess_{\text{ess},h}^{up} \right) \geq exp_{ii,i,h} \quad \forall i, ii, h \tag{7.6}$$

$$\sum_{g \in i} \left(resgen_{g,h}^{\text{down}} \right) + \sum_{ii \neq i} \left(reslink_{ii,i,h}^{\text{down}} \right) + \sum_{\text{ess} \in i} \left(resess_{\text{ess},h}^{\text{down}} \right) \geq Resdown_i^{\min} \quad \forall i, h \tag{7.7}$$

7.1.2.2.3 Charging and Discharging Limits of the ESS

Limits to the rate of charge and discharge of each unit are imposed by

$$0 \leq pdisch_{\text{ess},h} \leq \left(1 - \delta_{\text{ess},h}^{\text{char}}\right) \cdot Pdisch_{\text{ess}}^{\max} \quad \underset{x \to \infty}{\forall h, ess} \tag{7.8}$$

$$0 \leq pchar_{\text{ess},h} \leq \delta_{\text{ess},h}^{\text{char}} \cdot Pchar_{\text{ess}}^{\max} \quad \forall h, ess \tag{7.9}$$

Since an ESS cannot be charging and discharging at the same time, only one binary variable is necessary to indicate the charging and discharging of the unit.

7.1.2.2.4 Stored Energy of the ESS

The energy dynamics of the charge discharge of the ESS is modeled by

$$e_{\text{ess},h} = e_{\text{ess},h-1} + \eta_{\text{ess}} \cdot pchar_{\text{ess},h} - pdisch_{\text{ess},h} \quad \forall h, ess \tag{7.10}$$

It should be noted that round-trip efficiency of ESS has been taken into account by means of a constant η_{ess}, although efficiency might vary according to the ESS operation.

Limits to the energy level in each hourly period (for instance, to impose that no ess unit is completely discharged, but each can provide as up reserve the rated power) are written as

$$E_{ess}^{min} \leq e_{ess,h} \leq E_{ess}^{max} \quad \forall h, ess \tag{7.11}$$

7.1.2.2.5 Selection of Power System Applications

The constraints described by Equations 7.1 through 7.10 apply to the generic case in which the ESS can provide simultaneously primary reserve and peak-shaving generation. In the case of uniquely providing primary reserve, Equations 7.8 and 7.9 change to

$$pdisch_{ess,h} = 0 \quad \forall h, ess \tag{7.12}$$

$$pchar_{ess,h} = 0 \quad \forall h, ess \tag{7.13}$$

By contrast, if the ESS provides only peak-shaving service, Equations 7.2 and 7.3 change to

$$resess_{ess,h}^{up} = 0 \quad \forall h, ess \tag{7.14}$$

$$resess_{ess,h}^{down} = 0 \quad \forall h, ess \tag{7.15}$$

7.2 Case Study 1: Economic Assessment of ESS Providing Reserve and Peak Shaving in Island Power Systems

7.2.1 Description of Case Study

In this first case study, the savings of providing the primary regulation requirements and peak-shaving generation by ESS are investigated in two Spanish island power systems of different sizes and with different generation technologies: Gran Canaria (GC) and La Gomera (LG). GC is a large island power system with 1045 MW of thermal generation installed capacity and a peak demand of 577 MW, whereas LG is a small system with thermal installed capacity of 20 MW and peak demand of 12.1 MW. The GC power system is supplied by 22 generating units of a wide variety of generation technologies: combined-cycle gas, steam, gas, and diesel. The LG power system is supplied by 10 diesel generating units.

For each power system, five ESS scenarios were applied to 11 weekly scenarios, each corresponding to a different month of 2010 (in this way, the results of a weekly unit commitment model can be used to reach mid-term conclusions). In each weekly scenario, hourly demand and renewable production are considered as known input data. The five ESS scenarios are characterized by different nominal powers and storage capacities, which are analyzed and evaluated for the 11 weekly scenarios. It was assumed that the ESS had a cycling efficiency η_{ess} of 70%. Cycle efficiencies of 70% are typical, for instance, for pumped-storage plants, compressed-air energy storage, or lead-acid or flow batteries. It was also assumed that the ESS was in middle charge at the beginning of the week, and ended the week also at a middle level of charge (as is the usual weekly operation of typical pump-storage units).

The five ESS scenarios assessed in the GC island power system are shown in Table 7.1. It should be noted that the first scenario represents a power system with no ESS, and thus, the comparison of the rest of the scenarios with this first one will yield the benefits of different sizes of ESS in providing primary reserve and peak-shaving generation.

Similarly, the five ESS scenarios of La Gomera island are given in Table 7.2.

7.2.2 GC Island Results

First, the impact of providing primary reserve and peak shaving separately and then of providing them simultaneously was analyzed for one weekly scenario and for one particular ESS. Thereafter, the analysis was repeated for the 11 weekly scenarios and for the five ESS scenarios.

TABLE 7.1

ESS Scenarios of Gran Canaria

	ESS 1	ESS 2	ESS 3	ESS 4	ESS 5
Power (MW)	0	25	50	75	100
Capacity (MWh)	0	150	300	450	600

TABLE 7.2

ESS Scenarios of La Gomera

	ESS 1	ESS 2	ESS 3	ESS 4	ESS 5
Power (MW)	0	0.25	0.5	1	2
Capacity (MWh)	0	1.5	3	6	12

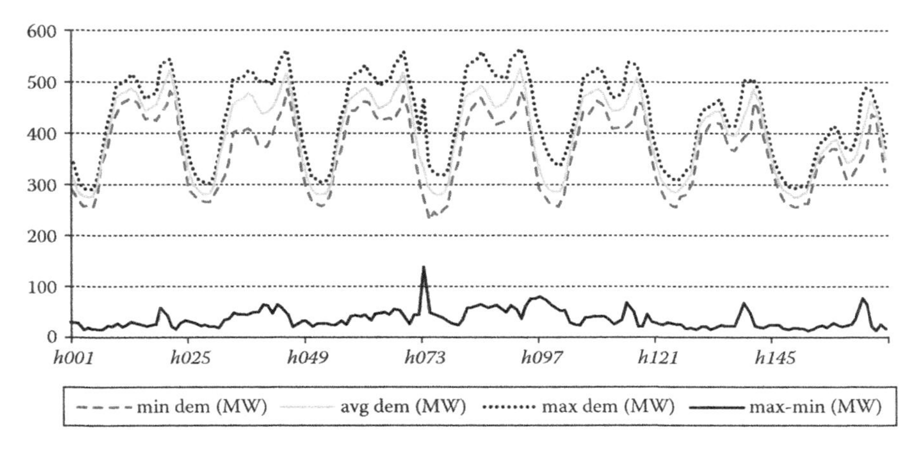

FIGURE 7.1
Average, maximum, and minimum demand in Gran Canaria scenarios.

Figure 7.1 shows the average, maximum, and minimum hourly demand in each hour of the 11 analyzed weeks (renewable generation in GC in 2010 was small compared with demand). Hours *h001–h120* correspond to Monday through Friday, while *h120–h168* represent the weekend. In addition, Figure 7.1 depicts the difference between the maximum and minimum hourly demand in the 11 weekly scenarios, which is under 100 MW most of the time, reaching a peak of 136 MW in hour *h073*.

Considering one of the simulated weeks of GC (corresponding to January 2010) and one of the simulated scenarios (EES5 scenario corresponding to ESS unit 100 MW, 600 MWh), Figure 7.2 shows the primary reserve provided by scenario ESS 5 during 1 week together with the evolution of the demand curve, in the case that the ESS device only provides primary reserve. It can be seen that on working days, the island power system requires the full up reserve (here 100 MW) of the ESS device during the day to diminish the dispatch cost. It's interesting to see that during the night (00:00–06:00), the island system does not require the available ESS reserve, since the thermal units, needed to satisfy the demand and operating close to the minimum technical limit, provide sufficient primary reserve. The spikes in the afternoon are due to the load reduction and consequent generation reduction.

Considering the same week (January 2010), in Figure 7.3, the demand, thermal conventional generation, and net ESS generation (discharge minus charge) of scenario ESS 5 are depicted when the ESS was uniquely providing peak-shaving generation. Figure 7.3 clearly shows how the charging and discharging of the ESS unit flattens out the thermal generation profile. The ESS unit charges during the night hours, and discharges during peak hours in daytime. This indicates that fewer conventional thermal generating units need to be started up or shut down.

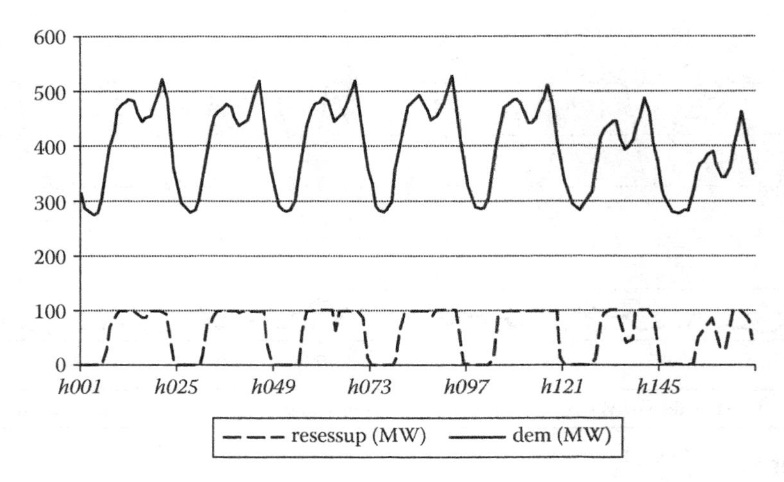

FIGURE 7.2
Gran Canaria demand and up reserve provided by ESS 5 unit (100 MW, 600 MWh)—ESS provides reserve only.

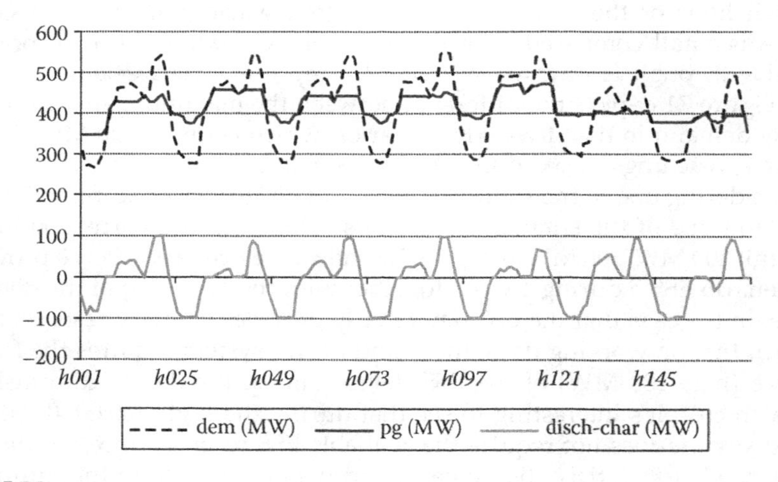

FIGURE 7.3
Gran Canaria demand, thermal conventional generation, and net ESS generation—ESS provides peak shaving only.

Figure 7.4 compares the thermal generation when unit ESS 5 provides reserve only (in this case, thermal generation equals demand minus renewable generation), when unit ESS 5 provides peak shaving only, and when the unit provides both services to the island power system of GC. When providing both services, again it can be seen that the generation profile is flattened out with respect to the reserve-only case, though to a lesser degree

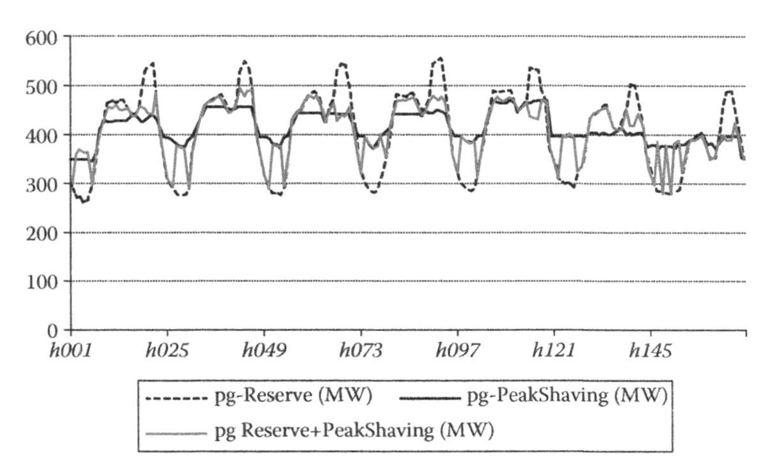

FIGURE 7.4
Comparison of Gran Canaria thermal generation when ESS provides reserve only, ESS provides peak shaving only, and ESS provides both.

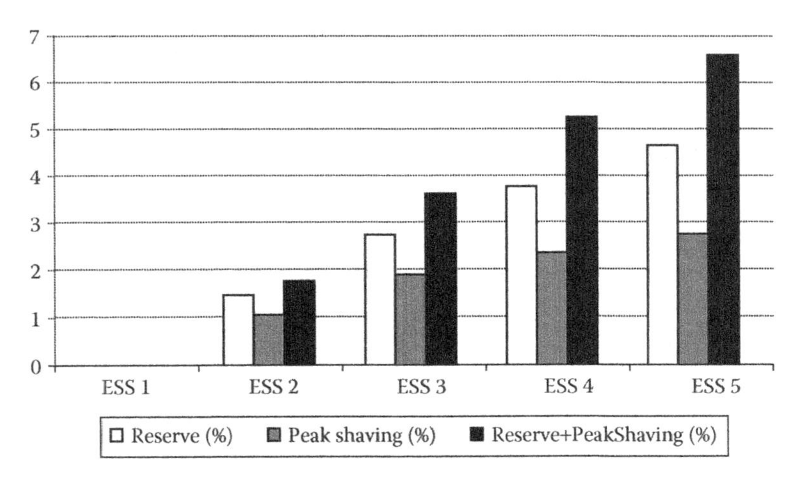

FIGURE 7.5
Gran Canaria mean annual cost reduction (%) of ESS scenarios when ESS provides reserve only, ESS provides peak shaving only, and ESS provides both.

than in the peak shaving only case, since part of the discharging capacity is employed to increase the island up reserve.

Figure 7.5 displays the mean annual cost reductions for GC of the five ESS scenarios when ESS provides reserve only, when ESS provides peak shaving only, and when ESS provides both. It can be seen that the provision of primary reserve and peak shaving reduces the island system operation costs with increasing size of the ESS. This reduction is nearly linear for the analyzed ESS scenarios; that is, multiplying the size by a factor instigates a reduction by the same factor. Note that this linearity doesn't necessarily apply for even larger ESS, since, for example, the maximum reserve provided by ESS 5 is already close to the largest connected unit (150 MW, corresponding to 50% of one of the combined-cycle power plants), and some units are always connected to cover the demand. In addition, the combined provision of primary reserve and peak-shaving generation improves the system operation costs relative to their separate provisions. In fact, this improvement due to combined provision ranges from 70% to 88% of the sum of the separate provisions. The reduction of the combined provision is not exactly the sum of the separate provisions, since when providing peak-shaving generation during peak hours, less reserve can be provided. However, the larger the ESS, the closer the combined provision comes to the sum of the separate provisions, since a larger ESS is more likely to be able to provide sufficient power for both primary reserve and peak-shaving generation.

According to the mean annual cost reductions, an ESS 5, providing primary reserve and peak-shaving generation, would yield a total cost reduction of 23.2 million €/year. By assuming an average total cost of 2800 €/kW, a typical lifetime of 15 years (EPRI 2010a), and constant cash flows, the installation of the ESS 5 in GC would yield an internal rate of return (IRR) of 7.25%.

Table 7.3 shows system cost savings for each of the 11 weekly scenarios when the ESS was simultaneously providing both peak-shaving generation and primary reserve. It can be inferred that the system cost savings for the 11 weekly scenarios are very similar, with a maximum difference of 1.2%. The results are therefore very similar for the weeks corresponding to a particular ESS scenario, and system savings increase linearly with ESS size.

Finally, Figure 7.6 compares the number of connected units for the week in January 2010 when there are no ESS devices, when scenario ESS 5 uniquely

TABLE 7.3

Comparison of the System Costs Savings for Gran Canaria among the Weekly Scenarios

	System Costs Savings (%)				
	ESS 1	**ESS 2**	**ESS 3**	**ESS 4**	**ESS 5**
Jan-10	0.00	1.79	3.72	5.41	7.01
Feb-10	0.00	1.75	3.73	5.04	6.30
Mar-10	0.00	1.29	3.47	5.25	6.80
Apr-10	0.00	1.73	3.65	5.24	6.23
May-10	0.00	2.09	3.88	5.53	7.13
Jun-10	0.00	2.27	3.80	5.30	6.33
Jul-10	0.00	1.68	3.09	4.88	5.63
Aug-10	0.00	1.63	3.58	5.46	6.62
Sep-10	0.00	1.88	3.78	5.32	6.96
Oct-10	0.00	1.64	3.49	5.16	6.72
Nov-10	0.00	1.89	3.53	5.14	6.64

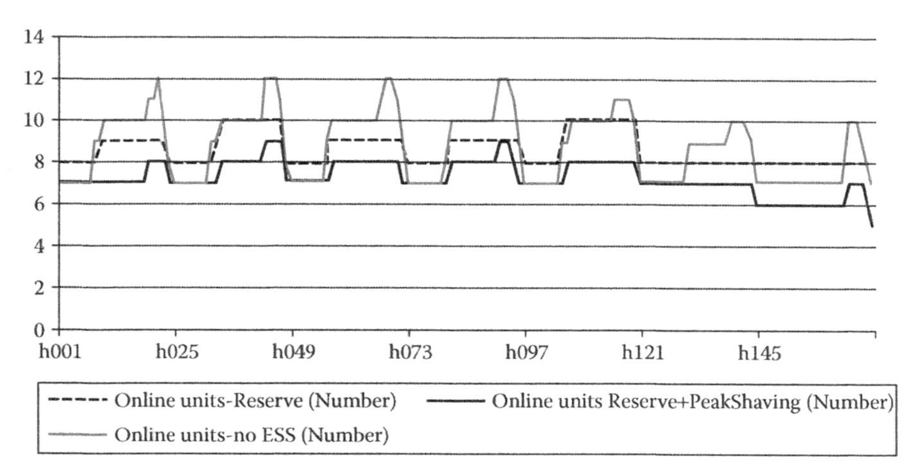

FIGURE 7.6
Comparison of the number of connected units in 1 week for Gran Canaria.

provides primary reserve, and when it provides both primary reserve and peak-shaving generation simultaneously. In the case without ESS, the largest number of online units is observed. The number of units online is reduced when the ESS is providing primary reserve only. When it is providing both primary reserve and peak-shaving generation, the number of units online is further reduced.

7.2.3 La Gomera Island Results

First, the impact of providing primary reserve and peak shaving separately and then of providing them simultaneously was analyzed for one weekly scenario and for one particular ESS device. Thereafter, the analysis was repeated for the 11 weekly scenarios. Figure 7.7 shows the average, maximum, and minimum hourly demand in LG among the 11 scenarios, and in addition, the difference between the maximum and minimum hourly values. No renewable generation exists within the system.

Considering one of the simulated weeks of LG island (corresponding to January 2010), and one of the simulated ESS scenarios (ESS 5, corresponding to unit 2 MW, 12 MWh), Figure 7.8 shows the ESS primary reserve together with the evolution of the demand curve, in the case when the ESS device only provides primary reserve. It can be seen that the island power system of LG requires ESS primary reserve, but the ESS full nominal power (2 MW) is only necessary for some hours as system up reserve. This indicates that less reserve must be provided by the conventional generation. However, the required ESS reserve is lower in LG than in GC (see Figure 7.2), mainly because the unit commitment needs to keep three units always online to guarantee system stability. This implies that there is always some reserve in

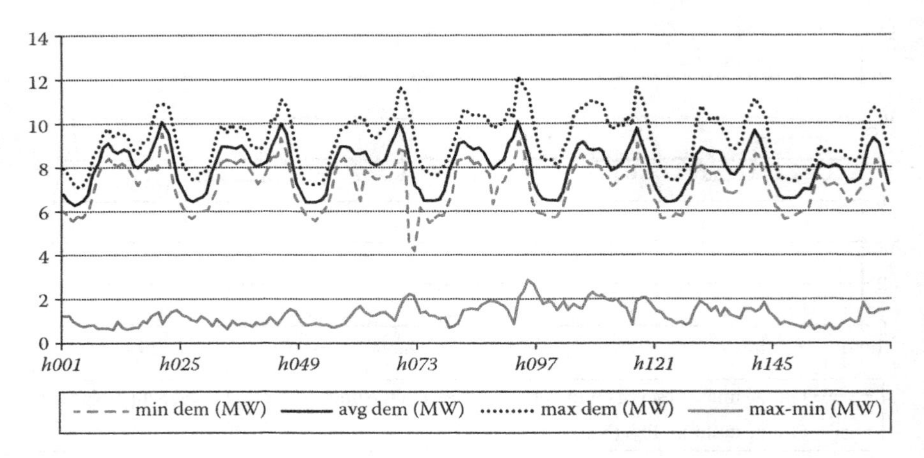

FIGURE 7.7
Average, maximum, and minimum demand in La Gomera scenarios.

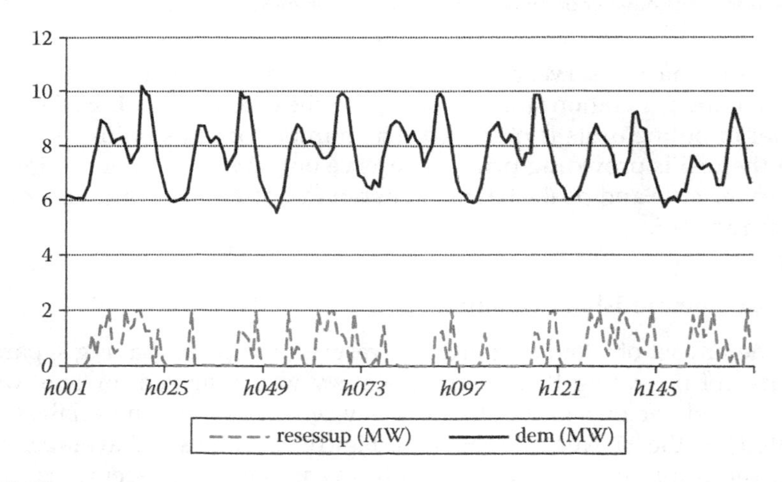

FIGURE 7.8
La Gomera demand and up reserve provided by ESS 5 unit (2 MW, 12 MWh)—ESS provides reserve only.

the system, large enough during valley hours. In fact, the ESS is mostly providing reserve during peak hours (08:00–12:00 and 20:00–24:00).

Figure 7.9 depicts the demand, thermal conventional generation, and net ESS generation (discharge minus charge) of scenario ESS 5 of LG when the ESS was uniquely providing peak-shaving generation. The thermal generation profile flattened out due to charging and discharging of the ESS, mainly during the afternoon and night peak hours. This indicates that fewer conventional generating units need to be started up during peak hours or shut down during afternoon hours. The peak-shaving effect is less significant than in GC (see Figure 7.3) due to the unit commitment decision to keep

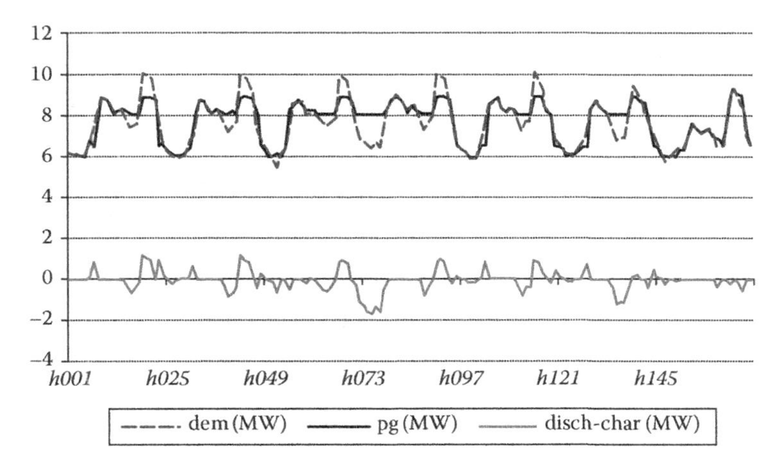

FIGURE 7.9
La Gomera demand, thermal conventional generation, and net ESS generation—ESS provides peak shaving only.

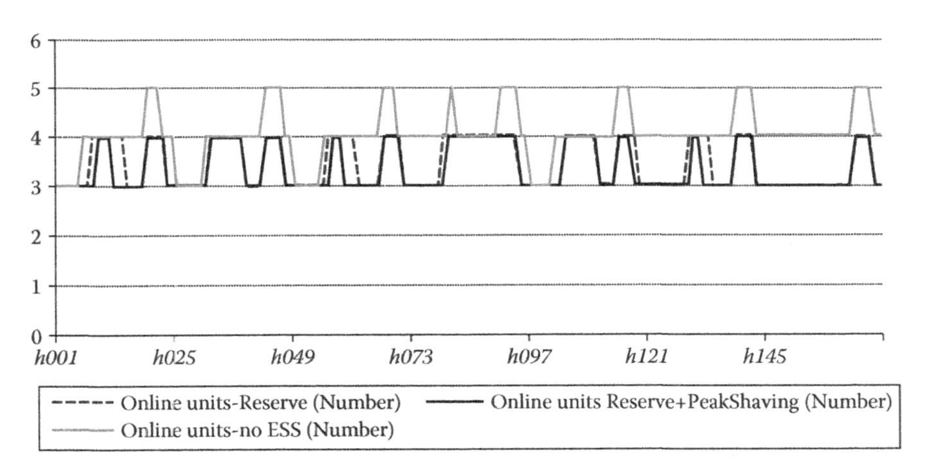

FIGURE 7.10
Comparison of the number of connected units in 1 week for La Gomera.

three units online during valley hours, as a result of which generating costs do indeed increase, but start-up costs can be reduced for units needed during noon or afternoon hours. In fact, start-up costs in LG are larger than in GC, and therefore it's cheaper to keep units online, which influences the impact of the ESS providing peak-shaving generation.

Figure 7.10 compares the number of connected units in LG for the week corresponding to January 2010 when there are no ESS devices, when scenario ESS 5 uniquely provides primary reserve, and when it is providing both primary reserve and peak-shaving generation simultaneously. As in the case of

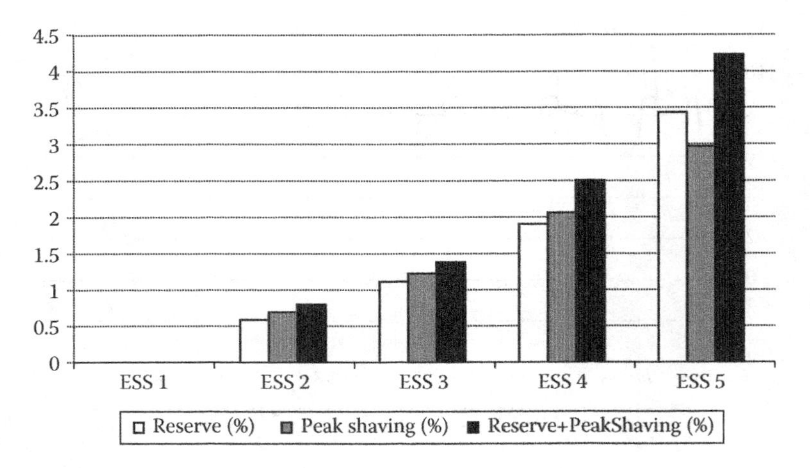

FIGURE 7.11
La Gomera mean annual cost reduction (%) of ESS scenarios when ESS provides reserve only, ESS provides peak shaving only, and ESS provides both.

GC, exploiting the use of ESS for both reserve and peak shaving induces a smaller number of units than when only ESS reserve is employed or when no ESS is used for system services.

Figure 7.11 displays the mean annual cost reductions for LG of the five ESS scenarios when ESS provides reserve only, when ESS provides peak shaving only, and when ESS provides both. It can be seen that the provision of primary reserve and peak shaving reduces the system operation costs with increasing size of the ESS. This reduction is nearly linear; that is, multiplying the size by a factor instigates a reduction by the same factor. In addition, the combined provision of primary reserve and peak-shaving generation improves the system operation costs relative to their separate provisions. However, and in contrast to the results for GC, the reduction by the simultaneous provision is nowhere near the sum of the separate provisions. This is mainly due to the three units kept online, the larger start-up costs, the smaller peak demand to valley demand ratio, and the fact that the size of the generating units with respect to the peak demand is larger.

Table 7.4 shows the annual system cost savings for each of the 1 weekly scenarios when the ESS was simultaneously providing both peak-shaving generation and primary reserve. It can be inferred that, as for GC, the system cost savings for the 1 weekly scenarios show a difference of about 1.3%.

According to the mean annual cost reductions, an ESS 5, providing primary reserve and peak-shaving generation, would yield a total cost reduction of 654,800 million €/year. By assuming an average total cost of 2800 €/kW, a typical lifetime of 15 years, and constant cash flow, the installation of the ESS 5 in LG would yield an IRR of 8%.

TABLE 7.4

Comparison of the System Costs Savings for La Gomera among the Weekly Scenarios

	System Costs Savings (%)				
	ESS 1	**ESS 2**	**ESS 3**	**ESS 4**	**ESS 5**
Jan-10	0.00	0.78	1.30	2.33	3.64
Feb-10	0.00	0.58	1.05	2.24	3.49
Mar-10	0.00	0.89	1.40	2.53	4.04
Apr-10	0.00	1.08	1.92	2.16	3.93
May-10	0.00	0.77	1.57	2.91	3.89
Jun-10	0.00	0.78	1.16	2.13	4.35
Jul-10	0.00	0.33	1.09	2.34	4.45
Aug-10	0.00	0.95	1.18	2.83	4.34
Sep-10	0.00	1.06	1.46	2.42	5.05
Oct-10	0.00	0.90	1.64	2.31	4.64
Nov-10	0.00	0.85	1.64	3.24	4.71

7.2.4 Conclusions of Case Study 1

The first case study has illustrated the use of a weekly unit commitment model to make economic assessments of real island power systems for the purpose of making operating and investment decisions about ESS devices. The application to two isolated Spanish power systems has shown that the provision of primary reserve and peak shaving by ESS could be an economical alternative to scheduling spare generation capacity in the online units. It has been shown for both systems that the combined provision of both primary reserve and peak-shaving generation improved island power systems' savings relative to the separate provisions. The results also indicate that the presence of ESS in the larger power system of GC is more beneficial because in the smaller system of LG, three units are always kept online by the unit commitment for stability reasons, and since start-up costs are higher, the peak demand to valley demand ratio is smaller, and the size of the generating units with respect to the peak demand is larger.

7.3 Renewable Energy Sources Modeling within Unit Commitment

7.3.1 Motivation

It is recognized worldwide that increasing the generation share of RES sources is a key target to diminish carbon emissions and increase sustainability. As already mentioned in Section 6.2 of Chapter 6, extensive research has tackled the influence of RES on power system reserve. Especially in island power systems, their intermittent behavior can affect stability if suitable constraints are not imposed on system operation. Examples of the effect of wind on island power systems can be given for the United Kingdom (Strbac et al. 2007), Ireland (Meibom et al. 2011), and Crete (Hansen and Papalexopoulos 2012).

However, according to the actual grid codes of certain island power systems (such as the Spanish grid code for isolated power systems [REE 2006]), the upward primary reserve should be larger than either any connected unit or the most probable loss of renewable power. However, this last statement is quite ambiguous, since actual Spanish grid codes do not specify the percentage of the predicted renewable power that should be considered reliable for that purpose. In this sense, not only should the possibility of curtailing RES be modeled in unit commitment models, but also the modeling of the quantity of RES that can be considered reliable to allocate reserve.

7.3.2 Mathematical Formulation

The modeling of RES curtailment and RES reliability within unit commitment models modifies the objective function, the demand balance equation, and the system reserve constraints. Only the new nomenclature and the equations of Chapter 6 that have to be updated are presented in this section. Parameters start with uppercase letters, while variables start with lowercase letters.

7.3.2.1 Nomenclature

7.3.2.1.1 Parameters

C_h^{rescurt} is the penalization cost of renewable curtailments in hour h (€/MW)

7.3.2.1.2 Variables

$prescurt_{i,h}$ is the curtailed RES generation for island i in hour h (MW)

7.3.2.2 Equations

7.3.2.2.1 Objective Function

A wind curtailment penalization has to be added to the objective function:

$$\min \sum_{g,h} \left(c_{g,h} + c_{g,h}^{\text{start-up}} + c_{g,h}^{\text{shut-down}} \right) + \sum_{i,h} \left(C_h^{\text{rescurt}} \cdot prescurt_{i,h} \right) \qquad (7.16)$$

Penalization of reserve curtailment should be established such that security criteria for reserve are always met; that is, it is preferable to curtail renewable generation rather than violate reserve criteria.

7.3.2.2.2 Demand Balance Constraint

RES curtailment effectively reduces the effective generation by RES:

$$\sum_{g\in i}(p_{g,h})+\sum_{ess\in i}(pdisch_{ess,h}-pchar_{ess,h})+(Res_{i,h}-prescurt_{i,h})$$
$$=D_{i,h}+\sum_{ii}(exp_{i,ii,h})\quad \forall h,i \tag{7.17}$$

7.3.2.2.3 System Reserve Constraints

According to the actual grid codes of isolated power systems, the upward primary reserve should be larger than either the largest connected unit or the most probable loss of RES. However, actual grid codes do not specify the percentage of renewables that should be considered reliable for that purpose. To analyze the effect, the parameter $Resloss_i$ was introduced in Chapter 6 to formulate the percentage of renewable power that may be lost and that needs to be included in the up reserve constraint of Equation 7.5. This equation must be updated to include possible RES curtailments:

$$\sum_{g\in i}(resgen_{g,h}^{up})+\sum_{ii\neq i}(reslink_{ii,i,h}^{up})$$
$$+\sum_{ess\in i}(resess_{ess,h}^{up})\geq Resloss_i\cdot(Res_{i,h}-prescurt_{i,h})\quad \forall i,h \tag{7.18}$$

Note that if RES curtailments are permitted, effective RES power should be considered here as potential downward reserve, and thus, the down reserve constraint needs to be updated:

$$\sum_{g\in i}(resgen_{g,h}^{down})+\sum_{ii\neq i}(reslink_{ii,i,h}^{down})+\sum_{ess\in i}(resess_{ess,h}^{down})+(Res_{i,h}-prescurt_{i,h})$$
$$\geq Resdown_i^{min}\quad \forall i,h \tag{7.19}$$

7.4 Case Study 2: Impact of Wind Reliability and ESS on Island Power System Costs

As an example of the use of the unit commitment model that includes the modeling of RES, this second case study tackles the problem of the integration of high quantities of wind energy within an island power system. The influence of both wind power reliability and ESS providing reserve and peak-shaving services on system costs is investigated for an actual small Spanish isolated power system (with peak demand around 700 MW) formed by a set of generating units of a wide variety of generation technologies. Current, as well as mid-term and long-term, demand scenarios are considered to derive significant conclusions. In addition, ESS scenarios of different sizes are used to assess the impact of these storage devices.

In a first step, the building of the actual (year 2010), mid-term (year 2020), and long-term (year 2030) scenarios is explained. Then, a detailed analysis of the actual scenarios (year 2010) is provided. This gives a first insight into the degree to which ESS influence system costs and wind power curtailments. Finally, system cost savings for different values of $Resloss_i$ and different sizes of ESS devices are analyzed.

7.4.1 Building of Scenarios

7.4.1.1 Power System Scenarios

Real operation data for 2 weeks, one corresponding to summer (July 12–18, 2010) and one corresponding to winter (January 11–17, 2010), have been used as a starting point to characterize the actual, mid-term, and long-term scenarios. Winter and summer scenarios are labeled starting with the letters *w* and *s*. Mid-term and long-term scenarios are built by scaling the demand and wind power profiles of January 11–17, 2010 and July 12–18, 2010, respectively.

Actual scenarios (year 2010) consider 0% of demand increment, mid-term scenarios (year 2020) consider a yearly 3% demand increase, representing a 35% increase with regard to the actual scenarios, and long-term scenarios (year 2030) consider a demand increase of 50% with regard to the actual scenarios. These demand scenarios are labeled *D0, D35,* and *D50,* respectively.

It is very difficult to forecast wind power generation in each scenario accurately. This is due not only to wind variability, but also to uncertainties with respect to the future capacity of installed wind power generation, depending, among other things, on the economic environment. To consider different wind penetration levels (low, high, and extreme penetration), real wind power profiles of past scenarios have been scaled to different values. Note that some of the scaled-up wind power scenarios are not necessarily realistic, but they can provide information on the existence of wind penetration limits due to reserve requirements. These wind power scenarios are labeled *WX,* where *X* is the percentage of weekly wind energy out of weekly energy demand.

Table 7.5 summarizes the main features of the 20 power system scenarios considered. In Table 7.5, Demand, Wind, and Wind Penetration represent weekly values, whereas Peak demand, Off-Peak demand, and MaxWind generation represent hourly values. Note that maximum wind power generation

TABLE 7.5

Power System Scenarios

Time Horizon	Name of Scenario	Dem (MWh)	Wind (MWh)	Wind Pen (%)	Peak (MW)	Off-Peak (MW)	Max Wind (%)	Max Wind (MW)
Actual	w-D0-W1	68715	354	0.5%	554	265	3.5%	12
Mid-Term	w-D35-W0	92765	372	0.4%	748	358	2.7%	13
Mid-Term	w-D35-W1	92765	620	0.7%	748	358	4.5%	21
Mid-Term	w-D35-W2	92765	1417	1.5%	748	358	10.3%	48
Long-Term	w-D50-W0	103072	372	0.4%	831	398	2.4%	13
Long-Term	w-D50-W1	103072	620	0.6%	831	398	4.0%	21
Long-Term	w-D50-W2	103072	2480	2.4%	831	398	16.1%	85
Long-Term	w-D50-W4	103072	3897	3.8%	831	398	25.4%	133
Actual	s-D0-W13	67375	8509	12.6%	523	282	19.2%	59
Mid-Term	s-D35-W10	90956	8935	9.8%	705	380	14.9%	62
Mid-Term	s-D35-W16	90956	14891	16.4%	705	380	24.9%	103
Mid-Term	s-D35-W23	90956	21273	23.4%	705	380	35.6%	148
Long-Term	s-D50-W9	101062	8935	8.8%	784	422	13.4%	62
Long-Term	s-D50-W15	101062	14891	14.7%	784	422	22.4%	103
Long-Term	s-D50-W21	101062	21273	21.0%	784	422	32.0%	148
Long-Term	s-D50-W51	101062	51056	50.5%	784	422	76.9%	354
Actual	w-D0-W7	68715	4783	7.0%	554	265	46.7%	163
Actual	w-D0-W13	68715	9211	13.4%	554	265	90.0%	314
Actual	s-D0-W22	67375	14891	22.1%	523	282	33.6%	103
Actual	s-D0-W32	67375	21273	31.6%	523	282	48.0%	148

TABLE 7.6
ESS Scenarios

	No EES	ESS 1	ESS 2	ESS 3	ESS 4	ESS 5	ESS 6	ESS 7	ESS 8
Power (MW)	0	20	80	200	20	80	200	10	10
Capacity (MWh)	0	6	6	6	10	10	10	2	4

is highest for summer scenarios, except for one actual and one long-term winter scenario.

7.4.1.2 ESS Scenarios

Nine ESS scenarios, characterized by different nominal powers and storage capacities, as shown in Table 7.6, are analyzed and evaluated for each of the 20 power system scenarios. The first scenario represents the case without ESS, and thus, the comparison of the rest of scenarios with this first one will yield the benefits of the ESS.

It is assumed that the ESS has a cycling efficiency η_{ESS} of 65%. It is further assumed that the ESS is halfway charged at the beginning of the week, and that it is also halfway charged at the end of the week, allowing ±5% of deviation.

7.4.1.3 Variability of Wind Scenarios

Three different values of $Resloss_i$, corresponding to 0%, 25%, and 50%, are used to analyze the influence of wind reliability on wind power curtailments and system costs. Clearly, a larger value of $Resloss_i$ will have a larger impact on the system costs, since more reserve might well be required.

7.4.2 Results of Actual Scenarios

First, a detailed analysis of results of the actual demand and wind power generation scenarios is provided. No increment of demand has been applied (*D0* scenarios in Table 7.5), and a $Resloss_i$ value of 50% has been considered.

7.4.2.1 Wind Behavior

Figure 7.12 shows the wind power curtailment as a percentage of the wind power generation as a function of the actual demand and wind power scenarios. It can be seen that wind power curtailments are only required for the extreme wind penetration scenarios. Note that, measured on a weekly basis, wind power curtailments are no higher than 0.35% when no ESS is installed, implying that curtailment is actually required only for a very few hours. This is mainly because a $Resloss_i$ factor of 50% is considered, and thus,

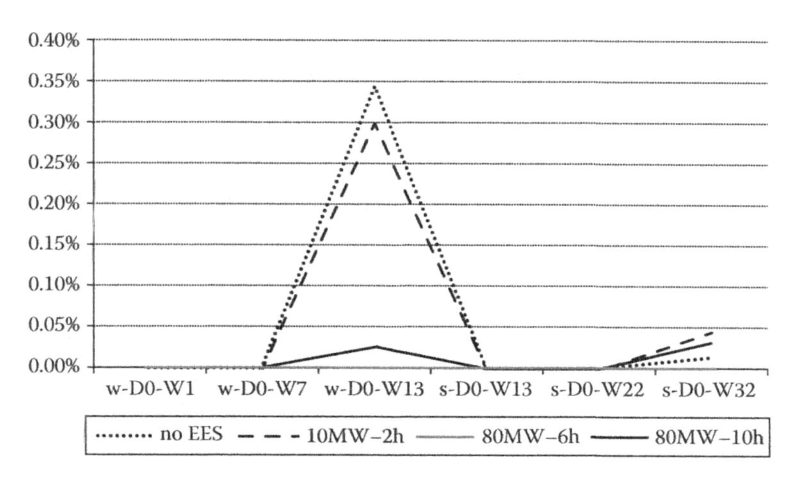

FIGURE 7.12
Wind curtailments of actual scenarios (%), considering $Resloss_i = 50\%$ and no ESS deployed.

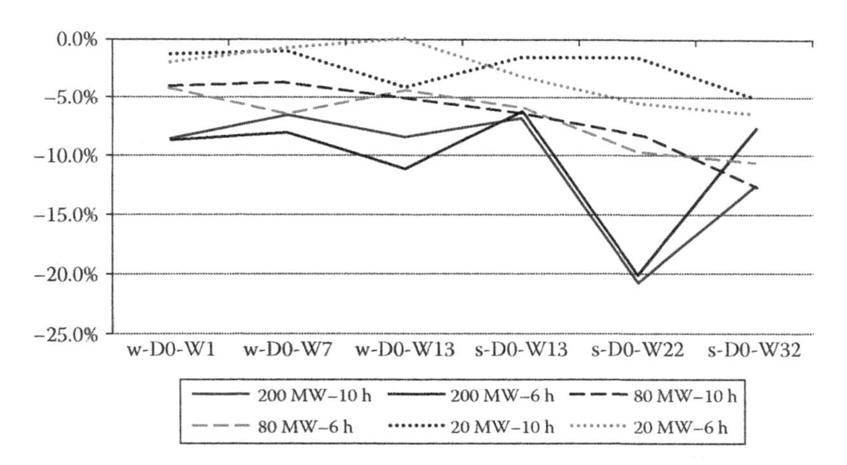

FIGURE 7.13
Relative system cost savings of installing ESS considering $Resloss_i = 50\%$.

the required upward reserve is dominated by the largest unit. Wind curtailments are decreased when a larger ESS is installed. In fact, ESS units of size greater than 80 MW–6 h, that is, devices of 80 MW–10 h, 200 MW–6 h, and 200 MW–10 h, completely avoid wind curtailments.

7.4.2.2 System Cost Analysis

Figure 7.13 shows the system cost savings relative to the case without ESS for each actual power system scenario and for different ESS scenarios. ESS 6 (200 MW–10 h) and ESS 3 (200 MW–6 h) achieve similar savings between

10% and 20% of the system cost. The same conclusion can be drawn for ESS 5 (80 MW–10 h) and ESS 2 (80 MW–6 h), obtaining savings between 5% and 10%. ESS devices of nominal power of 20 MW (ESS 1, ESS 4) achieve less significant savings. It seems, therefore, that system cost savings are dominated by the nominal power of the ESS device, and not so much by the ESS's capacity. One reason is that the provision of reserve does not require huge capacities. Another reason lies in the fact that the plant begins and ends with the same level of charge.

7.4.2.3 ESS Operation Profile

Figure 7.14 shows the evolution of the charge level of ESS 5 (80 MW–10 h) throughout the week for the different actual power system scenarios. The purpose of the figure is to show that the optimal operation profile of the ESS device significantly depends on the scenario. Since the ESS has a maximum energy capacity of 800 MWh, Figure 7.14 clearly depicts the operation assumption that the ESS begins and ends the week at medium charge (400 MWh). Further, two scenarios discharge power during the first 4 days, whereas the remaining scenarios charge power during that period.

7.4.3 Sensitivity Analysis of *Resloss*

To measure the impact of wind reliability, the influence of the parameter $Resloss_i$ on system costs is analyzed. For each value of $Resloss_i$, the absolute and per unit system cost savings (€/MW) that ESS devices achieve are investigated. Since the ESS 2, 3, 5, and 6 (200 and 80 MW devices) were shown to be most effective (see Figure 7.13), only these ESS devices (with 10 h of discharging capacity) are detailed further. Table 7.7 summarizes the results yielded by the weekly optimization model.

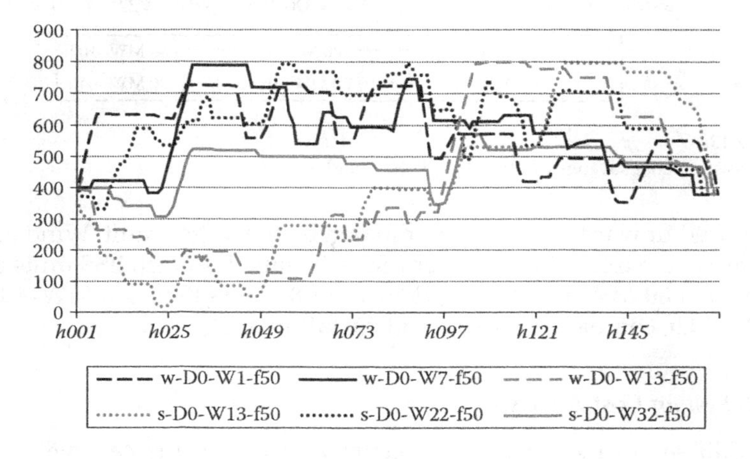

FIGURE 7.14
Energy operation profile of ESS 5 (80 MW–10 h) for the actual power system scenarios, considering $Resloss_i = 50\%$.

TABLE 7.7

Power System Cost Depending on *Resloss* and Energy Storage Systems

	System Cost (k€), *Resloss* = 50%			System Cost (k€), *Resloss* = 25%			System Cost (k€), *Resloss* = 0%		
Scenario	**No EES**	**200 MW-10 h**	**80 MW-10 h**	**No EES**	**200 MW-10 h**	**80 MW-10 h**	**No EES**	**200 MW-10 h**	**80 MW-10 h**
w-D0-W1	9,019	8,306	8,663	8,903	8,306	8,497	8,903	8,306	8,497
w-D0-W7	8,554	8,027	8,243	8,314	7,590	7,956	8,309	7,614	8,002
w-D0-W13	8,156	7,515	7,788	7,932	7,107	7,459	8,004	7,081	7,543
s-D0-W13	7,793	7,290	7,322	7,551	7,232	7,350	7,551	7,232	7,350
s-D0-W22	7,562	6,257	6,982	6,951	6,109	6,403	6,840	6,113	6,474
s-D0-W32	6,539	5,805	5,806	6,650	5,426	5,622	6,066	5,428	5,713
w-D35-W0	11,850	11,243	11,338	11,707	11,178	11,334	11,707	11,178	11,334
w-D35-W1	11,865	11,208	11,382	11,690	11,162	11,339	11,690	11,162	11,339
w-D35-W2	11,877	11,132	11,287	11,584	11,134	11,245	11,584	11,134	11,245
s-D35-W10	10,447	10,140	10,258	10,344	10,001	10,154	10,344	10,001	10,154
s-D35-W16	9,766	9,438	9,733	9,686	9,265	9,522	9,686	9,265	9,522
s-D35-W23	9,240	8,507	8,994	9,001	8,451	8,766	9,001	8,449	8,766
w-D50-W0	13,022	12,392	12,851	12,961	12,398	12,727	12,961	12,398	12,727
w-D50-W1	13,177	12,363	12,835	12,924	12,351	12,704	12,924	12,351	12,704
w-D50-W2	12,782	12,129	12,583	12,724	12,137	12,449	12,724	12,137	12,449
w-D50-W4	12,636	11,989	12,423	12,567	11,985	12,270	12,553	11,983	12,264
s-D50-W9	11,839	11,232	11,391	11,649	11,093	11,290	11,649	11,093	11,290
s-D50-W15	11,076	10,630	10,648	10,895	10,488	10,603	10,895	10,488	10,603
s-D50-W21	10,230	9,867	10,063	10,125	9,755	9,925	10,120	9,771	9,930
s-D50-W51	7,168	6,503	6,611	6,739	6,304	6,426	6,703	6,326	6,429
MEAN	10,230	9,599	9,860	10,045	9,473	9,702	10,011	9,475	9,717

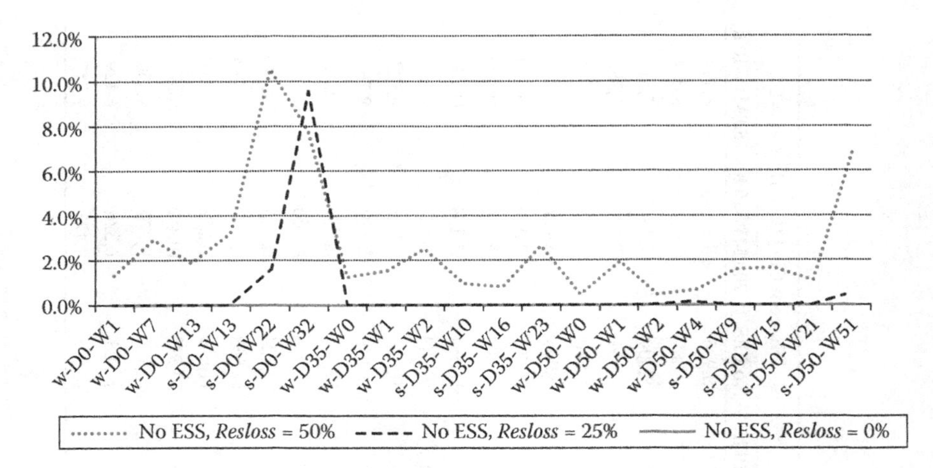

FIGURE 7.15
System cost increment (%) with respect to $Resloss_i = 0\%$, with no ESS.

Several drawings will be extracted from Table 7.7. First, in Figure 7.15, the impact of $Resloss_i$ is shown for all actual, mid-term, and long-term scenarios when no ESS is present. It can be seen that for most power system scenarios, system cost increments are below 2% with respect to the case where wind power variability has not been taken into account for reserve allocation ($Resloss_i = 0\%$). However, when wind power variability is taken into account by incrementing $Resloss_i$ to 25% and 50%, system cost increments can reach up to 10% in summer scenarios with high and extreme wind penetration levels.

By using ESS 6 (200 MW–10 h), and unlike the case without ESS, no significant cost increment occurs in the case of $Resloss_i$ of 25%, as can be seen in Figure 7.16. This means that the flexibility provided by ESS 6 can accommodate the increment in reserve requirements due to the increase in wind power penetration. However, if $Resloss_i$ increases to 50%, cost increments up to 7% persist.

To compare ESS plants of different capacity, per unit cost savings are shown in Figure 7.17 considering $Resloss_i = 50\%$. It can be deduced that for ESS 6 (200 MW–10 h), the average savings are 2900 €/MW, reaching a maximum value of 6000 €/MW. Figure 7.17 suggests that the smaller one (80 MW–10 h) reaches better per unit power system cost savings.

7.4.4 Conclusions of Case Study 2

The analysis of actual scenarios shows that under the assumption of weekly operation of ESS, their impact significantly depends on their nominal power and not so much on their capacity. Only ESS devices over 80 MW achieve significant impacts on system savings. Further, it can be concluded that wind power curtailments only occur for extreme wind penetration

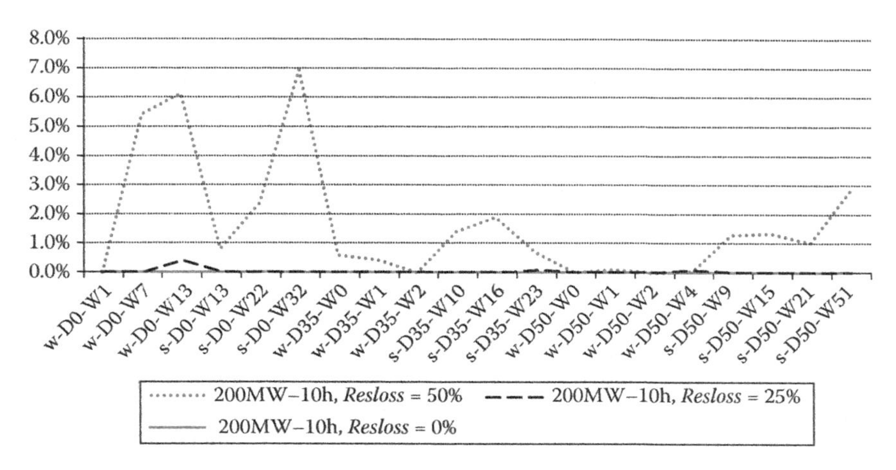

FIGURE 7.16
System cost increment (%) with respect to $Resloss_i = 0\%$, with ESS 6 (200 MW–10 h).

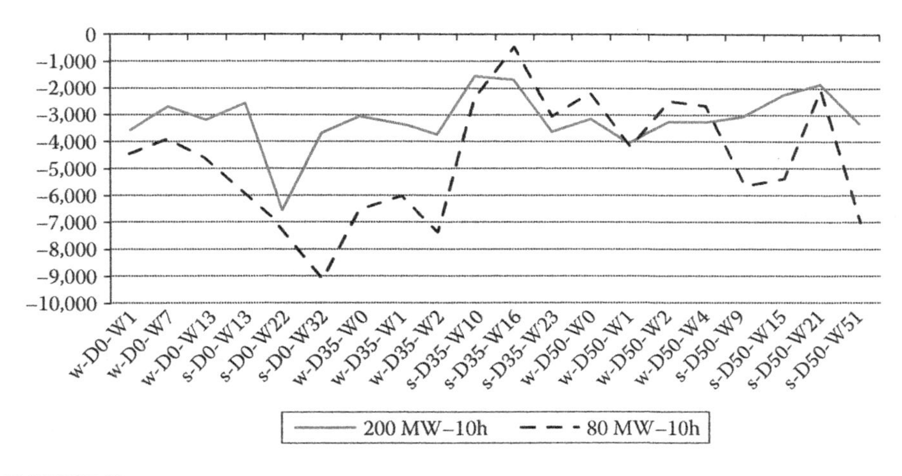

FIGURE 7.17
Comparison of system cost savings per unit of ESS installed power for plants ESS 5 (80 MW–10 h) and ESS 6 (200 MW–10 h), for $Resloss_i = 50\%$.

scenarios. ESS devices are able to reduce and, in the case of an ESS device of 200 MW, to eliminate wind power curtailments. Cost savings basically originate from the flexibility of operation provided by ESS: part of the thermal generation is substituted by the ESS, and less reserve needs to be allocated in thermal units. The increase in the level of wind that should be secured with reserve (parameter $Resloss_i$) can increase island power system cost significantly, although this is extremely dependent on the island system and the scenario considered. The use of ESS also reduces the influence of this reliability level.

7.5 Demand-Side Management and Electric Vehicle Modeling within Unit Commitment

The implementation of DSM measures and the promotion of EVs are both measures aimed at modifying the load profile that must be supplied by conventional thermal generation. In this section, the modeling of DSM measures and EV devices within the weekly unit commitment model is shown. Both DSM and EV are modeled on an aggregated level in each island.

First, the nomenclature of parameters (uppercase letters) and variables (lowercase letters) is shown. Then, the new equations and constraints corresponding to DSM and EV are formulated, and the demand balance constraint is updated to include these actions.

7.5.1 Nomenclature

7.5.1.1 Sets

d day

t charging/discharging time window of EV in each day

7.5.1.2 Parameters

$Ddisp_i^{\max}$ maximum dispatchable power demand in island i (MW)

$Ddisp_i^{\min}$ minimum dispatchable power demand in island i (MW)

$Ddispd_{i,d}^{\max}$ maximum dispatchable energy demand in day d and island i (MWh)

$Ddispd_{i,d}^{\min}$ minimum dispatchable energy demand in day d and island i (MWh)

$\delta Ddisp_{i,h}$ binary parameter indicating whether demand of island i is dispatchable in hour h (0/1)

$Pev_i^{\max}$ maximum charging/discharging EV power in island i (MW)

$Eev_i^{\min}$ minimum energy storage capacity of EVs in island i (MWh)

$Eev_i^{\max}$ maximum energy storage capacity of EVs in island i (MWh)

$Eevd_{i,d}^{\max}$ maximum daily EV energy level in day d and island i (MWh)

$Eevd_{i,d}^{\min}$ minimum daily EV energy level in day d and island i (MWh)

ηev round-trip efficiency of EV

$\delta ev_{h,i}$ binary parameter indicating whether EV charging/discharging is allowed in hour h and island i (0/1)

7.5.1.3 Variables

$ddisp_{i,h}^{up}$ upward variation of the dispatchable component of the power demand of island i in hour h (MW)

$ddisp_{i,h}^{down}$ downward variation of the dispatchable component of the power demand of island i in hour h (MW)

$pev_{i,h}^{char}$ charging of EV of island i in hour h (MW)

$pev_{i,h}^{disch}$ discharging of EV of island i in hour h (MW)

$resev_{i,h}^{up}$ up reserve used from the EVs of island i in hour h (MW)

$resev_{i,h}^{down}$ down reserve used from the EVs of island i in hour h (MW)

$eev_{i,h}$ actual energy storage capacity of EV of island i in hour h (MWh)

7.5.2 Modeling of DSM and EV within Unit Commitment

7.5.2.1 DSM Equations and Constraints

DSM is modeled on an aggregated level. First, hourly dispatchable DSM demands are constrained between minimum and maximum limits (which are set as a fraction of the energy demand):

$$Ddisp_i^{\min} \cdot \delta Ddisp_{i,h} \le ddisp_{i,h}^{up} \le Ddisp_i^{\max} \cdot \delta Ddisp_{i,h} \quad \forall h,i \tag{7.20}$$

$$Ddisp_i^{\min} \cdot \delta Ddisp_{i,h} \le ddisp_{i,h}^{down} \le Ddisp_i^{\max} \cdot \delta Ddisp_{i,h} \quad \forall h,i \tag{7.21}$$

DSM variations must be balanced during 1 day and during their availability slots. This condition is expressed by

$$\sum_{h \in d} ddisp_{i,h}^{down} = \sum_{h \in d} ddisp_{i,h}^{up} \quad \forall d,i \tag{7.22}$$

Finally, the amount of dispatchable demand is limited for a particular day:

$$Ddispd_{i,d}^{\min} \le \sum_{h \in d} ddisp_{i,h}^{up} \le Ddispd_{i,d}^{\max} \quad \forall d,i \tag{7.23}$$

7.5.2.2 EV Equations and Constraints

As in the case of DSM, EVs are modeled as dispatchable demands on an aggregated level. However, they can only be allocated in selected hours (set t) of every day d. In addition, EVs can also inject power into the system, acting as generators, and they can also provide reserve.

First, limits in the hourly values of EV charge and discharge are formulated. It should be noted that these limits depend on the number of available EV cars.

$$0 \leq pev_{i,h}^{\text{char}} \leq Pev_i^{\max} \cdot \delta ev_{i,h} \quad \forall h,i \tag{7.24}$$

$$0 \leq pev_{i,h}^{\text{disch}} \leq Pev_i^{\max} \cdot \delta ev_{i,h} \quad \forall h \tag{7.25}$$

Then, energy capacity limits should be applied:

$$Eev_i^{\min} \leq eev_{i,h} \leq Eev_i^{\max} \quad \forall h,i \tag{7.26}$$

EV energy dynamics during driving are not modeled, but an average state of charge is assumed. The energy dynamics of EV when connected to the grid are formulated as follows:

$$eev_{i,h} = eev_{i,h-1} + pev_{i,h}^{\text{char}} \cdot \eta ev - pev_{i,h}^{\text{dischar}} \quad \forall h \in t,i \tag{7.27}$$

Finally, the limit of available energy for EVs after the charging/discharging time window t of a day d is imposed:

$$Eev_{i,d}^{\min} \leq eev_{i,h}^{\min} \leq Eev_{i,d}^{\max} \quad \forall d,h \in t,i \tag{7.28}$$

In a similar fashion to ESS devices, EVs contribute to up and down island power system reserve. These up and down EV reserves are computed similarly to ESS devices:

$$resev_{i,h}^{up} = Pev_i^{\max} - pev_{i,h}^{\text{disch}} + pev_{i,h}^{\text{char}} \quad \forall h,i \tag{7.29}$$

$$resev_{i,h}^{\text{down}} = pev_{i,h}^{\text{disch}} + Pev_i^{\max} - pev_{i,h}^{\text{char}} \quad \forall h,i \tag{7.30}$$

7.5.2.3 Demand Balance Constraint

Both DSM actions and EV charge/discharge modify the overall demand balance constraint. This equation, including all the actions described in this chapter (EES, RES, DSM, and EV), is finally written as

$$\sum_{g \in i}(p_{g,h}) + \sum_{ess \in i}(pdisch_{ess,h} - pchar_{ess,h}) + (Res_{i,h} - prescurt_{i,h}) + pev_{i,h}^{\text{disch}} - pev_{i,h}^{\text{char}}$$
$$= D_{i,h} + ddisp_{i,h}^{up} - ddisp_{i,h}^{\text{down}} + \sum_{ii}(exp_{i,ii,h}) \quad \forall h,i \tag{7.31}$$

7.5.2.4 System Reserve Constraints

Taking into account the up and down EV reserve, all the total system reserve constraints written in Chapter 7, including ESS, RES, and EV, are finally updated in the following way:

$$\sum_{\substack{gg\in i\\ gg\neq g}}\left(resgen_{gg,h}^{up}\right)+\sum_{ii\neq i}\left(reslink_{ii,i,h}^{up}\right)+\sum_{ess\in i}\left(resess_{ess,h}^{up}\right)+resev_{i,h}^{up}\geq p_{g,h}\quad \forall i,g\in i,h \tag{7.32}$$

$$\sum_{g\in i}\left(resgen_{g,h}^{up}\right)+\sum_{ii\neq i}\left(reslink_{ii,i,h}^{up}\right)+\sum_{ess\in i}\left(resess_{ess,h}^{up}\right)$$
$$+resev_{i,h}^{up}\geq Resloss_i\cdot\left(Res_{i,h}-prescurt_{i,h}\right)\quad \forall i,h \tag{7.33}$$

$$\sum_{g\in i}\left(resgen_{g,h}^{up}\right)+\sum_{\substack{iii\neq i\\ iii\neq ii}}\left(reslink_{iii,i,h}^{up}\right)+\sum_{ess\in i}\left(resess_{ess,h}^{up}\right)+resev_{i,h}^{up}\geq exp_{ii,i,h}\quad \forall i,ii,h \tag{7.34}$$

$$\sum_{g\in i}\left(resgen_{g,h}^{\text{down}}\right)+\sum_{ii\neq i}\left(reslink_{ii,i,h}^{\text{down}}\right)+\sum_{ess\in i}\left(resess_{ess,h}^{\text{down}}\right)+resev_{i,h}^{\text{down}}$$
$$\geq Resdown_i^{\text{min}}\quad \forall i,h \tag{7.35}$$

7.6 Case Study 3: Assessment of Initiatives toward Smarter Island Power Systems

7.6.1 Overview

To increase island sustainability, a combination of several actions needs to be carried out, customized for specific islands, opportunities, and constraints (Eurelectric 2012). Actions can be basically separated into three categories: (1) generation side: use of natural gas and/or RES for power generation and use of ESS for reserve provision, (2) grid side: interconnection of island systems with other island systems or the continental system, and (3) demand side: use of ESS, implementation of DSM, and promotion of EV. The use of natural gas instead of oil for power generation is, however, affected by the availability of local resources and/or the existence of economies of scale in both gas pipelines and liquefied natural gas. Similarly, interconnection of an island system to the continent can be prevented by the presence of deep waters and the existence of economies of scale (in the case of both ac and HVDC transmission).

The main objective of this ambitious case study is to assess initiatives that can transform island power systems into smart ones. The study aims to yield significant valid worldwide conclusions. An initiative is understood as either a single action or a set of multiple actions. Different penetration levels of each action are considered. Since the shift from oil to gas and the interconnection of islands to a continent can depend on local factors and on economies of scale, in this study, RES (particularly wind and photovoltaic [PV] generation), ESS, DSM, and EV actions are investigated. The assessment consists of determining the impact of single-action and multiaction initiatives on the system operation costs of an island power system. The hourly unit commitment on a weekly basis, including the modeling of renewable energy and advanced control devices, is used to assess the impact of the initiatives on the system operation costs of an island. A set of five prototype island power systems is used to take into account the differences among island power systems. Then, the different investment costs of the initiatives are accounted for by determining their corresponding IRR throughout their lifetime.

7.6.2 Methodology

The methodology for assessing initiatives toward smarter and more sustainable island power systems comprises three consecutive steps: (1) classification of islands by means of clustering techniques, (2) unit commitment launching over a wide set of scenarios, and (3) economic assessment of initiatives by means of IRR computation. These three steps will next be briefly described.

7.6.2.1 Prototype Islands

Since system operation and operation costs depend on island features, the diversity of island power systems is taken into account by analyzing prototype islands determined by cluster techniques. The identification of prototype islands can be realized by means of several clustering techniques. Clustering refers to the partitioning of a data set into clusters, so that the data in each subset ideally share some common trait and differ from the data in other subsets. The K-Means clustering algorithm is used here, although the Fuzzy C-Means or Kohonen Self Organizing Map (KSOM) algorithm can be used as well with similar results.

The K-Means algorithm attempts to cluster N objects into K_N partitions. The main objective is to find K_N clusters such that the quadratic quantization error (QE) is minimal:

$$\min(\text{QE}) \tag{7.36}$$

where the quadratic error is expressed as the sum of the quadratic error of each cluster *c*:

$$\mathrm{QE} = \sum_{c=1}^{K_N} \mathrm{QE}_c = \sum_{c=1}^{K_N} \sum_{e \in c} \left\| \mathbf{x}_e - \mathbf{c}_c \right\|^2 \tag{7.37}$$

In general, the Euclidean distance is used as a distance measure between clusters and the input data vectors. The input vectors $\mathbf{x}_e$ belong to the learning set Ω and describe the features of an island power system. Meaningful variables describing island power systems include the energy demand D_{Energy}, the peak demand D_{Peak}, the installed capacity P_{gen}, the highest transmission and distribution (T&D) voltage level V_{TDmax}, the population n_{Pop}, the average system operation cost C_{Ave}, and the gross domestic product (GDP). The input vector of the *j*th island is then expressed as

$$\mathbf{x}_j = \begin{bmatrix} D_{\mathrm{Energy}} & D_{\mathrm{Peak}} & P_{\mathrm{gen}} & V_{\mathrm{TD\,max}} & n_{\mathrm{Pop}} & C_{\mathrm{Ave}} & \mathrm{GDP} \end{bmatrix}$$

$\mathbf{c}_c$ is the centroid, which is computed by averaging the $\mathbf{x}_e$ associated with the *c*th cluster. The real island power system (the input vector $\mathbf{x}_j$) closest to the centroid $\mathbf{c}_c$ is then a prototype island. The number of clusters K_N is a priori unknown and needs to be estimated. Furthermore, K-Means strongly depends on the initial solution due to its inherent gradient-based optimization algorithm to solve QE. Nevertheless, K-Means is a fast algorithm.

7.6.2.2 Weekly Unit Commitment

Each alternative is economically assessed by using the unit commitment formulation described in Chapter 6, which is extended in this chapter by ESS, RES, DSM and EV models. For each prototype island power system, the unit commitment is run for some representative weeks of a year to deduce the yearly system operation cost. For each representative demand scenario, all possible combinations of actions (i.e., wind, PV, ESS, DSM, and EV) are simulated. Each combination of actions constitutes an initiative, which needs to be economically assessed. Figure 7.18 shows the action tree of a prototype island for a particular demand scenario.

7.6.2.3 Economic Assessment of Initiatives

Finally, each initiative is assessed in terms of both the reduction of system operation costs and IRR. IRR depends on the investment parameters of each initiative. The net present value (NPV) is defined by

$$\mathrm{NPV} = \sum_{t=0}^{n_{\mathrm{years}}} \frac{C_t}{(1+r)^t} \tag{7.38}$$

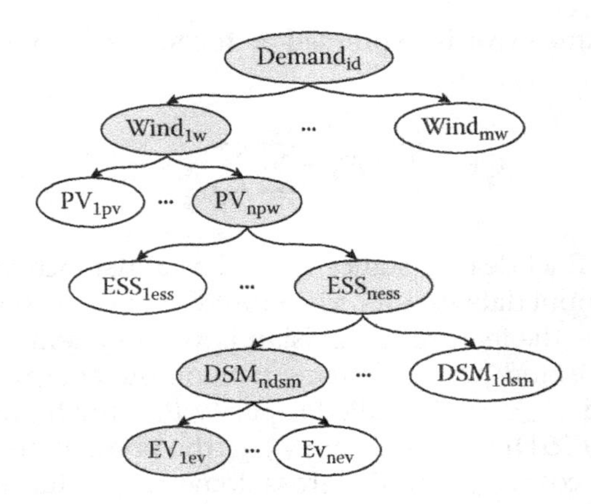

FIGURE 7.18
Action tree of a prototype island for a particular demand scenario.

where:
n_{years} is the expected lifetime
C_t is the cashflow of year t
r is the rate of return

An initiative is considered to be financially feasible only if the corresponding NPV is positive. The cashflow for $t>0$ is given by the difference of the yearly system operation cost of the ith initiative C_{tot,ini_i} with respect to the current yearly system operation cost without initiatives $C_{tot,current}$:

$$C_t = C_{tot,ini\_i} - C_{tot,current} \tag{7.39}$$

IRR is the rate of return for a zero NPV; that is, IRR provides a lower feasibility bound. Equation 7.4 can be rewritten in polynomial form, as shown in Equation 7.40:

$$0 = \sum_{t=0}^{n_{years}} \frac{C_t}{(1+\mathrm{IRR})^t} = C_0 + C_1 a_1 + \cdots + C_{n_{years}} a_{n_{years}} \tag{7.40}$$

The problem of finding the IRR is thus transformed into one of finding a real root p of this polynomial. Once the root has been found, the IRR becomes

$$\mathrm{IRR} = \frac{1}{p} - 1 \tag{7.41}$$

An initiative is viable if the IRR is higher than a predefined acceptance limit, which is equal to or higher than the nominal annual discount rate.

7.6.3 Results

This section presents the results of applying the proposed methodology to a set of 60 real power islands worldwide of different sizes and features. First, the identification of the prototype islands is described. Then, the unit commitment input data are outlined. Finally, the results of the economic assessment are presented.

7.6.3.1 Prototype Islands

Figure 7.19a shows the relation between the energy demand, the GDP, and the population of the nearly 60 island power systems used in the study. It can be seen that these features are quite well correlated, although some degree of dispersion exists. The larger the population, the larger the GDP and the

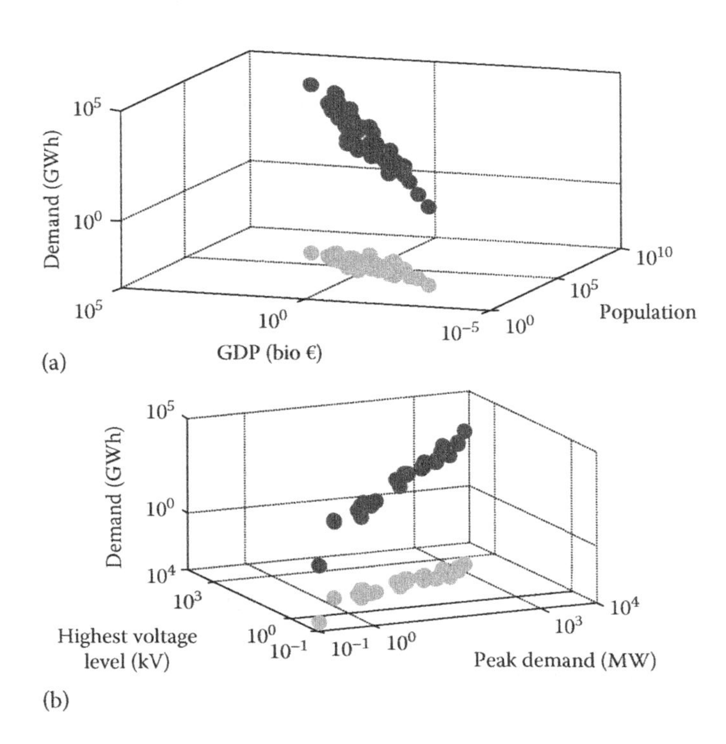

FIGURE 7.19
Relation between different island features: (a) energy demand, GDP, and population size, and (b) energy demand, peak demand, and highest T&D voltage level.

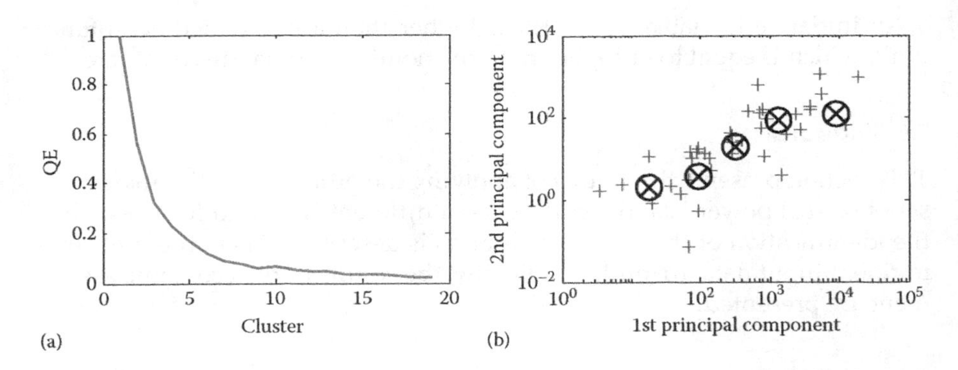

FIGURE 7.20
(a) Quadratic quantization error as a function of number of clusters and (b) principal component analysis of the clusters and the input data.

energy demand. Similarly, Figure 7.19b shows the relation between energy demand, highest T&D voltage level, and peak demand. The highest voltage level is determined by the distances between generation and load centers and the amount of power transmitted. Again, the three features seem to be correlated. This makes sense, since the larger the peak demand, the more power needs to be transmitted, favoring higher transmission voltage levels.

Figure 7.20 shows the results of applying the K-Means clustering algorithm. According to Figure 7.20a, which plots the quadratic QE as a function of the number of clusters, five clusters (prototype islands) seem to be sufficient to describe the nearly 60 island power systems. Figure 7.20b compares the five prototype and the nearly 60 island power systems by means of principal component analysis (PCA). PCA is mathematically defined as an orthogonal linear transformation that transforms the data to a new coordinate system such that the greatest variance by any projection of the data comes to lie on the first coordinate (called the *first principal component*), the second greatest variance on the second coordinate, and so on. The result is that the five prototype islands satisfactorily represent the diversity of all island power systems.

Finally, Table 7.8 summarizes some of the features of the five prototype islands in terms of annual energy demand, installed generation capacity, highest T&D voltage level, population, and average generation cost.

7.6.3.2 Unit Commitment Input Data

The weekly unit commitment is simulated for four representative weeks of the year of each prototype island, one per season. For each action, five different penetration levels have been defined. The penetration levels and cost figures of wind, PV, ESS, DSM, and EV actions are shown in Table 7.9. For

TABLE 7.8

Features of the Prototype Island Power Systems

Prototype	1	2	3	4	5
D_{Energy} (GWh)	10,627	1,446.7	309.8	90.5	16.6
P_{gen} (MW)	2,802.7	457.4	99.2	26.7	6.2
V_{TDmax} (kV)	220	66	66	32	32
n_{Pop}	2,550	244	88	33	5.5
C_{ave} (€/MWh)	81	118	101	146	140

TABLE 7.9

Penetration Levels and Cost Figures of Contemplated Actions

	Installed Power P_{in}	Energy Capacity	Investment
Wind	(0, 25, 50, 75, 100)% average power demand	—	1000 €/kW
PV	(0, 25, 50, 75, 100)% average power demand	—	2300 €/kW
ESS	(0, 2.5, 5, 7.5, 10)% average power demand	4 h × P_{in}	2800 €/kW
DSM	(0, 1,2, 3, 4)% average power demand	(0, 0.4, 0.8, 1.25, 1.7)% of daily average energy demand	650 €/kW
EV	(0, 5, 10, 15, 20)% number of cars	10 h × P_{in} (slow charging)	2400 €/kW

ESS, a round-trip efficiency of 80% has been assumed, and energy capacity has been set such that the ESS is able to operate at P_{in} for 4 h. ESS are operated on a weekly basis, whereas EVs are operated on a daily basis, assuming an initial state of charge of 50% (a 30 kWh EV consuming 25 kWh/100 km for an average driving distance of around 60 km [JRC 2012]). Investment costs have been extracted from a variety of sources: wind and PV (Schröder et al. 2013), EES (EPRI 2010b; IRENA 2015), DSM (Xcel Energy 2014; US Dept Energy 2014), and EV (C2ES 2015; Transport for London 2010).

Note that the penetration levels of all actions except EV are given as a percentage of the average power demand, whereas the penetration levels of EV actions are given as a percentage of the estimated number of cars of an island. The installed power of EV can be expressed as

$$P_{in,EV} = P_{EV} \cdot n_{cars} \cdot \%EV \tag{7.42}$$

where:

P_{EV} is the power demand of an EV (e.g., 3.3 kW)
n_{cars} is the number of cars
$\%EV$ is the percentage of EV of n_{cars}

The cost parameter of EV actions corresponds to the cost of the charging point infrastructure, excluding the cost of the EV itself. For a slow charging point of 3.3 kW EV, the EV cost figure C_{EV} in euros per kilowatt per car can be computed as

$$C_{EV} = \frac{C_{CP} \cdot n_{CP/EV}}{P_{EV}} \tag{7.43}$$

where:

C_{CP} is the cost of the charging point

$n_{CP/EV}$ is the number of charging points per EV

It must be mentioned that the assumed EV investment cost ignores other costs for utilities, such as promotion and marketing costs (US Dept Energy 2010). In the case of DSM programs, these costs might account for to 10% of the total program costs (Xcel Energy 2014).

7.6.3.2 Economic Assessment of Initiatives

The impact of all possible single and multiaction initiatives (3125 initiatives in total) on the system operation cost of each of the five prototype islands has been simulated for the four representative seasonal demand scenarios.

Figures 7.21 and 7.22 show the annual system operation cost reduction in the case of Prototype Island 2 (a larger island power system) and Prototype Island 4 (a smaller island power system) for all considered single and multiaction initiatives. The insets in Figures 7.21 and 7.22 allow the impact of different actions to be dissected. Arrows illustrate the direction of increasing sizes of actions. It can be inferred that larger reductions of system operation costs are obtained when actions are implemented at higher penetration levels. This conclusion is also valid for Prototype Islands 1, 3, and 5, and it makes sense, since the considered actions usually exhibit low variable costs in comparison with conventional power generation. Nonetheless, larger and smaller islands, as shown in Figures 7.21 and 7.22, exhibit certain differences. Whereas the increase of wind and PV penetration levels always leads to an increase in annual cost reduction in the case of a smaller island, the wind and PV penetration levels of a larger system clearly show a maximum cost reduction. The penetration level of PV generation with maximum cost reduction increases with decreasing wind penetration level.

The impact of a particular action also varies according to the prototype island. In the case of a larger island system (see Figure 7.21), wind and ESS actions have a higher influence on annual cost reduction (around €10–15 million) than PV, DSM, and EV actions (around €2–5 million). Both wind and PV actions have, however, a greater influence on the annual cost reduction of a smaller island system (around €1.5–3 million) than ESS, DSM, and EV

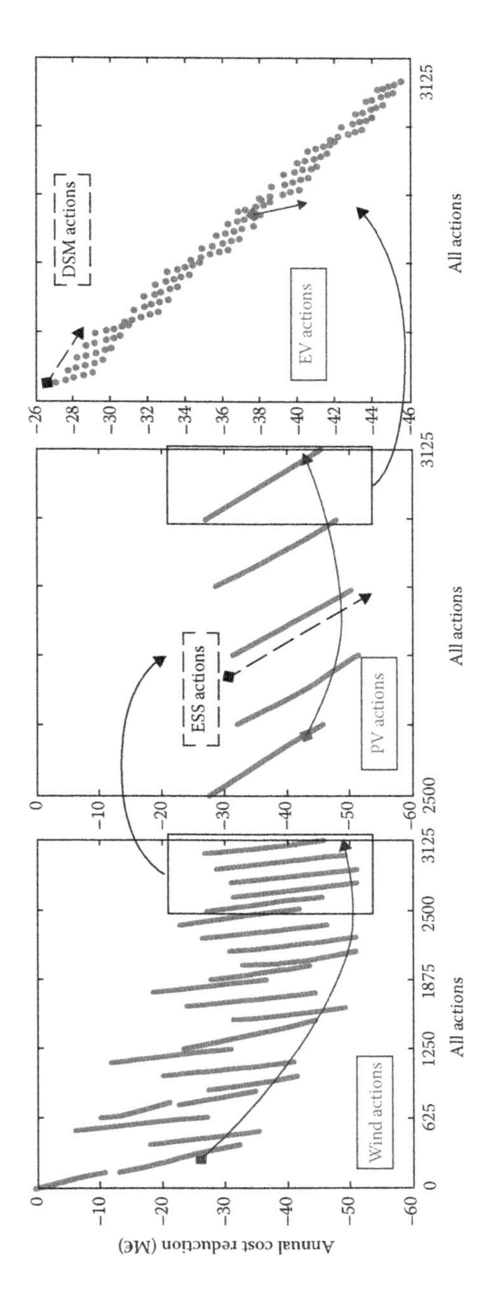

FIGURE 7.21
Annual system cost reduction of Prototype Island 2—aggregated impact of actions.

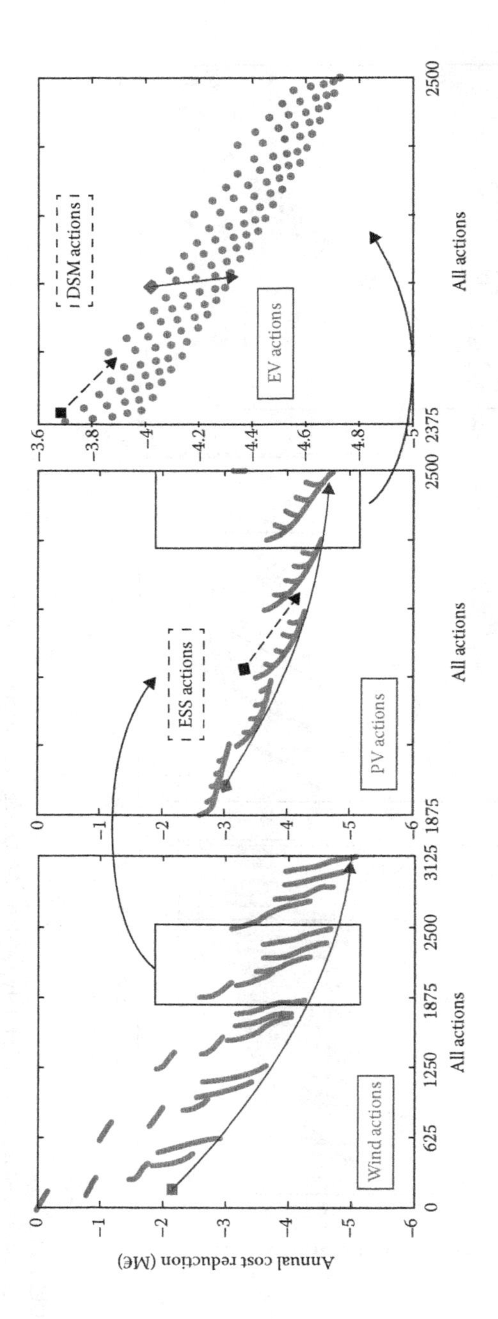

FIGURE 7.22
Annual system operation cost reduction of Prototype Island 4—aggregated impact of actions.

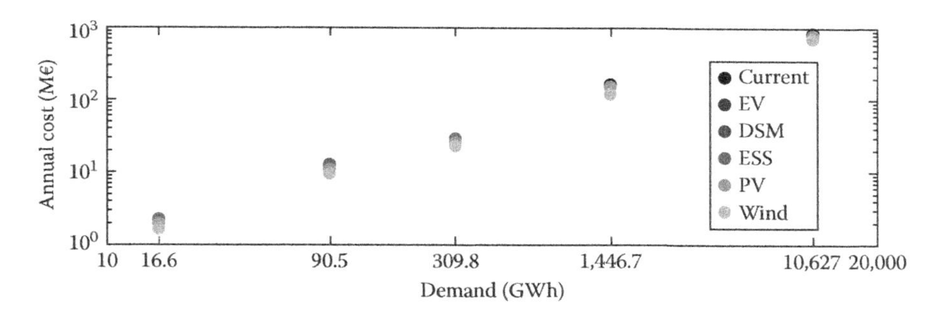

FIGURE 7.23
Annual system operation costs as a function of the annual demand of the five prototype islands with and without single-action initiatives implemented at their highest penetration levels.

actions. ESS in particular yields a lower cost reduction for a smaller system than for a larger one.

Figure 7.23 shows and compares the annual system operation costs as a function of the annual energy demand of the five prototype islands with and without single-action initiatives (e.g., only wind or only ESS actions) implemented at their highest penetration levels. It can be seen that annual system costs are correlated with the annual demand of the prototype islands. This finding is in line with those deduced from Figure 7.19.

The previous analysis of the impact of initiatives on annual system operation costs pinpoints that multiaction initiatives are most effective. However, this analysis does not consider the investment cost required to implement the initiatives. Congruently, the IRR of all possible single- and multiaction initiatives is computed on the basis of the investment costs shown in Table 7.9.

Figures 7.24 and 7.25 compare the five best initiatives in terms of annual cost reduction with the five best initiatives in terms of IRR for Prototype Islands 2 and 4, respectively. In contrast to the results in terms of annual cost reduction, single actions or a few multiple actions implemented at their lower penetration levels succeed in terms of IRR, since the investment costs of the considered actions are usually higher than the investment costs of conventional generation. The IRR over 15 years points to the use of smaller DSM and/or EV actions for Prototype Island 2, whereas mostly wind, together with DSM actions, dominates in the case of Prototype Island 4. It can be concluded that the same single-action and multiaction initiatives have different impacts in terms of relative size with regard to the prototype island. In other words, different island power systems benefit from different initiatives. Note also that EV actions do not contribute to annual cost reductions in the case of Prototype Island 4, since the reduction of both start-up costs and reserve costs does not compensate for the costs of increasing the demand.

Table 7.10 shows the best and the fifth best initiatives in terms of IRR for the five prototype islands. It can be inferred that single-action initiatives are the solutions with the highest IRR in three of the five islands. Multiple

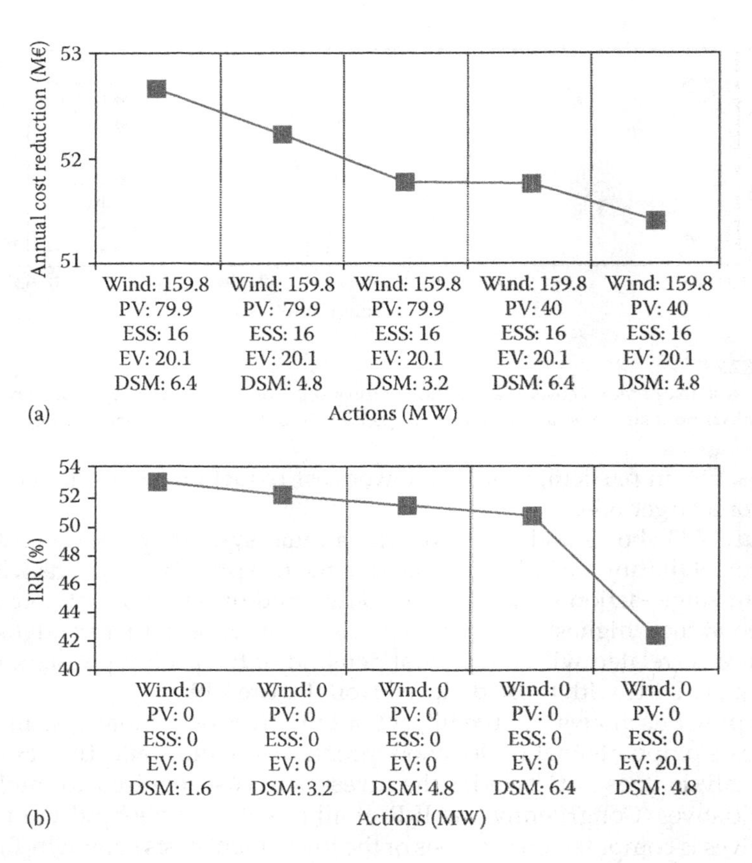

FIGURE 7.24
Five best initiatives of Prototype Island 2: (a) in terms of annual cost reduction and (b) in terms of IRR.

actions with highest IRR are usually based on lower penetration levels of wind generation and DSM. Again, it can be concluded that successful initiatives are those that implement actions at very low penetration levels. From Table 7.10, it can be further inferred that larger islands (Prototypes 1 and 2) benefit most from DSM and/or EV actions, whereas medium-sized to smaller islands (Prototypes 3–5) mostly benefit from wind power and solar PV power actions. In the case of smaller islands, wind power and solar PV power actions are accompanied by DSM actions, exhibiting, however, a slightly lower IRR than without the DSM actions. Finally, it is interesting to see that ESS actions are not present. The main reasons are the relatively low penetration levels of wind and PV actions, where conventional generation is still able to provide sufficient reserve, and the investment costs. Note that in Case Study 2 of this chapter, ESS providing reserve and peak-shaving services yielded an IRR around 8%.

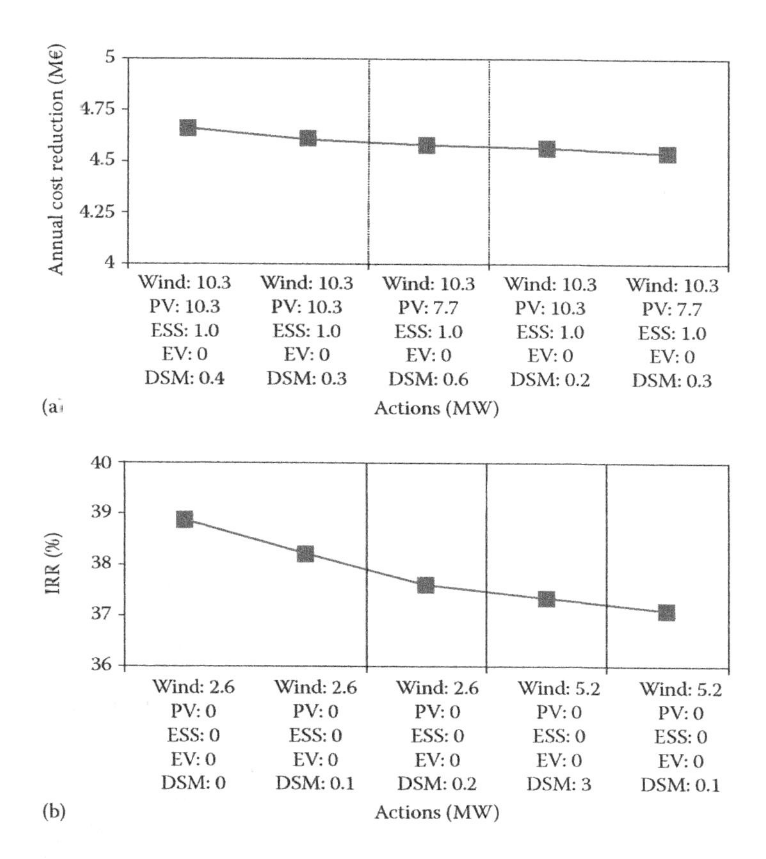

FIGURE 7.25
Best initiatives of Prototype Island 4: (a) in terms of annual cost reduction and (b) in terms of IRR.

TABLE 7.10
Best and Fifth Best Initiatives in Terms of IRR for the Considered Prototype Islands

Prototype Island	Initiative	Wind (MW)	PV (MW)	ESS (MW)	EV (MW)	DSM (MW)	IRR (%)
1	Best	0	0	0	0	12.1	46.7
1	5th best	303.3	0	0	0	12.1	21.4
2	Best	0	0	0	0	1.6	53.0
2	5th best	0	0	0	20.1	4.8	42.3
3	Best	8.8	8.8	0	0	0	67.3
3	5th best	8.8	8.8	0	0	1.4	65.4
4	Best	2.6	0	0	0	0	38.9
4	5th best	5.2	0	0	0	0.1	37.1
5	Best	0.5	0	0	0	0	34.2
5	5th best	0.5	0	0	0.	0.08	33.9

TABLE 7.11

Best and Fifth Best Initiatives in Terms of IRR for the Considered Prototype Islands with Increased EV and DSM Investment Costs

Prototype Island	Initiative	Wind (MW)	PV (MW)	ESS (MW)	EV (MW)	DSM (MW)	IRR (%)
1	Best	0	0	0	0	12.1	22.3
1	5th best	303.3	0	0	0	12.1	19.3
2	Best	79.9	0	0	0	1.6	29.4
2	5th best	79.9	0	0	6	6.4	29.2
3	Best	8.8	8.8	0	0	0	67.3
3	5th best	8.8	8.8	0	0	1.4	63.5
4	Best	2.6	0	0	0	0	38.9
4	5th best	5.2	0	0	0	0.2	36
5	Best	0.5	0	0	0	0	34.2
5	5th best	0.95	0	0	0	0.02	31.6

Although the investment costs of wind generation are rather well established, the investment costs of PV, ESS, and particularly EV and DSM might vary significantly. A further analysis of the IRR has been carried out by considering an investment cost of EV and DSM twice as high as assumed in Table 7.9. Table 7.11 shows again the best and the fifth best initiatives in terms of IRR for the five prototype islands. By comparing Tables 7.10 and 7.11, a major change can be detected for Prototype Island 2, where the best and fifth best initiatives now also include wind actions. It can also be seen that IRR is considerably reduced (except for the best initiatives of Prototype Islands 3 and 5, consisting only of wind and PV actions, which are not affected by DSM and EV costs).

ESS would become a competitive initiative (among the five best) only if DSM and EV actions were dramatically higher (four to five times the assumed value) or if ESS cost were to fall significantly. Actually, if a 50% reduction of ESS investment cost is applied, the results of Table 7.10 remain the same except for Prototype Island 1, where the fifth best initiative now involves a 30.3 MW–4 h ESS action and a 48.5 MW DSM action.

7.6.4 Conclusions of Case Study 3

Case Study 3 has shown how the weekly unit commitment model can be used to plan the deployment of single and multiple actions that can transform island power systems into smart ones. Initiatives fostering the deployment of wind and PV generation, ESS, DSM, and EVs are considered. The study is ambitious, since it has considered a total number of 60 real islands worldwide, grouped into five prototype islands. The economic assessment of the initiatives has been made in terms of their IRR. The assessment shows that islands of different sizes and features require different initiatives. Usually, single-action initiatives are most successful. Larger islands tend toward DSM-dominated initiatives, whereas smaller islands tend toward RES (wind and PV) initiatives. ESS actions also show a positive IRR, but far lower than that for other actions.

References

Anagnostopoulos, J. S., and D. E. Papantonis. 2012. Study of pumped storage schemes to support high RES penetration in the electric power system of Greece. *Energy* 45 (1):416–423.

Centre for Climate and Energy Solutions (C2ES. 2015. Business models for financially sustainable EV charging networks, Available at www.c2es.org/docUploads/business-models-ev-charging-infrastructure-03-15.pdf.

Chen, C., S. Duan, T. Cai, B. Liu, and G. Hu. 2011. Optimal allocation and economic analysis of energy storage system in microgrids. *IEEE Transactions on Power Electronics* 26 (10):2762–2773.

Chen, H., T. N. Cong, W. Yang, C. Tan, Y. Li, and Y. Ding. 2009. Progress in electrical energy storage system: A critical review. *Progress in Natural Science* 19 (3):291–312.

Daneshi, H., and A. K. Srivastava. 2012. Security-constrained unit commitment with wind generation and compressed air energy storage. *IET Generation, Transmission and Distribution* 6 (2):167–175.

Delille, G., B. Francois, and G. Malarange. 2012. Dynamic frequency control support by energy storage to reduce the impact of wind and solar generation on isolated power system's inertia. *IEEE Transactions on Sustainable Energy* 3 (4):931–939.

Dietrich, K., J. M. Latorre, L. Olmos, and A. Ramos. 2012. Demand response in an isolated system with high wind integration. *IEEE Transactions on Power Systems* 27 (1):20–29.

EPRI. 2010a. Electric Energy Storage Technology Options: A White Paper Primer on Applications, Costs and Benefits, available from http://large.stanford.edu/courses/2012/ph240/doshay1/docs/EPRI.pdf. Palo Alto, CA: EPRI.

EPRO. 2010b. *Electric Energy Storage Technology Options: A White Paper Primer on Applications, Costs, and Benefits*. Palo Alto, CA: EPRI, 1020676.

Eurelectric. 2012. EU Islands: Towards a sustainable energy future, D/2012/12.105/24, June 2012, available from www.eurelectric.org/.

Ghofrani, M., A. Arabali, M. Etezadi-Amoli, and M. S. Fadali. 2014. Smart scheduling and cost-benefit analysis of grid-enabled electric vehicles for wind power integration. *IEEE Transactions on Smart Grid* 5 (5):2306–2313.

Hansen, C. W., and A. D. Papalexopoulos. 2012. Operational impact and cost analysis of increasing wind generation in the Island of Crete. *IEEE Systems Journal* 6 (2): 287–295.

Hittinger, E., J. F. Whitacre, and J. Apt. 2012. What properties of grid energy storage are most valuable? *Journal of Power Sources* 206:436–449.

Hu, W., C. Su, Z. Chen, and B. Bak-Jensen. 2013. Optimal operation of plug-in electric vehicles in power systems with high wind power penetrations. *IEEE Transactions on Sustainable Energy* 4 (3):577–585.

IRENA. 2015. Battery storage for renewables: Market status and technology outlook, Available at www.irena.org/DocumentDownloads/Publications/IRENA_Battery_Storage_report_2015.pdf.

JRC. 2012. *Driving and Parking Patterns of European Car Drivers: A Mobility Survey*. Luxembourg: Publications Office. Available at https://setis.ec.europa.eu/sites/default/files/reports/Driving_and_parking_patterns_of_European_car_drivers-a_mobility_survey.pdf

Kottick, D., M. Blau, and D. Edelstein. 1993. Battery energy storage for frequency regulation in an island power system. *IEEE Transactions on Energy Conversion* 8 (3):455–459.

Leadbetter, J., and L. G. Swan. 2012. Selection of battery technology to support grid-integrated renewable electricity. *Journal of Power Sources* 216:376–386.

Lu, C.-F., C.-C. Liu, and C.-J. Wu. 1995. Effect of battery energy storage system on load frequency control considering governor deadband and generation rate constraint. *IEEE Transactions on Energy Conversion* 10 (3):555–561.

Margaris, I. D., S. A. Papathanassiou, N. D. Hatziargyriou, A. D. Hansen, and P. Sorensen. 2012. Frequency control in autonomous power systems with high wind power penetration. *IEEE Transactions on Sustainable Energy* 3 (2):189–199.

Meibom, P., R. Barth, B. Hasche, H. Brand, C. Weber, and M. O'Malley. 2011. Stochastic optimization model to study the operational impacts of high wind penetrations in Ireland. *IEEE Transactions on Power Systems* 26 (3):1367–1379.

Mercier, P., R. Cherkaoui, and A. Oudalov. 2009. Optimizing a battery energy storage system for frequency control application in an isolated power system. *IEEE Transactions on Power Systems* 24 (3):1469–1477.

Mufti, M. ud din, S. A. Lone, S. J. Iqbal, M. Ahmad, and M. Ismail. 2009. Supercapacitor based energy storage system for improved load frequency control. *Electric Power Systems Research* 79 (1):226–233.

Palensky, P., and D. Dietrich. 2011. Demand side management: Demand response, intelligent energy systems, and smart loads. *IEEE Transactions on Industrial Informatics* 7 (3):381–388.

Quintero J.M. 2013, Hybrid storage (pumping)-wind solutions for small islands: El Hierro island 100 % RES project.national strategy. GENERA 2013, Session on Island Power Systems, Madrid.

REE. 2006. Ministry of industry, tourism and commerce of Spain, resolution 9613 of 26 April 2006 establishing the operation procedures for insular and extra peninsular power systems (in Spanish), official bulletin of the state nº 219 of 31 May 2006, available from www.ree.es.

Roberts, B. 2009. Capturing grid power. *IEEE Power and Energy Magazine* 7 (4):32–41.

Roberts, B., and J. McDowall. 2005. Commercial successes in power storage. *IEEE Power and Energy Magazine* 3 (2):24–30.

Ross, M., R. Hidalgo, C. Abbey, and G. Joós. 2011. Energy storage system scheduling for an isolated microgrid. *IET Renewable Power Generation* 5 (2):117–123.

Rouco, L., J. L. Zamora, I. Egido, and F. Fernandez-Bernal. 2008. Impact of wind power generators on the frequency stability of synchronous generators. Cigré Session., Paper no. A1-203. Paris, France. 24-29 August 2008.

Schröder, A., F. Kunz, J. Meiss, R. Mendelevitch, and C. von Hirschhausen. 2013. *Current and Prospective Costs of Electricity Generation until 2050*. Data Documentation, DIW, no. 68. Berlin: DIW. Available online at https://www.diw.de/documents/publikationen/73/diw_01.c.424566.de/diw_datadoc_2013-068.pdf.

Sebastian, R. 2011. Modelling and simulation of a high penetration wind diesel system with battery energy storage. *International Journal of Electrical Power and Energy Systems* 33 (3):767–774.

Sigrist L., Egido I., Rouco L. 2012, A Method for the Design of UFLS Schemes of Small Isolated Power Systems, IEEE Transactions on Power Systems 27(2):951–958.

Sigrist, L., I. Egido, E. Lobato Miguélez, and L. Rouco. 2015. Sizing and controller setting of ultracapacitors for frequency stability enhancement of small isolated power systems. *IEEE Transactions on Power Systems* 30 (4):2130–2138.

Strbac, G., A. Shakoor, M. Black, D. Pudjianto, and T. Bopp. 2007. Impact of wind generation on the operation and development of the UK electricity systems. *Electric Power Systems Research* 77 (9):1214–1227.

Tan, X., Q. Li, and H. Wang. 2013. Advances and trends of energy storage technology in microgrid. *International Journal of Electrical Power and Energy Systems* 44 (1):179–191.

Transport for London. 2010. Transport for London, Guidance for implementation of electric vehicle charging infrastructure, April 2010. Available online at http://app.thco.co.uk/WLA/wt.nsf/Files/WTA-3/$FILE/EVCP-guidance-version-1-Apr10[1].pdf.

US Dept Energy. 2010. US Department of Energy and Federal Energy Regulatory Commission, framework for evaluating the cost-effectiveness of demand response. Available online at https://emp.lbl.gov/sites/all/files/napdr-cost-effectiveness_0.pdf.

US Dept Energy. 2014. US Department of Energy, automated demand response benefits California utilities and commercial and industrial customers, September 2014. Available online at https://www.smartgrid.gov/files/C6-Honeywell-final-draft-091814.pdf.

van-der-Linden, S. 2006. Bulk energy storage potential in the USA, current developments and future prospects. *Energy* 31 (15):3446–3457.

Wang, S. Y., and J. L. Yu. 2012. Optimal sizing of the CAES system in a power system with high wind power penetration. *International Journal of Electrical Power and Energy Systems* 37 (1):117–125.

Xcel Energy. 2014. Xcel energy, demand-side management annual status report, April 1, 2013.

Index

B

C

D

J

K

L

Z